T0122303

Lecture Notes in Networks and Systems

Volume 9

Series editor

Janusz Kacprzyk, Polish Academy of Sciences, Warsaw, Poland
e-mail: kacprzyk@ibspan.waw.pl

The series "Lecture Notes in Networks and Systems" publishes the latest developments in Networks and Systems—quickly, informally and with high quality. Original research reported in proceedings and post-proceedings represents the core of LNNS.

Volumes published in LNNS embrace all aspects and subfields of, as well as new challenges in, Networks and Systems.

The series contains proceedings and edited volumes in systems and networks, spanning the areas of Cyber-Physical Systems, Autonomous Systems, Sensor Networks, Control Systems, Energy Systems, Automotive Systems, Biological Systems, Vehicular Networking and Connected Vehicles, Aerospace Systems, Automation, Manufacturing, Smart Grids, Nonlinear Systems, Power Systems, Robotics, Social Systems, Economic Systems and other. Of particular value to both the contributors and the readership are the short publication timeframe and the world-wide distribution and exposure which enable both a wide and rapid dissemination of research output.

The series covers the theory, applications, and perspectives on the state of the art and future developments relevant to systems and networks, decision making, control, complex processes and related areas, as embedded in the fields of interdisciplinary and applied sciences, engineering, computer science, physics, economics, social, and life sciences, as well as the paradigms and methodologies behind them.

Advisory Board

Fernando Gomide, Department of Computer Engineering and Automation—DCA, School of Electrical and Computer Engineering—FEEC, University of Campinas—UNICAMP, São Paulo, Brazil
e-mail: gomide@dca.fee.unicamp.br
Okyay Kaynak, Department of Electrical and Electronic Engineering, Bogazici University, Istanbul, Turkey
e-mail: okyay.kaynak@boun.edu.tr
Derong Liu, Department of Electrical and Computer Engineering, University of Illinois at Chicago, Chicago, USA and
Institute of Automation, Chinese Academy of Sciences, Beijing, China
e-mail: derong@uic.edu
Witold Pedrycz, Department of Electrical and Computer Engineering, University of Alberta, Alberta, Canada and
Systems Research Institute, Polish Academy of Sciences, Warsaw, Poland
e-mail: wpedrycz@ualberta.ca
Marios M. Polycarpou, KIOS Research Center for Intelligent Systems and
Networks, Department of Electrical and Computer Engineering, University of Cyprus, Nicosia, Cyprus
e-mail: mpolycar@ucy.ac.cy
Imre J. Rudas, Óbuda University, Budapest Hungary
e-mail: rudas@uni-obuda.hu
Jun Wang, Department of Computer Science, City University of Hong Kong, Kowloon, Hong Kong
e-mail: jwang.cs@cityu.edu.hk

More information about this series at http://www.springer.com/series/15179

Durgesh Kumar Mishra · Malaya Kumar Nayak
Amit Joshi
Editors

Information and Communication Technology for Sustainable Development

Proceedings of ICT4SD 2016, Volume 1

Springer

Editors
Durgesh Kumar Mishra
Microsoft Innovation Centre
Sri Aurobindo Institute of Technology
Indore, Madhya Pradesh
India

Amit Joshi
Department of Information Technology
Sabar Institute of Technology
Ahmedabad, Gujarat
India

Malaya Kumar Nayak
Dagenham
UK

ISSN 2367-3370 ISSN 2367-3389 (electronic)
Lecture Notes in Networks and Systems
ISBN 978-981-13-5002-3 ISBN 978-981-10-3932-4 (eBook)
https://doi.org/10.1007/978-981-10-3932-4

© Springer Nature Singapore Pte Ltd. 2018
Softcover re-print of the Hardcover 1st edition 2018
This work is subject to copyright. All rights are reserved by the Publisher, whether the whole or part
of the material is concerned, specifically the rights of translation, reprinting, reuse of illustrations,
recitation, broadcasting, reproduction on microfilms or in any other physical way, and transmission
or information storage and retrieval, electronic adaptation, computer software, or by similar or dissimilar
methodology now known or hereafter developed.
The use of general descriptive names, registered names, trademarks, service marks, etc. in this
publication does not imply, even in the absence of a specific statement, that such names are exempt from
the relevant protective laws and regulations and therefore free for general use.
The publisher, the authors and the editors are safe to assume that the advice and information in this
book are believed to be true and accurate at the date of publication. Neither the publisher nor the
authors or the editors give a warranty, express or implied, with respect to the material contained herein or
for any errors or omissions that may have been made. The publisher remains neutral with regard to
jurisdictional claims in published maps and institutional affiliations.

Printed on acid-free paper

This Springer imprint is published by Springer Nature
The registered company is Springer Nature Singapore Pte Ltd.
The registered company address is: 152 Beach Road, #21-01/04 Gateway East, Singapore 189721, Singapore

Preface

The volume contains the best selected papers presented at the ICT4SD 2016: International Conference on Information and Communication Technology for Sustainable Development. The conference was held during 1 and 2 July 2016 in Goa, India and organized communally by the Associated Chambers of Commerce and Industry of India, Computer Society of India Division IV and Global Research Foundation. It targets state of the art as well as emerging topics pertaining to ICT and effective strategies for its implementation for engineering and intelligent applications. The objective of this international conference is to provide opportunities for the researchers, academicians, industry persons and students to interact and exchange ideas, experience and expertise in the current trend and strategies for information and communication technologies. Besides this, participants are also enlightened about vast avenues, and current and emerging technological developments in the field of ICT in this era and its applications are thoroughly explored and discussed. The conference attracted a large number of high-quality submissions and stimulated the cutting-edge research discussions among many academically pioneering researchers, scientists, industrial engineers and students from all around the world. The presenter proposed new technologies, shared their experiences and discussed future solutions for designing infrastructure for ICT. The conference provided common platform for academic pioneering researchers, scientists, engineers and students to share their views and achievements. The overall focus was on innovative issues at international level by bringing together the experts from different countries. Research submissions in various advanced technology areas were received, and after a rigorous peer-review process with the help of program committee members and external reviewer, 105 papers were accepted with an acceptance ratio of 0.43. The conference featured many distinguished personalities like Prof. H.R. Vishwakarma, Prof. Shyam Akashe, Mr. Nitin from Goa Chamber of Commerce, Dr. Durgesh Kumar Mishra and Mr. Aninda Bose from Springer India Pvt. Limited, and also a wide panel of start-up entrepreneurs were present. Separate invited talks were organized in industrial and academia tracks on both days. The conference also hosted tutorials and workshops for the benefit of participants. We are indebted to CSI Goa Professional Chapter, CSI Division IV, and Goa Chamber of Commerce for their immense support to make this conference possible in such a grand scale. A total of 12 sessions were organized as a part of ICT4SD 2016 including nine technical, one plenary, one inaugural and one valedictory session. A total of 73 papers were presented in nine technical sessions with high discussion insights. Our sincere thanks to all sponsors, press, print and electronic media for their excellent coverage of this conference.

Indore, India   Durgesh Kumar Mishra
Dagenham, UK   Malaya Kumar Nayak
Ahmedabad, India   Amit Joshi
July 2016

Organisation Committee

Advisory Committee

Dr. Dharm Singh, Namibia University of Science and Technology, Namibia
Dr. Aynur Unal, Standford University, USA
Mr. P.N. Jain, Add. Sec., R&D, Government of Gujarat, India
Prof. J. Andrew Clark, Computer Science, University of York, UK
Dr. Anirban Basu, Vice President, CSI
Prof. Mustafizur Rahman, Endeavour Research Fellow, Australia
Dr. Malay Nayak, Director-IT, London
Mr. Chandrashekhar Sahasrabudhe, ACM, India
Dr. Pawan Lingras, Saint Mary's University, Canada
Prof. (Dr.) P. Thrimurthy, Past President, CSI
Dr. Shayam Akashe, ITM, Gwalior, MP, India
Dr. Bhushan Trivedi, India
Prof. S.K. Sharma, Pacific University, Udaipur, India
Prof. H.R. Vishwakarma, VIT, Vellore, India
Dr. Tarun Shrimali, SGI, Udaipur, India
Mr. Mignesh Parekh, Ahmedabad, India
Mr. Sandeep Sharma, Joint CEO, SCOPE
Dr. J.P. Bhamu, Bikaner, India
Dr. Chandana Unnithan, Victoria University, Australia
Prof. Deva Ram Godara, Bikaner, India
Dr. Y.C. Bhatt, Chairman, CSI Udaipur Chapter
Dr. B.R. Ranwah, Past Chairman, CSI Udaipur Chapter
Dr. Arpan Kumar Kar, IIT Delhi, India

Organising Committee

Organising Chairs
Ms. Bhagyesh Soneji, Chairperson ASSOCHAM Western Region

Co-Chair
Dr. S.C. Satapathy, ANITS, Visakhapatnam

Members
Shri. Bharat Patel, COO, Yudiz Solutions
Dr. Basant Tiwari, Bhopal
Dr. Rajveer Shekhawat, Manipal University, Jaipur
Dr. Nilesh Modi, Chairman, ACM Ahmedabad Chapter
Dr. Harshal Arolkar, Assoc. Prof., GLS Ahmedabad
Dr. G.N. Jani, Ahmedabad, India
Dr. Vimal Pandya, Ahmedabad, India
Mr. Vinod Thummar, SITG, Gujarat, India
Mr. Nilesh Vaghela, Electromech, Ahmedabad, India
Dr. Chirag Thaker, GEC, Bhavnagar, Gujarat, India
Mr. Maulik Patel, SITG, Gujarat, India
Mr. Nilesh Vaghela, Electromech Corp., Ahmedabad, India
Dr. Savita Gandhi, GU, Ahmedabad, India
Mr. Nayan Patel, SITG, Gujarat, India
Dr. Jyoti Parikh, Professor, GU, Ahmedabad, India
Dr. Vipin Tyagi, Jaypee University, Guna, India
Prof. Sanjay Shah, GEC, Gandhinagar, India
Dr. Chirag Thaker, GEC, Bhavnagar, Gujarat, India
Mr. Mihir Chauhan, VICT, Gujarat, India
Mr. Chetan Patel, Gandhinagar, India

Program Committee

Program Chair
Dr. Durgesh Kumar Mishra, Chairman, Div IV, CSI

Members
Dr. Priyanka Sharma, RSU, Ahmedabad
Dr. Nitika Vats Doohan, Indore
Dr. Mukesh Sharma, SFSU, Jaipur
Dr. Manuj Joshi, SGI, Udaipur, India
Dr. Bharat Singh Deora, JRNRV University, Udaipur
Prof. D.A. Parikh, Head, CE, LDCE, Ahmedabad, India
Prof. L.C. Bishnoi, GPC, Kota, India

Mr. Alpesh Patel, SITG, Gujarat
Dr. Nisheeth Joshi, Banasthali University, Rajasthan, India
Dr. Vishal Gaur, Bikaner, India
Dr. Aditya Patel, Ahmedabad University, Gujarat, India
Mr. Ajay Choudhary, IIT Roorkee, India
Dr. Dinesh Goyal, Gyan Vihar, Jaipur, India
Dr. Devesh Shrivastava, Manipal University, Jaipur
Dr. Muneesh Trivedi, ABES, Gaziabad, India
Prof. R.S. Rao, New Delhi, India
Dr. Dilip Kumar Sharma, Mathura, India
Prof. R.K. Banyal, RTU, Kota, India
Mr. Jeril Kuriakose, Manipal University, Jaipur, India
Dr. M. Sundaresan, Chairman, CSI Coimbatore Chapter
Prof. Jayshree Upadhyay, HOD-CE, VCIT, Gujarat
Dr. Sandeep Vasant, Ahmedabad University, Gujarat, India

Contents

Editors and Contributors

About the Editors

Dr. Durgesh Kumar Mishra has received his M. Tech degree in Computer Science from DAVV, Indore, in 1994 and Ph.D. degree in Computer Engineering in 2008. Presently, he has been working as a Professor (CSE) and Director, Microsoft Innovation Center at Sri Aurobindo Institute of Technology, Indore, MP, India. He is also a visiting faculty at IIT-Indore, India. He has 24 years of teaching and 10 years of research experience. He has completed his Ph.D. under the guidance of late Dr. M. Chandwani on "Secure Multi-Party Computation for Preserving Privacy". Dr. Mishra has published more than 90 papers in refereed international/national journals and conferences including IEEE and ACM conferences. He has organized many conferences such as WOCN, CONSEG and CSIBIG in the capacity of conference General Chair and Editor of conference proceedings. He is a Senior Member of IEEE and held many positions such as Chairman, IEEE MP Subsection (2011–2012) and Chairman IEEE Computer Society Bombay Chapter (2009–2010). Dr. Mishra has also served the largest technical and professional association of India, the Computer Society of India (CSI) by holding positions as Chairman, CSI Indore Chapter, State Student Coordinator—Region III MP, Member–Student Research Board and Core Member–CSI IT Excellence Award Committee. He has been recently elected as Chairman CSI Division IV Communication at National Level (2014–2016). Recently, he has been awarded with "Paper Presenter at International Level" by Computer Society of India. He is also the Chairman of Division IV Computer Society of India and Chairman of ACM Chapter covering Rajasthan and MP State.

Dr. Malaya Kumar Nayak is a technology and business solution expert and an academia visionary with over 17 years of experience in leading design, development, and implementation of high-performance technology solutions. He has proven abilities to bring the benefits of IT to solve business issues while delivering applications, infrastructure, costs, and risks. Dr. Nayak has provided strategic direction to senior leadership on technology. He is skilled at building teams of top 1% performers as well as displaying versatility for the strategic, tactical, and management aspects of technology. He has managed a team of professionals noted for integrity and competency. He is a member of Program Committee of Ninth International Conference on Wireless and Optical Communications Networks (WOCN 2012); Program Committee of 3rd International Conference on Reliability, Infocom Technologies and Optimization (ICRITO 2014); Editorial Advisory Board of International Book titled "Issues of Information Communication Technology (ICT) in Education"; Editorial Advisory Board of International Book titled "Global Outlook on Education"; Editorial Board of International Journal titled "Advances in Management"; Editorial Advisory Board of International Journal titled "Delving: Journal of Technology and Engineering Sciences (JTES)"; Editorial Advisory Board of International Journal of Advanced Engineering Technology

(IJAET) and Journal of Engineering Research and Studies (JERS); and Editorial Advisory Board of ISTAR: International Journal of Information and Computing Technology. Dr. Nayak has also served as reviewer in international reputed conferences such as CSI Sixth International Conference on Software Engineering (CONSEG 2012); Tenth International Conference on Wireless and Optical Communications Networks (WOCN 2013); Eleventh International Conference on Wireless and Optical Communications Networks (WOCN 2014); and CSI BIG 2014 International Conference on IT in Business, Industry and Government.

Mr. Amit Joshi is a young entrepreneur and researcher who has completed his graduation (B. Tech.) in information technology and M.Tech. in computer science and engineering and pursuing his research in the areas of cloud computing and cryptography. He has an experience of around 6 years in academic and industry in prestigious organizations of Udaipur and Ahmedabad. Currently, he is working as an Assistant Professor in Department of Information Technology at Sabar Institute in Gujarat. He is an active member of ACM, CSI, AMIE, IACSIT Singapore, IDES, ACEEE, NPA, and many other professional societies. He also holds the post of Honorary Secretary of CSI Udaipur Chapter and Secretary of ACM Udaipur Chapter. He has presented and published more than 30 papers in national and international journals/conferences of IEEE and ACM. He has edited three books on diversified subjects including Advances in Open Source Mobile Technologies, ICT for Integrated Rural Development, and ICT for Competitive Strategies. He has also organized more than 15 national and international conference and workshops including International Conference ICTCS 2014 at Udaipur through ACM–ICPS. For his contribution toward the society, he has been given Appreciation Award by the Institution of Engineers (India), ULC, on the celebration of engineers, September 15, 2014, and by SIG-WNs Computer Society of India on the Occasion of ACCE 2012 on February 11, 2012.

Contributors

Laxmi Ahuja Amity University, Noida, Uttar Pradesh, India

Theyazn H.H. Aldhyani School of Computer Sciences, North Maharashtra University, Jalgaon, India

M. Anusha Department of Computer Science and Engineering, K L University, Vaddeswaram, Andhra Pradesh, India

Anustha Electronics and Communication Engineering, Amity University, Noida, Uttar Pradesh, India

Ankit Arora Mindfire Solutions, Noida, India

Preeti Arora Bhagwan Parshuram Institute of Technology, New Delhi, India

Rajesh Kumar Arora Information Technology Applications Group, National Institute of Industrial Engineering (NITIE), Mumbai, India

Bhushan Atote Computer Department, MAEER's MIT, Pune, India

Sharad Awatade Department of Computer Engineering, Flora Institute of Technology, Pune, India

Pattanaik Balachandra Faculty of Engineering and Technology, Department of Electrical and Computer Engineering, Mettu University, Mettu, Ethiopia

Soham Banerjee Department of Computer Science and Engineering, Amity School of Engineering and Technology, Amity University, Noida, Uttar Pradesh, India

Mangesh Bedekar Computer Department, MAEER's MIT, Pune, India

Santi Kumari Behera Veer Surendra Sai University of Technology, Burla, India

R. Belwal AIT, Haldwani, India

Bharti Suri USICT, Guru Gobind Singh Indraprastha University, Dwarka, New Delhi, India

Jayeeta Chakraborty Computer Engineering Department, National Institute of Technology, Kurukshetra, India

Pankaj Chandre Department of Computer Engineering, Flora Institute of Technology, Pune, India

Ashwani Chaudhary C.B.P. Government Engineering College, New Delhi, India

Himadri Chaudhary GTU PG School, Ahmedabad, Gujarat, India

Sunita Chaudhary Department of Computer Science and Information Technology, Jagannath University, Jaipur, India

Maxwell Christian MCA Department, GLS University, Ahmedabad, Gujarat, India

Bijal N. Dalwadi Department of Information Technology, Birla Vishvakarma Mahavidyalaya, Anand, India

Meenu Dave Department of Computer Science and Information Technology, Jagannath University, Jaipur, India

Pooja Deshmukh Department of Computer Engineering, D.Y. Patil Institute of Engineering & Technology, Pimpri, Pune, India

Jenish Dhanani Sarvajanik College of Engineering and Technology, Surat, India

Nitika Vats Doohan Computer Science Engineering Department, Jaipur National University, Jaipur, India

Ritu Garg National Institute of Technology, Kurukshetra, Haryana, India

Sudhanshu Gautam Department of Computer Science and Engineering, Jaypee Institute of Information Technology, Noida, India

Sakshi Goel Amity School of Engineering and Technology, Amity University, Noida, Uttar Pradesh, India

Rahul S. Goradia Department of Electronics and Communication, G. H. Patel College of Engineering and Technology, Anand, India

Kavita Gupta Maharishi Markandeshwar University, Mullana, India

C.R. Jadhav Department of Computer Engineering, D.Y. Patil Institute of Engineering & Technology, Pimpri, Pune, India

Urvashi Jadon Department of Electronics and Communication Engineering, ITM University, Gwalior, Madhya Pradesh, India

V.K. Jain COER School of Management, Roorkee, India

Rutvij H. Jhaveri Department of Computer Engineering, Shri S'ad Vidya Mandal Institute of Technology, Bharuch, India

Abhimanyu Jindal Department of IT, University Institute of Engineering and Technology, Panjab University, Chandigarh, India

Bansidhar Joshi Department of Computer Science and Engineering, Jaypee Institute of Information Technology, Noida, India

Manish R. Joshi School of Computer Sciences, North Maharashtra University, Jalgaon, India

Bintu Kadhiwala Department of Computer Engineering, Sarvajanik College of Engineering and Technology, Surat, India

Nara Kalyani G. Narayanamma Institute of Technology and Science, Hyderabad, Telangana, India

Manjeet Kantak Computer Science and Engineering Department, Goa College of Engineering, Ponda, Goa, India

Pallavi Khatri Department of Computer Science and Engineering, ITM University, Gwalior, Madhya Pradesh, India

E. Kughan Velammal Engineering College, Chennai, India

Lalit Kulkarni Department of Information Technology, Maharashtra Institute of Technology, Pune, India

Vaishnavi Kulkarni Sandip Institute of Technology and Research Center, Nashik, India

Amit Kumar Central University of South Bihar, Patna, India

Ashish Kumar Department of Mathematics and Statistics, Manipal University Jaipur, Jaipur, Rajasthan, India

Narander Kumar Department of Computer Science, Babasaheb Bhimrao Ambedkar University (A Central University), Lucknow, India

Somayya Madakam Information Technology Applications Group, National Institute of Industrial Engineering (NITIE), Mumbai, India

Saurabh Malgaonkar Department of Computer Science, Whitacre College of Engineering, Texas Tech University, Lubbock, TX, USA

Mani USICT, Guru Gobind Singh Indraprastha University, Dwarka, New Delhi, India

Pattnaik Manjula Faculty of Business and Economics, Department of Accounting & Finance, Mettu University, Mettu, Ethiopia

Manoj Kumar Ambedkar Institute of Advanced Communication Technologies and Research, Geeta Colony, New Delhi, India

Princy Matlani Computer Science and Engineering Department, Guru Ghasidas University, Bilaspur, Chhattisgarh, India

Deepti Mehrotra Amity School of Engineering and Technology, Amity University, Noida, Uttar Pradesh, India

Latika Mehrotra Computer Science Engineering Department, Jaipur National University, Jaipur, India

Sourav Mishra Electronics and Communication Engineering, Amity University, Noida, Uttar Pradesh, India

Vivekanand Mishra Department of Electronics and Communication Engineering, SVNIT, Surat, Gujarat, India

Monika Department of IT, University Institute of Engineering and Technology, Panjab University, Chandigarh, India

Renuka Nagpal Amity School of Engineering and Technology, Amity University, Noida, Uttar Pradesh, India

Hiroshama Nain Department of Electronics and Communication Engineering, ITM University, Gwalior, Madhya Pradesh, India

Lincolin Nhapi Department of Computer Science and Engineering, ITM University, Gwalior, Madhya Pradesh, India

Dhiraj Nitnaware Electronics and Telecommunication Department, Institute of Engineering and Technology, DAVV, Indore, Madhya Pradesh, India

Nishant Painter Department of Computer Engineering, Sarvajanik College of Engineering and Technology, Surat, India

Suja Panicker Computer Department, MAEER's MIT, Pune, India

Devanshi P. Patel Department of Computer Engineering, Shri S'ad Vidya Mandal Institute of Technology, Bharuch, India

Dhrumil M. Patel Department of Computer Engineering, Shri S'ad Vidya Mandal Institute of Technology, Bharuch, India

Kalyani Patel K S School of Business Studies, Ahmedabad, India

Neel N. Patel Department of Computer Engineering, Shri S'ad Vidya Mandal Institute of Technology, Bharuch, India

Sagar M. Patel Department of Information Technology, Charotar University of Science and Technology, Changa, India

Sandip Patel Department of Information Technology, Charotar University of Science and Technology, Changa, India

Prajakta Patrikar Sandip Institute of Technology and Research Center, Nashik, India

Gayatri Phade Sandip Institute of Technology and Research Center, Nashik, India

Fruitwala Pranav Institute of Technology, Nirma University, Ahmedabad, India

P. Prittopaul Velammal Engineering College, Chennai, India

Rashmi Priya TMU, Moradabad, India

Joshua Reginald Pullagura Vignan University Guntur, Guntur, Andhra Pradesh, India

Keyur Rana Sarvajanik College of Engineering and Technology, Surat, India

Kritika Rani Department of Computer Science and Engineering, Jaypee Institute of Information Technology, Noida, India

Prabhat Ranjan Central University of South Bihar, Patna, India

Nidhi Rehani National Institute of Technology, Kurukshetra, Haryana, India

Ravitej Singh Rekhi Electronics and Communication Engineering, Amity University, Noida, Uttar Pradesh, India

Nishant Sahni Computer Engineering Department, Mukesh Patel School of Technology Management & Engineering, NMIMS University, Mumbai, India

Chandan Kumar Sahu Sambalpur University, Sambalpur, India

Monika Saini Department of Mathematics and Statistics, Manipal University Jaipur, Jaipur, Rajasthan, India

Prashant Salunke Sandip Institute of Technology and Research Center, Nashik, India

Amit Sanghi Department of Computer Science and Engineering, Marudhar Engineering College, Bikaner, India

Hari Bhaskar Sankaranarayanan Amadeus Software Labs, Bangalore, India

M.V.S. Santhosh Velammal Engineering College, Chennai, India

Prashant Sahai Saxena Computer Science Engineering Department, Jaipur National University, Jaipur, India

Swati Saxena Department of Computer Science, Babasaheb Bhimrao Ambedkar University (A Central University), Lucknow, India

Prabira Kumar Sethy Sambalpur University, Sambalpur, India

Prerak S. Shah Department of Computer Engineering, Shri S'ad Vidya Mandal Institute of Technology, Bharuch, India

R. Sharath Velammal Engineering College, Chennai, India

Mugdha Sharma Amity University, Noida, Uttar Pradesh, India

Sneha Birendra Tiwari Sharma Computer Science and Engineering Department, Goa College of Engineering, Ponda, Goa, India

Manish Shrivastava Computer Science and Engineering Department, Guru Ghasidas University, Bilaspur, Chhattisgarh, India

Katta Shubhankar Reddy Chaitanya Bharathi Institute of Technology, Hyderabad, Telangana, India

Diksha Shukla Department of Computer Science, Babasaheb Bhimrao Ambedkar University (A Central University), Lucknow, India

Hansraj Sidh Department of Computer Science and Engineering, Marudhar Engineering College, Bikaner, India

Aarti Singh Guru Nanak Girls College, Yamuna Nagar, Haryana, India

E. Sivasankar National Institute of Technology, Tiruchirapalli, Tamilnadu, India

Kailash Srinivasan Computer Engineering Department, Mukesh Patel School of Technology Management & Engineering, NMIMS University, Mumbai, India

Devesh Kumar Srivastava Department of Information Technology, Manipal University Jaipur, Jaipur, Rajasthan, India

Vekariya Subhadra Institute of Technology, Nirma University, Ahmedabad, India

Sanket B. Suthar Department of Information Technology, Charotar University of Science and Technology, Changa, India

Birju Tank GTU PG School, Ahmedabad, Gujarat, India

Rajneesh Tanwar Department of Information Technology, Amity University, Noida, India

Vyas Tarjni Institute of Technology, Nirma University, Ahmedabad, India

Sanjeev Thakur Department of Computer Science and Engineering, Amity School of Engineering and Technology, Amity University, Noida, Uttar Pradesh, India

Siddharth Tripathi Marketing Management Group, National Institute of Industrial Engineering (NITIE), Mumbai, India

Nimish Ukey Department of Information Technology, Maharashtra Institute of Technology, Pune, India

Kapadiya Urvashi Institute of Technology, Nirma University, Ahmedabad, India

N. Usha Rani Department of Computer Engineering, D.Y. Patil Institute of Engineering & Technology, Pimpri, Pune, India

M. Usha Velammal Engineering College, Chennai, India

Naman Vaishnav GTU PG School, Ahmedabad, Gujarat, India

Karan Vala Computer Engineering Department, Mukesh Patel School of Technology Management & Engineering, NMIMS University, Mumbai, India

Srikanth Vemuru Department of Computer Science and Engineering, K L University, Vaddeswaram, Andhra Pradesh, India

Dhulipalla Venkata Rao Narasaraopet Institute of Technology, Guntur, Andhra Pradesh, India

Shanti Verma L J Institute of Computer Applications, Ahmedabad, India

Vijay Verma Computer Engineering Department, National Institute of Technology, Kurukshetra, India

Nagaraj Vernekar Computer Science and Engineering Department, Goa College of Engineering, Ponda, Goa, India

J. Vijaya National Institute of Technology, Tiruchirapalli, Tamilnadu, India

Deepali Virmani Bhagwan Parshuram Institute of Technology, New Delhi, India

Garima Vyas Electronics and Communication Engineering, Amity University, Noida, Uttar Pradesh, India

Arun Kumar Yadav Department of Computer Science and Engineering, ITM University, Gwalior, Madhya Pradesh, India

Saniya Zahoor IT Department, NIT Srinagar, Jammu & Kashmir, India

Internet of Things Applications @ Urban Spaces (Tel Aviv Smart City: A Case Study)

Somayya Madakam, Siddharth Tripathi and Rajesh Kumar Arora

Abstract The Internet of Things (IoT) is one of the evolving concepts in different continents including Europe, USA, Australia, Africa, and Asia. Diverse continents are using this concept for different applications in medical, pharmacy, manufacturing, business, education, mining, entertainment, logistics, forest, agricultural, research, and development sectors to name a few. Some of the studies also show that these IoT technologies are essential in bringing Quality of Life (QoL) to urban citizens. This chapter is based on exploratory study using secondary data. This investigation will explain the Internet of Things phenomena. Research manuscript also inspects on "Tel Aviv Smart City." Moreover, this explores the Internet of Things applications in bringing the operational efficiency of Tel Aviv Smart City. Different IoT applications in city axes and collaborative works are exemplified in crystal clear.

Keywords Smart Cities · Brown field cities · Tel Aviv Smart City
Internet of Things · IoT · Collaboration · Quality of Life · 6As

S. Madakam (✉) · R.K. Arora
Information Technology Applications Group, National Institute
of Industrial Engineering (NITIE), Mumbai, India
e-mail: somu4smart@gmail.com

R.K. Arora
e-mail: arorarajesh84@gmail.com

S. Tripathi
Marketing Management Group, National Institute of Industrial
Engineering (NITIE), Mumbai, India
e-mail: siddharthtripathi.nitie@gmail.com

© Springer Nature Singapore Pte Ltd. 2018   1
D.K. Mishra et al. (eds.), *Information and Communication Technology
for Sustainable Development*, Lecture Notes in Networks and Systems 9,
https://doi.org/10.1007/978-981-10-3932-4_1

1 Introduction

Konrad Zuse is the first person in inventing the first freely programmable computer in the year 1936. That means the computer device usage started way back in an isolated manner to do just some simple mathematical calculations. Later, many computer scientists, researchers, and visionary people used computers for sharing data via ARPANET (The Advanced Research Projects Agency Network) project. ARPANET was one of the world's first operational packet switching networks. Later, as usage increased tremendously in our daily life, industries, manufacturing, education, health care, and business areas, researchers, academicians, and programmers started working more on data communication and networking technologies. Finally, Tim Berners-Lee was the first man leading in the development of the World Wide Web (WWW), HTML, HTTP, and URLs which is nothing but the Internet. Now, it reached to the latest HTML5 in which real-time dynamic data can be floated in the Web site in 24×7 in the form of text, picture, audio, and video. This data may be a structured, semi-structured, and even unstructured format on the Internet. The real-world objects and physical things will also be connected to advanced Future Internet (FI). This is named as the Internet of Things (IoT), in which physical objects can also be connected to the Internet through various technologies for human life. I got surprised, when my neighbor (Ms. Arpana Rai) was talking about the koubachi sensor in which it completely disseminates about her rose plant's health conditions through IoT technologies and applications in her smartphone intimation from a sensor with respect to the requirement of water, humidity, sunny conditions, and pesticide. I have also gone through in one of the blogs about Chinese cities that RFID, GIS, GPS, and sensors are deployed in the school buses and connected to the Internet in order to provide security updates about children to their parents. In some of the megacities like Amsterdam, Masdar cities are trying to use power in efficiently by applying natural renewal energy resource methods and deploying IoT technologies like smart grid and ZigBee to Internet. PlanIT Valley city is using Urban OS for its city operational efficiency.

2 Internet of Things

IoT is the most hyped concept nowadays in this Information Technology (IT) arena. The Internet of Things (IoT) becomes an attractive research topic, in which the real entity in physical world becomes a virtual entity in the cyber world, and both physical and digital entities are enhanced with sensing, processing, and self-adopting capabilities to perform interaction through special addressing scheme [1]. Even in International Telecommunication Union (ITU) report, the authors mentioned that a new dimension has been added to the world of ICTs (Information and Communication Technologies): From anytime, anyplace, connectivity for anyone, we will now have connectivity for anything; connections will multiply and create an entirely new dynamic network of networks which is nothing but an Internet of Things or Internet of Objects. There are a lot of confusions about what does mean by the Internet of Things? Because of no standard definition for the Internet of Things at a global level, defining the Internet of Things is like describing an "elephant" by the six blind people in six different ways. Some of the working definitions on "Internet of Things", contextually various by authors wise.

The Internet of Things (IoT), also called as the Internet of Objects, refers to a wireless network between objects; usually, the network will be wireless and self-configuring, such as household appliances. In the World Summit on Information Summit—2005 held at Tunis by International Telecommunication Union (ITU), the Internet of Things is defined as that by embedding short-range mobile transceivers into a wide array of additional gadgets and everyday items, new forms of communication are enabled between people and things, and between things themselves. Even though the authors [2] say that the Internet of Things assumes that objects have digital functionality and can be identified and tracked automatically, in the Internet of Things, everything real becomes virtual, which means that each person and thing has a locatable, addressable, readable counterpart on the Internet. According to IERC (European Research Cluster on Internet of Things), the Internet of Things (IoT) is nothing but [Anything] can be accessed at [Any time] for [Any service] from [Anywhere] through [Any network] for any service by [Anybody], so we can say that the Internet of Things can be called as 6As. The 6As are shown in Fig. 1.

Fig. 1 Internet of Things (6As)

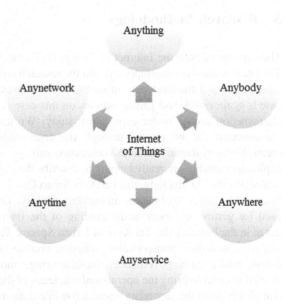

Some people called IoT as the technological God because of the (omnipresent) presence of real objects everywhere including all continents in which the Internet connection + embedded technology is penetrated via wireless, wired, and remote sensing technologies. It is also called as omnipotent in which it has supremacy over handling any kind of situations even in natural disaster situations by ambient intelligence and other allied technologies. Last but not least, it is also to be called as omniscient by knowing everything by cloud computing technologies and taking better intelligent decisions to handle incidents with the help of big data analytic software. There is no field left without the Internet of Things applications including disaster management, transportation, health care, infrastructure, governance, education, infrastructure, home appliances. That's why all the IT corporates, developers, designers, governments, and end users are running behind the IoT technologies in its immense applications and services. By bringing the Internet connectivity to everything, IoT promises a number of benefits to the human being: resource optimization, more accurate sensing of our environment, and cost-efficient tracking of industrial processes. Even many corporates giants, software developing companies started designing, developing, and deploying these IoT technologies in city operational efficiency and for effectiveness. IBM, Cisco, Microsoft, Hitachi, HP, Wipro are too in the same race in the constitution of Smart Cities. Financial institutions, banks, foreign direct investments (FDIs), and urban citizens in that particular city also are coming together in building Smart Cities. They believe that IoT is playing a major role in bringing the Quality of Life in conversion from Brown to Smart City like Tel Aviv. Let us see Tel Aviv Smart City with IoT applications.

3 Research Methodology

The concept of both the Internet of Things (IoT) and Smart Cities are totally new. The phenomenon is contemporary, and the research articles' availability on the both topics increased tremendously in recent years. However, the Smart City Tel Aviv case is again novel, and getting articles on this case is limited. Hence, the research phenomenon comes under exploratory study. We know that exploratory studies phenomenon can be proved through the case study/in-depth interviews/survey methods/observational methods/sometimes mix of some or all of these. The exploratory studies are really helping to describe the object/thing/event/case in detail in a 360° view. In this light, the Tel Aviv Smart City is taken to look into deeply the Internet of Things Applications in building the city. The case study methodology is used for getting the more understanding of the Internet of Things technological usage in the building the Tel Aviv of Urban Spaces. These technologies in different city axes including transportation, tourism, security, environment, business, education, health care, governance, manufacturing, and other city dimensions are studied in order to bring the operational efficiency of the city. So the data is collected completely from the secondary sources (on-line database). The data was in different

Table 1 Research methodology

Research type	Exploratory
Research method	Case study
Data collection	Online surfing
Data types	Text, picture, audio, video
Cases	Single
Keywords	Smart City, Tel Aviv, IoT, Internet of Things
Databases	Ebsco, KNIMBUS, Google Scholar, Google database
Analysis	Thematic/content analysis

formats including research articles (pdf), blogs, white papers, pictures, and videos. In quest of this, the keywords used are "Smart Cities," "Tel Aviv Smart City," "Internet of Things Technologies," "IoT," and "Quality of Life." The databases are Ebsco, KNIMBUS, and Google Scholar and open access articles in line with Google Search Engine. The data is collected from 3/3/2015 to 2/2/2016, i.e., around one year. Most of the observations in the data are that a tremendous work has been done on IoT and Smart Cities, except in Tel Aviv Smart City.

Now, the vital task is that-analysis part in which I did through the old narration method, i.e., thematic narration. This is a method, in which again most of the exploratory studies done because of data is qualitative in nature. Since data is in qualitative and semi-structured, structured, and unstructured formats in the subdivision of text, pictures, audio, and video, content analysis is done. The Web portals, in which Tel Aviv content is displayed, helped me in shaping this article. The limitation of this study is that it is unable to go for the primary data/survey/observation method due to the geographic constraint. However, the exploratory study and case studies admit to the validity of the article via secondary data. In a nutshell, the research methodology used in this article is shown in Table 1.

4 Tel Aviv Smart City

The efforts toward a novelty and an impressive array of technology deployment to city residents that include: citywide Wi-Fi (Wireless Fidelity) access in Tel-Net, location-based smartphone technology to help visitors get around the city, and active measures to engage residents through public round table policy discussions and a collaborative budget made by the Israeli municipality was recognized. This is about the Tel Aviv Smart City in Israel, the winner of the World Smart City Award-2014 under the best Smart City category by beating out almost 250 metropolises around the globe. The municipality of Tel Aviv-Yafo has dozens of nicknames for itself including the start-up city, the white city, and the nonstop city. Now, the city by the sea of technologies deployment for operational efficiency, so called as the world's smartest city [3]. The municipality officials of Tel Aviv city

Fig. 2 Wi-Fi access in public spaces

initially felt that there is a need for strengthening the link between people and city by creating a resident-oriented development. So they focused on identifying the main issues, gaps, and challenges existing in the city and targeted them with new products and solutions. The main thrust areas in which government focused were as follows: (1) social equality, (2) efficient governance, (3) creation of appealing urban environment, and (4) an economic and cultural center for opportunities not only through the technology-led solutions but also changing the approach toward the policies. The city persistently acts to create a climate that facilitates the formation of collaborations between residents, business establishments, third-sector organizations, and the municipality, while the making use of cutting-edge Internet of Things technologies enables learning, creativity, sharing to achieve social and economic prosperity [4] (Fig. 2).

4.1 Brown to Smart City

The Israel Government [5] has allotted financing for research and development on "Smart Cities," including pilot projects in some of the cities primarily of Tel Aviv City for (1) transportation, (2) smart lighting, (3) city energy monitoring and management, (4) wide bandwidth-free Wi-Fi, (5) an information data center, (6) smart grid, (7) smart water cycle, (8) socio-economic academic research, (9) adopting international standards, (10) promoting Smart Cities' audit specifications, and (11) local, as well as imported technologies that are being used to name a few. Tel Aviv-Yafo Smart City included successful planning phase in coordination with the Tel Aviv-Yafo municipal engineering department; a launch of hundreds of bicycles, tailor-made for city-use requiring only a simple regular maintenance; design and implementation of a sophisticated communication system which communicates with databases via a computerized rental terminal in order to identify subscribers and their ongoing activities; unique docking stations including a terminal that can present advertising, controlled from HQ; an efficient call center training for the project's employees. Tel Aviv Smart City considers itself as a leading Internet of Things technology hub, and as such, it has developed advanced many new products and solutions for urban administration and civic engagement for their better life. This Smart City actively involves residential people in the city experience and development. More emphasizing on engagement in city government decision-making processes and wisdom of the crowd as a means for Smart Municipal Corporation in the new age. Digi-Tel Residents' Club and city apps as key projects are personalized Web and mobile communication platforms which provide residents with individually tailored, location-specific information and services. As a part of the Smart City project, only the residents of the specific neighborhood are approached via online, asking to suggest ideas and have an impact on improving the Quality of Life in their neighborhood, for example, where and which trees to plant, where to install sitting benches, whether the sidewalks need to be repaired, water shortage problem intimation, garbage evacuation, and street lights to be mend. At the end of the process, the inhabitants are informed about the verdicts that were made, and now, they can reap the results. The mobile apps will help in intimating the complaint about resolving immediate urban issue. The mobile apps designed in city web site portal are completely Graphical User Interface (GUI) with rich colorful pictures to easily understand for any citizen. Besides, Kiosks for crowd-sourcing for complaints and suggestions of citizen services in the city.

4.2 Collaboration

"Tel Aviv Non-Stop City" project is a massive urban laboratory where more than 200 start-ups collaborate, innovate, and share the vision of this Smart City. As a part of the Tel Aviva Smart City notion and vision, Digital City Program (DCP) is focusing on building an open data infrastructure and tools to enable diversity of new on-line services, upgrading and optimization of existing services, personalization, engagement, sharing and all aspects of message and dialogue between city hall and the people of the city-citizens, business owners, labors, visitors and tourists by TSG, a subsidiary of Ness Group, is a global provider of these services [5]. In Tel Aviv Smart City, M/s. Motorola Company supplies the communication products and services, in which information gathering, data analytics and City Command and Control Centre services. The company is in technological support of critical infrastructure even like ponds, lakes, sea port, roads, forest, which is crucial to creating a safe and secure urban environment. Associating with Tel Aviv municipality, Motorola Company provides the end-to-end project management, multicity dimension integration, implementation, maintenance, and support that delivers the value promised by Smart City Tel Aviv. Israel Government is the home of many advanced smart solutions provider including Parko, Pango, Anagog, Hi-Park, Spaceek, and Polly the Parking Fairy [6, 7]. The Smart Parking solution was a smart combination of LPR systems in an automatic parking system, where the discount option for payment by Tel Aviv residents is based on the LPR input of approximately 300,000 inhabitants in the municipal system database. A feature for special events, concerts, and sporting events was also included whereby after payment on POF (Pay On Foot) machines, the transaction closes via the license plates while exiting the parking, with no need to insert the ticket at the exit. M/s. HTS company is providing Smart Parking solutions. Microsoft CityNext is a people-first approach in Tel Aviv Smart City to innovation that empowers government, businesses, and citizens to shape the future of this city. People-first means harnessing all the ideas, energy, and expertise of a Tel Aviv city's residential people as they create a healthier, safer, more sustainable place to live. Microsoft is uniquely equipped to enable this people-first approach in not only Tel Aviv Smart City but also across the world. Like this, many multinational companies are started their living laboratory centers for the innovative smart solutions for the city.

4.3 IoT Applications

The smart applications of Tel Aviv Smart City includes Social Media (Social media employ mobile and web based technologies to create highly interactive platforms via individuals and communities share, co-create, discuss, and modify user-generated content [8]), Wi-Fi, Intelligent Buildings (Intelligent buildings require integration of a variety of computer-based building automation and control system products that are usually made by different manufacturers. The exchange of information among these devices is critical to the successful operation of the building systems [9]), Open access data through mobiles, smart lighting systems, Smart Parking and mobile Digi-Tel apps are really helping in public services in Tel Aviv (Fig. 3).

4.4 Quality of Life

Tel Aviv city offers visitors and inhabitants a whole world of facilities such as citywide Wi-Fi, with more than 200 Access Points (AP) in 80 zones, counting the beach, boulevards, squares, and public gardens. The service is free for all, and unlike with other free Wi-Fi services, no registration is necessary, thus protecting

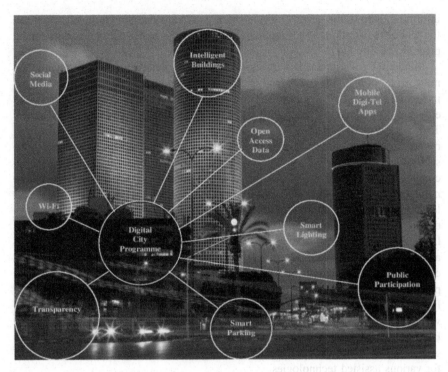

Fig. 3 IoT applications in Tel Aviv Smart City

the user's secrecy. Tel Aviv Smart City provides a unique service named "Digi-Tel" to its citizens allowing them to access public utility services via Wireless Fidelity (Wi-Fi) and information via e-mail, text message, social media, or a personalized web site that can be customized according to location, preferences, marital status, and many more apps. City's Digi-Tel program is a comprehensive, collaborating service for every Tel Aviv tenant 13 years and older who wants to join and enjoy its facilities. Tel Aviv has incorporated various digital resources into its urban ecosystem, including the city's interest and location-based Digi-Tel pass, through which citizens can pay water, electricity, gas, and municipal tax bills, land bills, issuing of date of birth, and death certificates, order parking permits, and send photographs of potholes or broken park and GPS (Global Position System)-based smartphone applications. In order to delight in the offerings of Digi-Tel, city residents need to enroll in program. The system software creates an individual profile with that person's interests, hobbies, and desired services on a priority basis. Then, the central control system sends targeted alerts to the concerned person's smart device/smartphone about promotions, deadlines to apply for particular municipal services, last-minute discounted tickets, coupons to cultural, festival, and sports events, etc. [10]. The platform facilitates a direct and holistic connection between the city and residents, from alerting residents to neighborhood road works or informing them of the nearest bicycle-sharing station to sending targeted reminders for school registration which facilitates access to the many cultural events taking place in the city [11].

Municipal government is very closer to citizens than central governments and more aware of and able to respond to their needs and expectations in (24 × 7) and in 365 days. The city government is trying to provide as much as citizen services efficiently for residents Quality of Life using Internet of Things (IoT) technologies. They not only provide basic public utility services such as good housing facilities, round the clock water and power supply, gas pipelines, and other infrastructures such as education, health, sanitation, roads, and transportation but also they do compulsory technology embeddedness in all the city dimensions. Hence, the municipality government can easy to automate the citizen services from the central location like City Command Control Centre (C4) to automate, monitor, and control the city services. City data center is fully mounted with a big number of computers, laptops, smart mobiles, and other electronic computational device to do urban analytics. The data generated through various citizens for various services from entire Tel Aviv urban and suburban places. The huge amount of data in which the process easily has been done for immediate decision making through Urban Analytics software. Urban Analytics places better information in the hands of citizens as well as government officials to empower people to make more informed choices. Even the IP cameras are installed in all the public places helping the municipality to give complete civic security. The data can be easily captured and stored in the C4 for various assisted technologies.

5 Conclusions

We knew that every city is facing with many urban issues such as traffic jams, hanging to local trains and buses, unhealthy journeys, queues for ticketing, theft in journeys, and inefficient transport infrastructure. Sudden rural migration and natural births also promote high densities. Beside, power shortcuts, non-potable water, applying for gas cylinder and waste dumps in cities, Illegal slums and insufficient housing, and less theme parks, Irresponsible urban citizens & difficulties for city mayors to provide services in round the clock. These are all the issues in which urban people are facing anomalies. However, urban dwellers are seriously looking for better solutions in order to come out of all these matters. So it is the eleventh hour that every urban citizen needs to wake up and collectively work for converting existing cities into Smart City along with city governments. Trying to deploy the IoT for the Smart Urban causes livable place like Tel Aviv Smart City. The mobile apps will help for Smart Life. The study intensifies on the importance of the collaborative work.

References

1. Ning H, Liu H (2012) Cyber-physical-social based security architecture for future internet of things. Adv Internet Things 2(01):1
2. Yong W et al (2006) A survey of security issues in wireless sensor networks. Commun Surv Tutorials IEEE 8:2–23
3. Madhya Pradesh Urban Infrastructure Investment Programme (MPUIIP) (2015) Report on Case Studies of Smart Cities International Benchmark. http://mpurban.gov.in/SmartCity/pdf/ReportonInternationalCaseStudies.pdf. Accessed 2 Dec 2015
4. Sofge M, Sharon Z, Shechter L, Oren H, Uziely G (2014) Tel Aviv Global, Smart City Tel Aviv, World Smart Cities Awards Report (2014). http://webmail.itongadol.com/noticias/val/83072/tel-aviv-wins-world-smart-cities-award.html
5. Bet Hazavdi E Director of department of energy conservation and smart cities. The Israeli governmental roll to promote smart cities in Israel—energy in the cities. http://sabadellsmartcongress.com/wp-content/uploads/Ponencia-Eddie-Bet-Hazavdi.pdf
6. Tel Aviv Non Stop City (2015). http://www.tel-aviv.gov.il/eng/GlobalCity/Documents/SMART%20CITY%20TEL%20AVIV.pdf. Accessed 24 Dec 2015
7. It takes a smart country to design smart cities (2015). http://www.israel21c.org/it-takes-a-smart-country-to-design-smart-cities/. Accessed 31 Dec 2015
8. Kietzmann JH, Hermkens K, McCarthy IP, Silvestre BS (2011) Social media? Get serious! Understanding the functional building blocks of social media. Bus Horiz 54(3):241–251
9. Bushby ST (1997) BACnet TM: a standard communication infrastructure for intelligent buildings. Autom Constr 6(5):529–540
10. Tel Aviv Municipal Corporation–Smart City-Wide Parking Management for the Bene-t of Urban Residents (2015). http://www.htsol.com/Images/uploaded/hts%20case%20study%20tel%aviv%20municipal%20parking.pdf. Accessed 30 Dec 2015
11. Tel Aviv Wins Best City Award (2016). http://www.ubmfuturecities.com/author.asp?section_id=407&doc_id=526900. Accessed 1 Jan 2016

Impact of Consumer Gender on Expenditure Done in Mobile Shopping Using Test of Independence

Shanti Verma and Kalyani Patel

Abstract In India, a number of mobile users have grown exponentially in the last decade. Now people spend more time with smartphones rather than personal meeting. Mobile commerce is the growing area of research nowadays. Most of the companies provide better pricing in mobile applications so they involve customer in mobile shopping. In this paper, the authors conduct an online survey on smartphone users in India. They try to find out that the gender of customer is dependent on the amount spent in mobile shopping. The authors analyze 258 data sets and perform test of intendance both parametric and nonparametric. The result of survey gives p value = 0.373 for parametric test and p value = 0.386 for nonparametric test. These results show that customer gender and expenditure done in mobile shopping are related to each other.

Keywords Mobile shopping · Mobile commerce · Chi-square test
Fisher's exact test

1 Introduction

Data mining (DM) is a technique to find meaningful information from the huge sets of data. This information is helpful to take business decisions. Data mining is also used to predict future patterns of customers' behavior with the help of given sets of data [1]. The various mechanisms of data mining are abstractions, aggregations, summarizations, and characterizations of data [2]. In this paper, we used summarization and transformation of data sets to do the analysis of data. Nowadays, all

S. Verma (✉)
L J Institute of Computer Applications, Ahmedabad, India
e-mail: verma.shanti@gmail.com

K. Patel
K S School of Business Studies, Ahmedabad, India
e-mail: patelkalyani05@gmail.com

© Springer Nature Singapore Pte Ltd. 2018
D.K. Mishra et al. (eds.), *Information and Communication Technology for Sustainable Development*, Lecture Notes in Networks and Systems 9,
https://doi.org/10.1007/978-981-10-3932-4_2

things are available online so there is a need to check how demographics of users effect their online shopping behavior.

Mobile commerce is a new emerging technology with greater scope which gives the ability to do commerce using mobile device [3, 4]. Shopping online via computer or mobile (M-shopping) makes the business process completely different than traditional business process. Products are same but the customer behavior changes as the business process changes. So there is a need to identify the customer judgment, perception, and behavior changes with respect to online or mobile shopping. Traditional predicting approaches are no longer applicable for M-business situations as the use of the Internet is rapidly spreading as an information gateway all around the world. In the current scenario, it is extremely difficult for marketing managers to deal with customer's buying patterns but also there is a need to deal with costumers regularly changing patterns [2]. As information technologies grow exponentially, now companies can store huge sets of data sets that are used to take decision about various offerings and fulfill customers' needs and expectations [5].

To collect primary data sets of customers' demographics and smartphone and mobile shopping usage, the authors used online survey method (Google forms). They collect 335 Indian customers' data sets from which after filtering 258 data sets are used for the study.

This paper is organized as follows: Introduction is provided in Sect. 1, the objective of the study is defined in Sect. 2, the literature review of mobile commerce, customer behavior, and test of independence is discussed in Sects. 3, 4 discusses survey description, Sect. 5 discusses the findings of experiment using R tool, and conclusion is provided in Sect. 6.

2   Objectives of Study

This section outlines the understanding of the mobile commerce and consumer buying behavior of the same as the core objectives of the study. To elaborate upon this theme, the main objectives of the study are identified as to test that customer gender is significantly independent of expenditure done in mobile shopping at once at 95% confidence. This study helps the companies to provide gender-based offers and find gender-centric behavior which helps to increase business profit of companies.

3   Literature Survey

3.1   Related Work: Test of Independence

A chi-square test is a statistical method to assess the goodness of fit between a set of observed values and expected values. There are two varieties of chi-square tests:

goodness of fit and test of independence. In this paper, the author used test of independence to determine whether the observed value (recorded) of one variable depends on the observed value (expected) of a different variable. Various authors used chi-square test to mine different sets of data. Some are as follows:

- Maryam Mahdavian and Fahimeh Mostajeran used the analysis of variance and chi-square test results to study key users skill of ERP system through a compressive skill measurement model [6].
- Badri et al. proposed a model that scrutinizes expected e-learners' intentions to offer e-learning programs efficiently and effectively using chi-square test values and structural equation modeling [7].
- Mr. Li used chi-square test and chi-square table analysis in customer satisfaction and empirical analysis [8].
- Rodney Graeme Duffett tries to find what effect does Facebook advertising have on the cognitive attitudinal component of Generation Y in an emerging country such as SA using ANOVA and chi-square test [9].
- Giannakos et al. performed statistical tests such as correlation coefficient and structural modeling and chi-square test values to investigating teachers' confidence on technological pedagogical and content knowledge: an initial validation of TPACK scales in K-12 computing education context [10].

3.2 Related Work—Mobile Shopping

In the current scenario, data mining can be used in various fields. E-commerce and M-commerce is one of the most upcoming fields where data mining is used. According to report of eMarketer, 2014, mobile phone Internet user growth in APAC by country India had the highest 38.3% growth in 2014. In future perspective, this report states that by 2018, India will have 11.7% growth in mobile phone Internet user. From this report, I conclude that there is highest growth in India toward mobile phone Internet usage and this leads to the use of mobile commerce growth in India.

- Ali Gohary and Kambiz Heidarzadeh Hanzaee performed a study that examines the relationship between Big Five personality traits with shopping motivation variables consisting of compulsive and impulsive buying, hedonic and utilitarian shopping values [11].
- Ceyda Aysuna Turkyilmaza, Sakir Erdema, and Aypar Uslua use factor analysis results for the personality traits and Kaiser-Meyer-Olkin test of sampling adequacy. They use variables personality traits (internal factor) and Web site quality (external factor) to check their effect on online impulse buying behavior [12].
- S. Muthukumar and Dr.N. Muthu prove from the study that India is the second largest cellular market in the world after China, with a massive subscriber base of 867.80 million, as of March 2013. This shows that in India there is a major role of mobile commerce for the growth of Indian economy [13].

- Ioannis Boutsis, Stavroula Karanikolaou, and Vana Kalogeraki presented PRESENT, our middleware that exploits the social behavior of the human crowd to identify group attendance behaviors and predict the next event for a user to attend [14].

4 Description of Survey

In this study, the authors used online survey method (Google forms) and collect data from all over the India.

4.1 Data Selection

- Out of 335 responses after filtering, 258 data sets are used for this study.
- Parameters of study are expenditure done in mobile shopping having five values (1) less than Rs. 1000, (2) Rs. 1000 to Rs. 3000, (3) Rs. 3000 to Rs. 5000, (4) Rs. 5000 to Rs. 10000, and (5) more than 10000 and gender of customer having two values male and female.

4.2 Data Collection

- Primary data collection through questionnaire: The mode of filling this questionnaire is through Google forms. The target audiences are smartphone sets who do mobile shopping.

4.3 Nature of Questionnaire

- A combination of multiple-choice questions was used in the questionnaires depending upon the complexity as well as the objective of the issues involved in the question. Cross-tabulations have been used in the questionnaires in order to simultaneously record the responses across more than one variable/response sets for meaningful analysis of the concerned issues.

4.4 Tool Used for Analysis

  i. Data mining tool "R"
 ii. Techniques: chi-square test and Fisher's exact test

5 Findings of Survey and Discussion

The author applied the chi-square technique to test that customer gender is significantly independent of expenditure done in mobile shopping at once at 95% confidence. Here, independent variable is gender of customer and expenditure done at once in mobile shopping is dependent variable.

5.1 Observed Frequency Table

table(Expenditure, Gender)

Expenditure	Gender	
	Female	Male
Less than Rs.1000	27	43
More than Rs. 10,000	3	8
Rs. 1000 to Rs. 3000	43	61
Rs. 3000 to Rs. 5000	15	26
Rs. 5000 to Rs. 10,000	4	17

The observed frequency table shows the original data obtained from survey conducted. In the above table for each value of expenditure, frequencies are given for male and female.

5.2 Expected Frequency Table

chi = chisq.test(tab)
chi$expected

Expenditure	Gender	
	Female	Male
Less than Rs.1000	26.072874	43.927126
More than Rs. 10,000	4.097166	6.902834
Rs. 1000 to Rs. 3000	38.736842	65.263158
Rs. 3000 to Rs. 5000	15.271255	25.728745
Rs. 5000 to Rs. 10,000	7.821862	13.178138

For the calculation of chi-square value, we have to calculate observed frequencies of each value that is present in observed frequency table. Here, we see that there are some differences between observed and expected frequencies.

5.3   Chi-square Value

chisq.test(tab, correct = T)
  Pearson's Chi-squared test data: tab
  X-squared = 4.2519, df = 4, p-value = 0.373
  Here, we see that at four degrees of freedom, chi-square value is 4.2519 and p value is 0.373. We know that if p value is less than 0.05 for 95% confidence, null hypothesis is accepted. Here, p value is more than 0.05 and so null hypothesis is rejected which clearly shows that customer gender is significantly dependent on expenditure done in mobile shopping at once at 95% confidence.

5.4   Fisher's Exact Test

fisher.test(tab, conf.int = T, conf.level = 0.99)
  Fisher's Exact Test for Count Data   data: tab

$$p - value = 0.3861$$

  alternative hypothesis: two-sided
  For the Fisher's exact test at 99% confidence, we see that p value is 0.386 which is greater than 0.05 and so null hypothesis is rejected.
  In above both test of independence, we see that p value is more than 0.05 which concludes that gender and expenditure are dependent on each other.

6   Conclusion

In the conducted survey, various demographics are taken, for example, age group, salary, family size, gender. In this paper, the author only used one demographic for study, i.e., gender. The results of study show that gender is dependent on the expenditure done in mobile shopping. These results are useful for the M-commerce companies to provide customers' gender-centric offers to increase the profit of their organization. In future, the authors try to test the effect of all demographics of users on the expenditure done in mobile shopping and also check which demographic factor effect is more to expenditure.

References

1. Jacinth Evangeline S, Subramanian KM, Venkatachalam K (2013) Survey on personal mobile commerce pattern mining and prediction. Int J Adv Res Comput Eng Technol (IJARCET) 2 (12):3163–3167
2. Moghadam AD, Jandaghi A, Safavi SO (2015) The probability of predicting e-customer's buying pattern based on personality type 4(1):18781–18785
3. Dhanalakshmi D, KomalaLakshmi J (2014) A survey on data mining research trends. Int J Eng Comput Sci 3(10):8911–8919
4. Ghode MPP (2014) Survey on personal mobile commerce pattern mining and prediction. Int J Technol Res Eng 1(8):538–540
5. Wei GT, Kho S, Husain W, Zainol Z (2015) A study of customer behaviour through web mining. J Inform Sci Comput Technol 2(1):103–107
6. Mahdavian Maryam, Mostajeran Fahimeh (2013) Studying key users' skills of ERP system through a comprehensive skill measurement model. Int J Adv Manuf Technol 69(9-12):1981–1999
7. Badri M et al (2014) Students' intention to take online courses in high school: a structural equation model of causality and determinants. Educ Inf Technol 1–27
8. Yong L (2009) Applications of Chi-square test and contingency table analysis in customer satisfaction and empirical analyses. In: IEEE computer society international conference on innovation management, pp 105–107
9. Duffett RG (2015) The influence of Facebook advertising on cognitive attitudes amid Generation Y. Electron Commer Res 15(2):243–267
10. Giannakos MN et al (2015) Investigating teachers' confidence on technological pedagogical and content knowledge: an initial validation of TPACK scales in K-12 computing education context. J Comput Educ 2(1):43–59
11. Gohary A, Hanzaee KH (2014) Personality traits as predictors of shopping motivations and behaviors: a canonical correlation analysis. Arab Econ Bus J 9(2):166–174
12. Turkyilmaz CA, Erdem S, Uslu A (2015) The effects of personality traits and website quality on online impulse buying. Procedia-Soc Behav Sci 175:98–105
13. Muthukumar S, Muthu N The Indian kaleidoscope: emerging trends in M-commerce
14. Boutsis I, Karanikolaou S, Kalogeraki V (2015) Personalized event recommendations using social networks. In: 2015 16th IEEE international conference on mobile data management (MDM), vol. 1. IEEE
15. Nithya J, Geetha R (2012) Agent-based data mining in mobile commerce: an overview. IJCSET l2(4):1065–1068

Detecting Phishing Websites Using Rule-Based Classification Algorithm: A Comparison

Sudhanshu Gautam, Kritika Rani and Bansidhar Joshi

Abstract In today's time, phishy website detection is one of the important challenges in the field of information security due to the large numbers of online transactions going through over the websites. Website phishing means stealing one's personal information over the Internet such as system backup data, user login credentials, bank account details or other security information. Phishing means creation of phishy or fake websites which look like legitimate ones. In this research paper, we use the associative classification data mining approach that is also named as rule-based classification technique by which we can detect a phishy website and thereby identifying the better detection algorithm which has a higher accuracy detection rate. The algorithms used are Naïve Bayes and PART algorithms of associative classification data mining approach. Moreover, we classify the websites into a legitimate website or a phishy website from the collected datasets of websites. The implementation will be done on the datasets of 1,353 websites which contain phishy sites as well as legitimate sites. At the end, results will show us the higher accuracy detection rate algorithm, which will more correctly identify phishing or legitimate websites.

Keywords Phishy website · Naïve Bayes algorithm · PART algorithm

S. Gautam (✉) · K. Rani · B. Joshi
Department of Computer Science and Engineering, Jaypee
Institute of Information Technology, Noida, India
e-mail: isudhanshugautam@gmail.com

K. Rani
e-mail: kritika.rani17@gmail.com

B. Joshi
e-mail: bansidhar.joshi@jiit.ac.in

© Springer Nature Singapore Pte Ltd. 2018
D.K. Mishra et al. (eds.), *Information and Communication Technology
for Sustainable Development*, Lecture Notes in Networks and Systems 9,
https://doi.org/10.1007/978-981-10-3932-4_3

1   Introduction

Website phishing is an Internet scam in which the user is unknown of the fact that the user is being targeted. A phisher targets the online user rather than the computer and shares his valuable information. Nowadays, website phishing has been the main security concern of online users due to the increased demand of using e-commerce websites. Here, we use associative classification (AC) in data mining which has a higher accuracy detection rate that may effectively detect phishy websites [1]. Phishing attacks are those in which victims are hacked by spoofed emails and fraud websites into giving up their personal information. Phishing will redirect the user to a different website through the user click within the website, email attachments, fake link, instant messages, spyware, etc. Also, the phishy attacker offers illegitimate websites to the user to fill up their personal information. Phisher mainly targets online users over the Internet. Associative classification is a method to define the classes from the trained datasets and classify the new class label in the predefined class or make new class groups. It is to get them trained from given train datasets and based on that gives the prediction results. The rule classification algorithms also called separate and conquer method which means it is an iterative process which means creating a rule subset in the trained datasets [2]. This paper will be followed as related works in Sect. 2, the overall system approach in Sect. 3, implementation and analysis in Sect. 4, results in Sect. 5, conclusion in Sect. 6, future works in Sect. 7 and lastly, the references.

2   Related Works

We have studied that the associative classification (AC) is known to extract classifiers which contain simple "if-then" rules with a high degree of predicting detection accuracy [1]. The algorithms that are studied for associative classification in data mining are PART and Naïve Bayes. These algorithms are based on the rule which works on learning the data and then classified the trained datasets. There are many methods which are used for detecting phishing websites such as black list, data mining approach, and machine learning technique [1, 3]. We surveyed phishing website indicators which can easily identify the phishy websites [1, 3]. We have collected the datasets from UCI machine learning repository, PhishTank as well as Millersmiles [4].

2.1   Website Features for Detecting Phishy Websites

Website features or attributes are used for identifying the websites from phishy or legitimate. This features classify the datasets into some group sets. We had collected a

Fig. 1 Phishing datasets

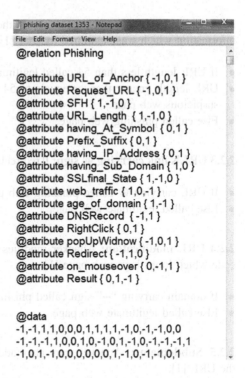

total of 1,353 phishing or legitimate or suspicious website datasets from free community sites [2] for analysing each feature with the classification approach of Naïve Bayes and PART algorithm. After analysing, we have 548 legitimate websites, 702 phishy websites and 103 suspicious websites datasets. We have classified 16 different features which will be used for detecting phishy or legitimate websites [1, 3].

From Fig. 1, the datasets of websites are having binary value with the use of following attributes which carry either phishy or legitimate status. The websites datasets of collected attributes hold categorical values with the numerical values such as "legitimate as 1", "suspicious as 0", and "phishy as −1", respectively.

2.2 Features/Attributes for Analysing the Performance and Their Corresponding Rules

2.2.1 IP address: URL carrying IP address is an indication of someone is trying to get enter in your personal information [1].

- If Internet Protocol address present in URL called phishing web page.
- Else called legitimate web page.

2.2.2 Long URL: An attacker conceals the bug code in URL to relocate the information which was submitted by user [1].

- If URL length less than 54 called legitimate web page.
- URL length greater than or equals to 54 and less than or equals to 75 called suspicious web page.
- Else called phishing web page.

2.2.3 URL's link carrying @ sign: It redirects the web page to another web page [1].

- If URL carrying @ called phishing web page.
- Else called legitimate web page.

2.2.4 URL modification: Attacker modifies the URL by adding some malicious code which looks legit web page [1].

- If domain carrying "−" sign called phishing web page.
- Else called legitimate web page.

2.2.5 Subdomains: Attacker can also add malicious code in subdomain of the URL [1].

- If less than 3 dots present in the domain then we call it a legitimate web page.
- If else equal to 3 than called suspicious web page.
- Else called phishing web page.

2.2.6 Fake HTTPS/SSL: Attacker may use fake HTTPs protocol by which online user thought they are connected with an authenticate website [1].

- If https protocol, trusted issuer, age greater than or equals to 2 years called legitimate web page.
- If else used https and issuer is not trusted called suspicious web page.
- Else called phishing web page.

2.2.7 Someone requesting their URL web page: An attacker modifies the web page with text, images and videos [1].

- If attacker request URL in the percentage of 22 called legitimate web page.
- If else request URL in greater or equals to 22% and less than with 61% called suspicious web page.
- Else called phishy web page.

2.2.8 URL anchor: URL of anchor means when online user type in the address bar and going to the different domain of web page because an attacker already set the malicious code within the web page which is connected to a different domain [1].

- If URL anchors percentage in less than 31% called legitimate web page.
- If else URL anchor greater or equals to and less than or equals to 67% called suspicious web page.
- Else called phishing web page.

2.2.9 SFH: SFH means server form handler. It is the submission of user information which passes the information from web page to a server [1].

- If SFH is about to blank, empty called phishing web page.
- If else SHD redirects to different domain called suspicious web page.
- Else called legitimate web page.

2.2.10 Abnormal URL: The record of a websites is not found in the database of WHOIS site [1].

- If no hostname in record database called phishing web page.
- Else called legitimate web page.

2.2.11 Pop-up window: Authenticate websites do not come up with the pop-up window [1].

- If right click disabled called phishing web page.
- If else right click showing alert sign called suspicious web page.
- Else called legitimate web page.

2.2.12 Redirect page: An online user tries to click on a link which automatically redirects to a phishing page [1].

- If redirect web page carrying #s greater than or equals to 1 called legitimate web page.
- If else redirect web page #s greater and less than 4 called suspicious web page.
- Else called phishing web page.

2.2.13 DNS record: A website not having a DNS record in a database comes into a phishing web page [1].

- If no DNS record called phishing web page.
- Else called legitimate web page.

2.2.14 Hiding the links: An attacker hides the genuine link with displaying a fraud link [1].

- If change of status bar on mouse over called phishing web page.
- If else no change called suspicious web page.
- Else called legitimate web page.

2.2.15 Website traffic: Legit websites have the high traffic as user visit regularly [1].

- If website traffic less than 1,50,000 called legitimate web page.
- If else website traffic greater than 1,50,000 called suspicious web page.
- Else called phishing web page.

2.2.16 Age of domain: Websites carrying online presence of below 1 year could be called as suspicious [1].

- If website age comes less than or equals to 6 months called legitimate web page.
- Else called phishing web page.

3 Overall System Approach

The proposed methodology is applying the Naïve Bayes and PART algorithms to classify the phishy, legitimate and suspicious websites based on phishing indicator attributes. We have used class association rules that means searching for a missing relation among the feature sets value and the group sets in the trained data and converted them as class association rule [1]. We have designed the processes for solving the phishing website detection using the rule-based classifier as shown in Table 1.

3.1 Naïve Bayes and PART Algorithms

Associative classification can effectively detect phishy websites with higher accuracy. The algorithms used here for associative classification in data mining are Naïve Bayes and PART algorithm. In PART algorithm, rules are generated in Weka data mining tool on the training dataset and classification is implemented in Java. And the other algorithm used is Naïve Bayes algorithm which will be implementing in Java.

Table 1 System design for identifying phishing websites with higher accuracy rate algorithm

1 Training datasets acquisition
Legitimate website data
Phishy website data
2 Rules discovery
Hidden correlations among the attribute values and classes are found (using CARs)
3 Applying the classification algorithm
Naïve Bayes and PART algorithm
4 Class assignment
Classifier is tested on test data
Class assigned to the website is tested according to the rules of classifier
5 Results
It will classify the higher accuracy detection rate algorithm

3.2 Naïve Bayes Classification Algorithm

A Naïve Bayes classifier follows the probabilistic classifier with the use of Bayes' rules [5]. In Naïve Bayes algorithm, we have used conditional probability and Bayes' rules. Bayesian classification, which predicts the categorical class value, classifies the dataset from the trained datasets and the class labels assigned from given attribute sets and uses it in classifying new data. We aimed to construct a rule which will allow us to assign future objects to a new class, given only the vectors of features describing the next objects sets. It is the scheme process which does not require any iterative parameter estimation where it can be used for identifying true-positive instances in huge datasets [6]. In this classifier, we used the probability rule.

3.2.1 Understanding Naïve Bayes Rules

Assume that we are having the samples of data X where class label is unknown. We make some hypothesis H in which X belongs to some class C. Then, the probability of hypothesis P (H|X) carries the data which observed in sample data X. The probability of posterior P (X|H) would be the H which is having the condition on X.

From Bayes' rule, we are having posterior probability.

$$\text{Posterior}\, P(H/X) = \frac{\text{Likelihood}(X/H) * \text{Prior}\, P(H)}{\text{Evidence}\, P(X)}$$

Therefore, we can calculate the posterior probability with the help of likelihood of datasets, prior knowledge of datasets and evidence of datasets. Similarly, we set the phishing datasets into trained and test data.

3.2.2 Pros of Naïve Bayes Algorithm

- It works on two phases:
  - Learning phase and test phase.
- It is a very easy to program, fast to train and very useful to use as a classifier, text classification and spam filtering.

3.3 PART Classifier Algorithm

PART stands for Projective Adaptive Resonance Theory. The concept of PART algorithm is to generate our own condition which will classify and gives the prediction output. PART classifier generates rules from training dataset and applies these rules on the testing dataset in the form of "if-then-else" statements to classify the class label of predictive class [2]. PART algorithm works on the generated own rules, and then based on that rules, it is assigned a class to the test data. Here, we have generated the many condition by taking the feature value in if-then-else format which makes the class of groups and classified the data into test.

3.3.1 Pros of PART Algorithm

- PART generates own rule condition and based on that rule a class is assigned to test the data.
- PART rules are formed from the trained datasets and these rules are applied on test data in the form of if-then-else statement to classify the class label of predictive class.
- Accuracy or free error rate classifier.

4 Implementation and Analysis

4.1 Approach Followed

We have used NetBeans software to implement the logic of Naïve Bayes and PART algorithms in Java language. Then, we collect the feature value and coded with both the classifier algorithms. We have connected the database for uploading and get read from the train datasets by using MySQL. First, we run the NetBeans software and used MySQL for the connection of datasets. We have divided our datasets in two parts, first trained data and test data. We have used the data of 1353 websites

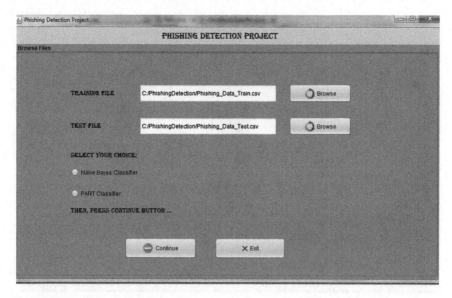

Fig. 2 Display after running the project

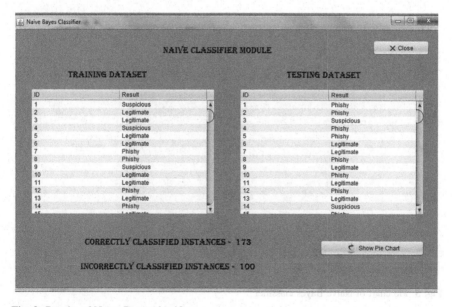

Fig. 3 Results of Naïve Bayes classifier

from which 1080 are used to train the data and 273 for testing the data. Hence, we load the trained data file and test data file. Afterwards, we can select any one classifier algorithm between Naïve Bayes and PART algorithm for evaluating the test results (Fig. 2).

Now, we will perform the classifier algorithms. First, we select Naïve Bayes classifier and thus have the following output (Fig. 3).

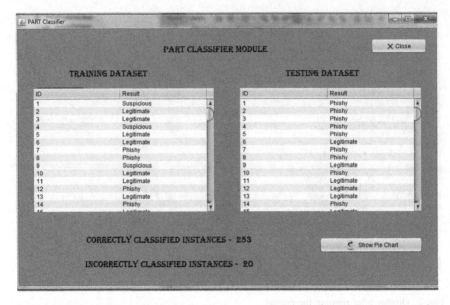

Fig. 4 Results of PART classifier

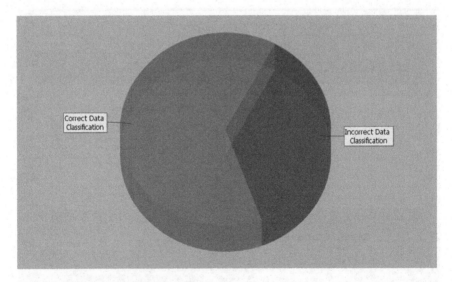

Fig. 5 Pie chart of Naïve Bayes classifier

Secondly, we select PART classifier and have the following output (Fig. 4).

Therefore, the classification algorithms have been implemented. We have the Naïve Bayes classifier which has been classified 173 correctly and 100 incorrectly identified instances. Similarly, PART classifier has been classified 253 correctly and 20 incorrectly identified instances. So, the PART algorithm is having the higher

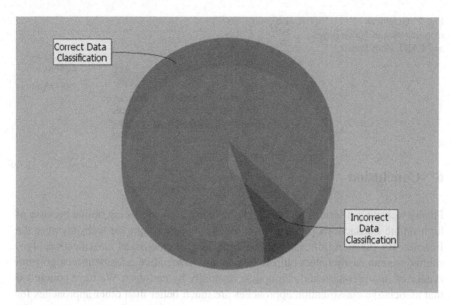

Fig. 6 Pie chart of PART classifier

Table 2 Results of Naïve Bayes and PART algorithm

Rule-based classification algorithm	Correctly classified instances	Incorrectly classified instances	% accuracy of correctly identified instances
Naïve Bayes algorithm	173	100	63.36
PART algorithm	253	20	92.67

accuracy detection rate in identifying the true-positive instances. Moreover, we can also see the pie chart of both the algorithms in Figs. 5 and 6.

5 Results

The result depends upon the accuracy and error rate of classifier. It shows the accuracy of detection in Naïve Bayes algorithm which holds the 63.36% accuracy in correctly identified instances and PART holds the 92.67% accuracy in correctly identified instances.

From Table 2 and Fig. 7, we have found that the accuracy of correctly identified instances is the PART algorithm which has the higher accuracy detection rate than the Naïve Bayes algorithm.

Fig. 7 Graphical
representation of Naïve Bayes
and PART algorithms

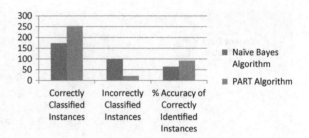

6 Conclusion

Phishy website is a serious problem that the many users are facing online because of high online transactions performed on a daily basis through the websites over the Internet. By given significant attributes and use of rule-based classification algorithms, we can easily detect phishing websites. Associative classification algorithm is ease of interpretation and can update manually by end-user. We have concluded that associative classification approaches are much better than other approaches for detecting phishing instances because of their simple rule transformation. It finds the missing information from the classified group sets which help us to minimize the true-negative scenarios. Naïve Bayes works on the probability theory and gives the prediction results. And the PART algorithm classified better results due to its own generated rule conditions. Therefore, from our result analysis, we have found that PART algorithm has a higher accuracy detection rate than the Naïve Bayes algorithm.

7 Future Works

Till now, the accuracy of used algorithm is not up to the mark. So, the work would be extended with the cons of used algorithms for enhancing these algorithms. Also, we would like to increase the feature value and perform the analysis with other classifier algorithms for more highly true-positive detection rate.

References

1. Abdelhamid N, Ayesh A, Thabtah F (2014) Phishing detection based associative classification data mining. Expert Syst Appl 41. Elsevier
2. Mahajan A, Ganpati A (2014) Performance evaluation of rule based classification algorithms. Int J Ad Res Comput Eng Technol (IJARCET)
3. Taalohi M, Langari N, Tabatabaee H (2015) Identifying phishing websites by techniques hyper heuristic and machine learning. ISSN Sci Int. Lahore

4. Datasets of phishing and legitimate websites from the sites as Phishtank, Millersmiles, and UCI machine learning repository site. https://archive.ics.uci.edu/ml/datasets.html
5. Naïve Bayes algorithm. https://en.wikipedia.org/wiki/Naive_Bayes_classifier
6. Wu X, Kumar V, Quinlan JR, Ghosh J, Yang Q, Motoda H, McLachlan GJ, Ng A, Liu B, Yu SY, Zhou ZH, Steinbach M, Hand DJ, Steinberg D (2007) Top 10 Algorithm in data mining. Springer Verlag London Limited published

Mobile Applications Usability Parameters: Taking an Insight View

Sakshi Goel, Renuka Nagpal and Deepti Mehrotra

Abstract Nowadays, mobiles are like the general purpose computers with inbuilt sensors, constant access to Internet and a huge variety of applications. Different applications are categorized in such a way that they can perform their task in the best possible manner. Usability of mobile applications is the ability of an individual to use the application for its intended purpose without getting frustrated. In this paper, the attention draws on the major usability factors of different applications. After finding out the factors, we are trying to give the brief introduction of various methodologies used to rank the factors and the structural relationships among these parameters are modeled. Major techniques among them are interpretive structural modeling (ISM) approach, analytical hierarchal approach (AHP) and DEMETAL (decision-making trial and evaluation technique). These methodologies are used to identify parameters affecting mobile applications, and the structural relationships between these parameters are modeled.

Keywords Usability · Mobile applications · Parameters · ISM
DEMATAL · AHP · TOPSIS

1 Introduction

The versatility that is emerging nowadays in mobile phones opens the doors for many new opportunities in the mobile world [1]. Mobile devices are becoming like a blessing to the users, and today millions of users are using it without any hindrance. The advancement that is seen in today's scenario with respect to mobile technology enabled a huge range of applications used by the population while they move [2]. Developers sometimes pay less attention toward the fact that the users are more interested in using these devices while they are moving. The key concern of

S. Goel (✉) · R. Nagpal · D. Mehrotra
Amity School of Engineering and Technology, Amity University,
Noida, Uttar Pradesh, India
e-mail: goel.sakshi.aries@gmail.com

© Springer Nature Singapore Pte Ltd. 2018
D.K. Mishra et al. (eds.), *Information and Communication Technology
for Sustainable Development*, Lecture Notes in Networks and Systems 9,
https://doi.org/10.1007/978-981-10-3932-4_4

using mobile phones is its limited screen size, huge power consumption and limited connectivity that lacks it behind when compared with desktops [3]. Major among all is the context in which we are using them. After all mobile devices are new PC nowadays, consumers are rapidly shifting toward smartphones and tablets instead of those bulky computers to access the wide range of services and products [4]. In 2010, it was the first time that smartphones are sold much higher than the PC's.

Mobile applications are software applications that are specifically designed to be used on the small, computing devices, wireless computing devices rather than laptops or desktops [5]. Mobile applications are categorized in three categories [6].

- *Native apps*—these applications are created particularly for a designed platform.
- *Web based*—these applications are dependent on the Web. They need Internet access every time in order to use them.
- *Hybrid apps*—these applications will combine the features of both the Web-based and native applications.

There are different applications in the smartphones that influence the life of an individual in either way. Different applications have different significance, and when we talk about usability, there are many factors that need to be considered at the time of defining its usability.

In this paper, we identified the factors for the usability of different applications. Apart from this, in this work the main emphasis draws on the studies carried out in decision making of the various usability factors. We try to make the user experience wonderful while using the mobile applications. In this regard, we identify the major usability factors of various applications and rank them in a decreasing order. Section 2 describes the related work that has been carried out in the field of usability. Section 3 of the paper describes the various applications and the usability factors associated with it and gives the brief idea of the techniques we can use to rank them and categorize them efficiently. Section 4 concludes the paper.

2   Related Work

Han et al. (2001), Kwahk and Han (2002) make use of usability evaluation framework consisted of two layers: usability formation and evaluation of usability [7]. Nigel Bevan et al. define the study for measuring the usability as a part of user-centered design process [8]. Rachel Harisson et al. define the PACMAD (people at the center of mobile application development) model of usability. This methodology brings different attributes of different usability factors together to develop a extraordinary comprehensive model [9]. The usability of the mobile applications can be measured by three criteria. They are efficiency, effectiveness and satisfaction. Some other attributes like cognitive load are overlooked, even though they are very prominent for the success or failure of a particular application. Cognitive load is the amount of cognitive processing needed by a particular user to

Fig. 1 PACMAD usability
evaluation framework

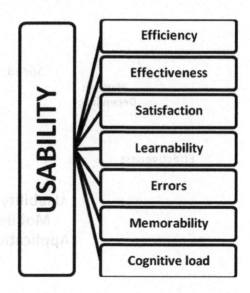

see the applications [10]. In order to overcome with this difficulty, PACMAD (people at the center of mobile development application) was introduced. The PACMAD gives a model that adds four attributes other than the three attributes, i.e., learnability, errors, cognitive load and memorability. In order to overcome the present model, PACMAD depends upon three factors. They are task, user and context of use [11] (Fig. 1).

Scott Gerber [12] gives the usability considerations that are highly responsible for the proper functionality of the mobile applications [13]. Fateh nayebi et al. present the state of art for the evaluation and measurement of the mobile usability applications [14]. Constantinos K. Coursaris et al. design a framework for the evaluation of the usability. A meta-analytical review is conducted of so many usability studies to draw the final conclusion [15].

3 Proposed Work

We have identified the mobile applications namely e-commerce applications, gaming applications, social applications, banking applications, books and references and news applications. We draw a table for the factors that are valuable for the usability of these applications. However, there are some common usability parameters that can be suited with any of the above applications. Basically the study shows the categorization of different usability applications along with their usability factors that are very basic while using the mobile applications. They are shown in Fig. 2 (Table 1).

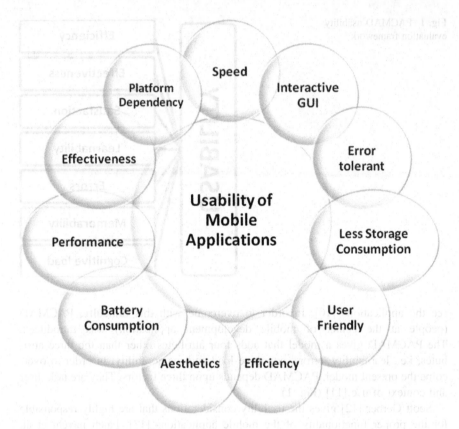

Speed

Platform
Dependency

Interactive
GUI

Effectiveness

Error
tolerant

Usability of
Mobile
Applications

Performance

Less Storage
Consumption

Battery
Consumption

User
Friendly

Aesthetics   Efficiency

Fig. 2 Usability factors of mobile applications

Proposed technology

In order to make the decision between the choices available for the attributes of e-commerce applications and other proposed applications, various techniques can be used. They are shown as follows

3.1 MCDM (Multiple Criteria Decision Making)

Multiple criteria decision making (MCDM) is a decision-making technique when their present multiple but conflicting choices. This technique is used in day to day life [16]. MCDM deals with choosing and identifying values on the basis of preferences of the person who is taking decisions. Making the decisions shows that some alternative choices are also present and the decision maker is considering them and in such situation our responsibility is to not only identifying the number of choices present but to choose the best one among all which can fits with the

Table 1 Usability parameters of different mobile applications

S. no	Gaming applications	Social applications	Banking applications	News application	Books and references	E-commerce applications
1	Varying screen sizes	High data rate	Security	High readability on every operating system	Adjustable font size	Hardware configuration
2	Swiping, tilting and replacement options	Content	E-transfer of money	Audio and video options along with text display	Less scrolling	Platform dependency
3	Battery consumption	Battery consumption.	Login credentials at every login	Quick updates	Interactive GUI	Secure payment gateway
4	Size of RAM	Security	Smooth backend	The pop ups that appear while browsing the news must be avoided	Efficiency	Network connectivity
5	User setting option (volume, language control, sound)	Platform dependency for new features	Ease of use	Low space consumption	Esthetics	Ease of use
6	Saved functionalities	Notifications at the lock screen	Voice recognition system	News must be available in all languages	Easily convertible in any format	Content
7	Level of game must be specified	Offline features	ATM locators application	GUI must be interactive	Efficient	Response time
8	Avoid hang problem	Interactive GUI	Accessing balance info without login	Content	High resolution of images and tables	Informational retrieval performance
9	Platform supportive	Efficiency	Network connectivity	Smooth navigation	Quick searching while entering keywords	Navigation
10	High resolution	Strong esthetics	Ease of use	Zooming without disturbing actual functionality	Consume less size in memory when downloaded	Optimization

model [17]. This is the methodology that comes under operational research which is categorized into two methods, namely multi-attribute decision making (MADM) and multi-object decision making (MODM) [18].

- *Multiple attribute decision making (MADM)*—it selects the "best" alternative from the predescribed alternatives with respect to multiple attributes.
- *Multiple objective decision making (MODM)*—it deals with the designing of the alternatives which actually optimize the multiple objectives of the person who is making the decisions.

3.2 AHP (Analytical Hierarchy Process)

It stands for analytical hierarchy process. It is a decision-making method that gives ratio scales of different parameters. In order to take input, it can be measurable (height, weight) or subjective (feeling, preference, satisfaction) opinion. The ratio scales are derived from Eigen values, and parameters are derived from Eigen vectors. AHP works on the human mentality who by nature clusters the things in their mind by their complexities and characteristics. It took both quantitative and qualitative factors into considerations [19]. The foremost importance of this method is its ease to solve multiple attribute problems. Moreover, AHP is quite easy to understand and deals with qualitative and quantitative data at a time. One reason of ease is the less involvement of mathematics. It only performs the pairwise comparison, generation of vectors and synthesis [20]. It breaks the problem into smaller and smaller parts and guides the decision maker with the help of pairwise comparison to give the relative intensity or the relative strength of the elements in the hierarchy.

It is that simple that there is no need of providing the formal training and they can understand and take participation actively. Satty [21] found one common behavior among various examples of the trend of dealing with complexity by humans—that is the hierarchical complexity structuring into the homogeneous clusters.

3.3 MAUT (Multi-Attribute Utility Theory)

MAUT is again a decision-making technique. It is a structured methodology that is build to handle the variations among various objectives. This technology was first introduced at Mexico airport in early 1970s to find the alternative locations for new airport at Mexico City [22]. MAUT is a compensatory strategy. This theory states that the preference of an individual between the alternative solutions for a particular problem can be expressed in terms of "utility function" which allots numbers to show the degree of desirability [23]. Multiple attributes are compared on the basis of their weights to find the best optimal solution [24]. The desirability is expressed such that the high number correlates with the higher desirability and the lower number with the low desirability [25].

3.4 Interpretive Structural Modeling (ISM)

It is a process of converting the poorly articulated model into a well-defined model that is helpful for many purposes [26]. The foremost focus of using ISM is to identify the directly and indirectly related elements. It identifies relationship among various sets of items that defines a particular problem. While using a system that is complex in nature, the user gets frustrated and does not want to spend much time on that system [27]. ISM provides the better understanding of the system by identifying the directly or indirectly related elements and to identify the structure within the system. It also changes the poorly defined attributes into the set of well-defined attributes. The very first step of ISM is to identify the variables. After choosing the contextually related elements, structural self-interaction matrix (SSIM) is developed. After finding the transitivity the levels are identified and finally with the help of MICMAC analysis, dependent are driving powers are identified [28].

3.5 TOPSIS (Technique for Order Preference by Similarity to Ideal Solution)

It is a technique that works with multi-attributes or with MCDM (multi-criteria decision making) problems. It provides the ease to the decision maker to manage the problem in the way that it will be solved and carry out the analysis and ranking of the different attributes after comparing them. Traditionally TOPSIS was introduced by yoon and hwang for solving MCDM. The concept behind this is that the alternatives so chosen must have shortest Euclidian distance form PIS (positive ideal solution) and farthest from NIS (negative ideal solution). Positive ideal solution is the solution that has minimum cost criteria and maximum benefit criteria. On the contrary, negative ideal solution has maximum cost criteria and minimum benefit criteria [29].

3.6 DEMATEL

Decision-making trial and evaluation laboratory (DEMATEL) is used in analytical network process (ANP), fuzzy set method and multi-criteria decision making (MCDM), etc. for enhancing these old methodologies into some new kind of applications for many hybrid methods. DEMATEL can sum up with many other techniques such as initial direct relation matrix. DEMATEL was come into light with the prior belief that the correct use of scientific research method will enhance the understanding of certain problems that are critical in nature. DEMATEL is applied to handle problems with regard to some crucial features for the problems and help in finding the best possible decisions. Some scientists use this method to change the application of the attributes and evaluation for the problems. DEMA-TEL determines the constraining and interdependent relations depend on some features.

4 Conclusion

In this work, we basically make an attempt to introduce the different mobile applications usability factors that are commonly used and after that we suggest some of the techniques that are helpful in drawing the decision and creating relationship among various factors. Among the different applications, we choose e-commerce as one of the applications where we will be applying interpretive structural modeling (ISM) along with the DEMATEL in our future work. With the parameters suggested above, we will try to create the relationships among the parameters and create a matrix on that basis.

References

1. Hardy R, Rukzio E (2009) Exploring expressive NFC-based mobile phone interaction with large dynamic displays. In: First international workshop on near field communication
2. Brodt A (2012) A mobile data management architecture for interoperability of resource and context data. In: 12th IEEE (international conference on mobile data management
3. Forman G, Zahorjan J (1994) The challenges of mobile computing. IEEE Comput
4. Hazarika P (2014) Recommendations for webview based mobile applications on android. In: IEEE international conference on advanced communication control and computing technologies (ICACCCT)
5. Constantinos K (2012) Coursaris: a meta-analytical review of empirical mobile usability studies. JUS J 6(3):117–171
6. Definition of mobile applications. https://www.nngroup.com/articles/mobile-native-apps/
7. Kwahka J, Han SH (2002) A methodology for evaluating the usability of audiovisual consumer electronic products. Elsevier science
8. Bevan N, Curson I Methods for measuring usability
9. Harrison R, Flood D, Duce D (2013) Usability of mobile applications: literature review an d rationale for a new usability model. J Interact Sci
10. Saleh A, Isamil RB, Fabil NB (2015) Extension of PACMAD model for usability evaluation metrics using goal question metrics (gqm) approach. J Theor Appl Inf Technol 79 (1):1992–8645
11. Patel N, Dalal P (2013) Usability evaluation of mobile applications. Int J Eng Res Technol (IJERT) 2(11):2278–018

12. Saleh AM, Ismail RB Usability evaluation frameworks of mobile application: a mini-systematic literature review
13. Scott gerber. http://thenextweb.com/apps/2015/08/28/10-usability-considerations-mobile-app/#gref
14. Nayebi F, Desharnais JM, Abran A The state of the art of mobile application usability evaluation
15. Coursaris CK, Kim DJ (2011) A meta-analytical review of empirical mobile usability studies. J Usability Stud 6(3):117–171
16. Xu L, Yang JB (2011) Introduction to multi-criteria decision making and the evidential reasoning approach
17. Gavade RK (2012) Multi-criteria decision making: an overview of different selection problems and methods: (IJCSIT). Int J Comput Sci Inf Technol 5(4):5643–5646
18. Triantaphyllou E, Shu B, Sanchez SN, Ray T (1998) Multi-criteria decision making: an operations research approach. In: Webster JG (ed) Encyclopedia of electrical and electronics engineering, vol 15. Wiley, New York, NY, pp 175–186
19. Arabameri A (2014) Application of the analytic hierarchy process (AHP) for locating fire stations: case study Maku city. Merit Res J Art, Soc Sci Humanit 2(1):001–010, ISSN 2350–2258
20. Abu-Sarhan Z (2011) Application of analytic hierarchy process (AHP) in the evaluation and selection of an information system reengineering projects. IJCSNS Int J Comput Sci Netw Secur 11(1)
21. Saaty TL (2008) Decision making with the analytic hierarchy process: Int J Serv Sci 1(1)
22. Definition of multi attribute utility theory. http://www.hsor.org/what_is_or.cfm?name=mutli-attribute_utility_theory
23. Definition of MAUT. http://www.prismleadership.com/300/MAUT.html
24. Guerrero-Baena MD, Gómez-Limón JA, Fruet Cardozo JV Are multi-criteria decision making techniques useful for solving corporate finance problems? pp 60–79
25. Kabassi K, Virvou M (2014) Multi-attribute utility theory and adaptive techniques for intelligent web-based educational software: Instr Sci 34(2):313–158
26. Sushil Interpreting the interpretive structural model. Global J Flex Syst Manage 0972–2696
27. George JP, Pramod VR (2014) An interpretive structural model (ISM) analysis approach in steel rerolling mills (SRRMs). IMPACT: Int J Res Eng Technol (IMPACT: IJRET) 2:161–174, ISSN 2321–8843
28. Jayalakshmi B (2014) Interpretive structural modeling of the inhibitors of wireless control system in industry. In: International conference on industrial engineering and operations management, Bali, Indonesia, January 7–9
29. Panda BN, Biswal BB, Deepak BBLV (2014) Integrated AHP and fuzzy TOPSIS approach for the selection of a rapid prototyping process under multi-criteria perspective. In: 5th international & 26th all india manufacturing technology, design and research conference (AIMTDR 2014) IIT Guwahati, Assam, India, December (2014)

Segmentation of Musculoskeletal Tissues with Minimal Human Intervention

Sourav Mishra, Ravitej Singh Rekhi, Anustha and Garima Vyas

Abstract Non-invasive methods of detection of diseases is very important in the medical domain. Imaging modalities such as MRI are usually employed and present the state-of-the-art. As of now, it is very widely used in the prognosis of heart diseases where tissue distribution is taken into account. This work exhibits multi-modal MRI to enable segmenting tissues in limb, which happens to be a crucial first step in analysis.

Keywords Musculoskeletal MRI · Segmentation · Region growing
Tissue classification

1 Introduction

Systematic segmentation allows for well-ordered post-processing of the acquired images. The quality of diagnosis has a high correlation with the ability to identify and measure tissue attributes. Recently, limb tissue segmentation in medical images has gained traction in the quest to find determinants of cardiovascular and geriatric diseases [1]. Pollack et al. have studied reactive hyperemia in the case of peripheral

This project has been partially founded by NIH grants 5R37AG018915-12 and 5P30AG021332-10 awarded to Wake Forest University School of Medicine.

S. Mishra (✉) · R.S. Rekhi · Anustha · G. Vyas
Electronics and Communication Engineering, Amity University, Noida, Uttar Pradesh, India
e-mail: smishra23@amity.edu

R.S. Rekhi
e-mail: ravitejsingh11@gmail.com

Anustha
e-mail: anushtha212@gmail.com

G. Vyas
e-mail: gvyas@amity.edu

© Springer Nature Singapore Pte Ltd. 2018
D.K. Mishra et al. (eds.), *Information and Communication Technology for Sustainable Development*, Lecture Notes in Networks and Systems 9,
https://doi.org/10.1007/978-981-10-3932-4_5

45

arterial diseases and found circulation to be lacking in arms [2]. Kitzman et al. have been trying to utilize a similar methodology in calf region for studying diastolic heart failure in the elderly population [3, 4]. In achieving this objective, an automated technique is relevant to quickly segment the images with minimal human intervention.

This paper exhibits a technique where calf region segmentation has been achieved by multi-modal MRI images. Information from several modalities have been used in tandem to delineate regions.

2 Segmentation of Limb: An Overview

A quantitative image can provide us with a region of interest (ROI) by mapping certain attributes. However, there are challenges since the intricacies of our body lead to various complications in the smallest of regions. Limb segmentation for clinical studies has been previously attempted by other research groups such as Karampinos et al., who investigated this pathway to investigate effects of Type-2 Diabetes [5]. A similar process has been used in the detection of weakness of muscles in respiratory disease patients by Andrews et al. [6]. The status quo involves some degree of human intervention in the processing anatomical datasets.

3 Acquisition and Preprocessing

The following steps were employed for isolating the information of interest.

3.1 Choice of Image Type

The semi-automated technique employs two major class of images, namely T2-weighted and spectral fitted fat & water fraction maps.

In T2-weighted images, the areas with water content appear bright whereas the areas with high fat content appear dark. All ^{1}H atoms (hydrogen atoms) have a fundamental property of nuclear spin. A characteristic resonant frequency called Larmor frequency exists which is related to the strength of the main magnetic field. Under such a field, the spins precess with resolvable horizontal and vertical components. Radiofrequency excitations, which disturb the spin equilibrium, makes it easy for researchers to distinguish tissues, since different materials have different rates of relaxation. T2-weighted technique exploits the vast difference of relaxation rates in the transverse plane of the spins. The question of fat and water maps employs this fundamental property of difference in chemistry and differential rates of spin evolution. Equipped with this know-how, we can design pulse sequences which will clearly separate the two tissue species.

3.2 Data Acquisition

A 3T MRI (Siemens Skyra^{TM}, Malvern, PA) system was used to acquire the required images of the subjects. Subjects were made to lie feet-first on the scanner. A 16-channel knee-coil was placed around the part of the calf where the girth was highest.

T2-weighted images were acquired with a long scan, where the Echo time (TE) and Repetition time (TR) were kept at 51 ms and 2120 ms, respectively. Field of view (FOV) was kept at 512 × 512 with 14 slices. The pixel dimensions were set at 0.5 mm × 0.5 mm × 2.4 mm for the X-Y-Z axes. A representative image of a T2 image from our dataset is given in Fig. 1.

The Fat and Water images have been obtained by the Spectral fitting technique wherein, 6 echoes were captured for different phases of water and $-CH_2$ (methylene) spin overlaps with TE and TR of 1.4 and 12 ms respectively, at 4° flip angle. By suitable polynomial fitting of signals at different spin-overlaps, Fat-only and Water-only images were extracted. A representative in-Phase image has been shown in Fig. 2.

The Fat and Water images are registered to a common space by the high-resolution T2 images. Using the common T2-space greatly simplifies the process of segmentation. An open source tool such as the FSL (Department of Neurobiology, Oxford University) has been used for 3D slice-by-slice registration.

Fig. 1 T2 image

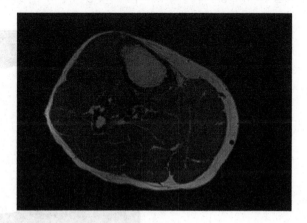

Fig. 2 In-phase image from
spectral estimation technique

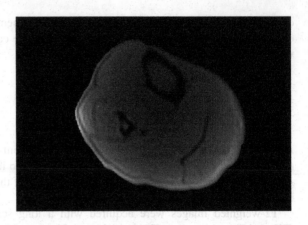

4 Methodology

4.1 Identification of Tissues

The limb analyzed is the human calf region. The axial slice of the limb exhibits four
main regions- the subcutaneous fat, the skeletal muscle, the bone cortex of the two
bones tibia and fibula, along with their respective marrows.

4.2 Tissue and Skin Mask

Using a segmentation script, a tissue mask was created out of the registered in-phase
image as illustrated in Fig. 3. This helps in retaining relevant information and nul-
ling the background for resampled images, as conveyed through Figs. 4 and 5.

Fig. 3 Tissue mask (*from T2 image*)

Fig. 4 Water fraction map

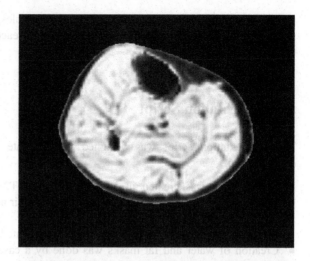

Fig. 5 Fat image, with background subtraction

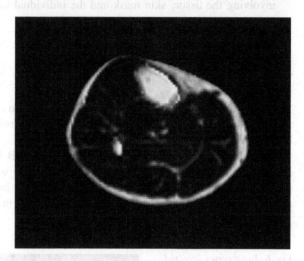

Tissue mask is eroded by morphological operations to produce skin mask- an outline of the tissue mask beneficial in registering perfusion images. This removes any bleed-through artifact.

4.3 Isolating Individual Component Tissues

Otsu's threshold method involves iterating through all the possible threshold values and calculating a measure of spread for the pixel levels each side of the threshold.

Due to the different chemical composition of molecules, fat fraction (Fig. 4) and water fraction maps (Fig. 5) are complementary to each other making the Otsu threshold easier to establish.

5 Segmentation Script

The segmentation script was designed to segregate the different regions as demonstrated.

- Tissue mask was created in MATLAB out of T2-weighted high-resolution image and in-phase image of water and fat after their normalization. Any gaps or holes were filled by morphological operations.
- By eroding the tissue mask, a skin mask was created.
- Creation of water and fat masks was done by a cascade of logical operations involving the tissue, skin mask and the individual fat and water fractions.

5.1 Uniqueness of Bone Cortex

The bone cortex is the only tissue which is non-uniform in shape & dimensions, and varies in location among human subjects. A region growing technique creates the bone-cortex by appropriate marker placing.

This output can be fed directly into the MATLAB based pipeline, for further accurate delineation. A representative output which was obtained from region-growing bone-cortex by seed-placing has been shown in Fig. 6. The display labels are finally assigned and saved as a file automatically in the parent folder.

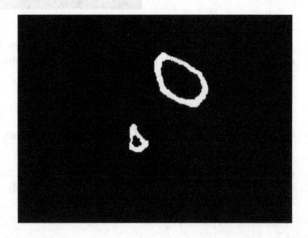

Fig. 6 Bone cortex (*created from region growing*)

6   Results and Discussion

A total of 27 calf MRI datasets were processed by this technique. A few representative results from the middle slices of the datasets have been provided in Figs. 7, 8 and 9. The generated segmentation overlay of the datasets can be efficiently viewed on a third-party tool like ITK-SNAP. This semi-automated technique has proved to be effective in segmenting the calf region with minimal human intervention.

This tool is crucial in cardiovascular research where absence of adequate tissue maps have led clinical assistants to manually paint different regions in limb slices. These segmented regions are very crucial to further calculating region ratios and statistics, which are quoted widely nowadays in literatures pertaining to Heart failure, Peripheral arterial diseases (PAD), and obesity research [7].

Fig. 7 Labels of sample dataset I

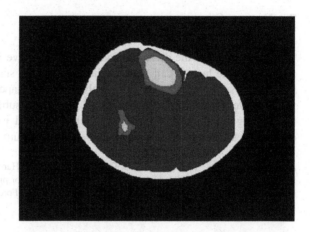

Fig. 8 Labels of sample dataset II

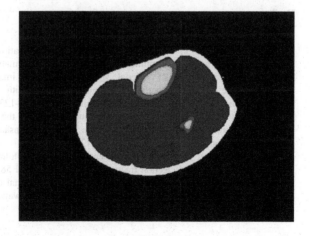

Fig. 9 Labels of sample
dataset III

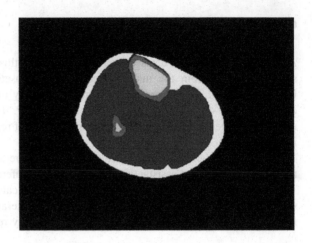

7 Future Scope

The future scope of the given method includes effective identification of the fibula
and tibia in the human calf region. The shape of the smaller bone- Fibula, varies
from being smooth to becoming roughly diamond-shaped at its ends. Its location
varies between human subjects and hence, it has been subjected to position-marking
in the program flow. Automation of identifying and isolating this sub-part will
enable quick processing of several datasets in minimum time.

Acknowledgements The authors thank Dr. RA Kraft, Dr. CA Hamilton of Virginia Tech and Dr.
Kitzman, MD of Wake Forest University School of Medicine for providing insights and feedback
into this solution. The authors also express gratitude to Wake Forest University School of Medicine for kindly allowing to use the MRI datasets.

References

1. Hernández-Vela A et al (2012) Human limb segmentation in depth maps based on
   spatio-temporal graph-cuts optimization. J Ambient Intell Smart Environ 4(6):535–546
2. Pollak AW et al (2012) Arterial spin labeling MR imaging reproducibly measures
   peak-exercise calf muscle perfusion: a study in patients with peripheral arterial disease and
   healthy volunteers. JACC: Cardiovasc Imaging 5(12):1224–1230
3. Haykowsky MJ et al (2012) Effect of endurance training on the determinants of peak exercise
   oxygen consumption in elderly patients with stable compensated heart failure and preserved
   ejection fraction. J Am Coll Cardiol 60(2):120–128
4. Haykowsky MJ et al (2011) Determinants of exercise intolerance in elderly heart failure
   patients with preserved ejection fraction. J Am Coll Cardiol 58(3):265–274
5. Karampinos DC et al (2012) Characterization of the regional distribution of skeletal muscle
   adipose tissue in type 2 diabetes using chemical shift-based water/fat separation. J Magn Reson
   Imaging 35(4):899–907

6. Andrews S et al (2011) Probabilistic multi-shape segmentation of knee extensor and flexor muscles. In: Medical image computing and computer-assisted intervention–MICCAI 2011. Springer, pp 651–658
7. Kitzman DW et al (2001) Importance of heart failure with preserved systolic function in patients ≥65 years of age. Am J Cardiol 87(4):413–419
8. Kittler J, Illingworth J (1985) On threshold selection using clustering criteria. IEEE Trans Syst Man Cybern 5:652–655

Implementation of Modified ID3 Algorithm

Latika Mehrotra, Prashant Sahai Saxena and Nitika Vats Doohan

Abstract Data classification algorithms are very important in real world applications like- intrusion classification, heart disease prediction, cancer prediction etc. This paper presents a novel decision tree based technique for data classification. Basically it is an enhanced variant of ID3 algorithm. ID3 is a popular and common decision tree based technique for data classification. in this paper, an upgraded version of ID3 is proposed. This version calculates information gain in a different way by giving more weightage to more important attribute instead of an attribute which is having more different values. The fundamentals of data classification are also discussed in brief. The experimental results have proven that the accuracy of the presented method is better.

Keywords IDS · NIDS · HIDS · SBIDS · Data mining

1 Introduction

Because of large volumes of security audit data as well as complex and dynamic properties of intrusion behaviors, the optimization of the performance of IDS becomes an important open problem that is receiving more and more attention from the research community. Uncertainty to explore if certain algorithms perform better for certain attack classes constitutes the motivation for the reported herein (Fig. 1).

L. Mehrotra (✉) · P.S. Saxena · N.V. Doohan
Computer Science Engineering Department, Jaipur National University, Jaipur, India
e-mail: latika19mehrotra@gmail.com

P.S. Saxena
e-mail: sahai.prashant@gmail.com

N.V. Doohan
e-mail: nitika.doohan@gmail.com

© Springer Nature Singapore Pte Ltd. 2018   55
D.K. Mishra et al. (eds.), *Information and Communication Technology for Sustainable Development*, Lecture Notes in Networks and Systems 9, https://doi.org/10.1007/978-981-10-3932-4_6

Fig. 1 Traditional IDS framework

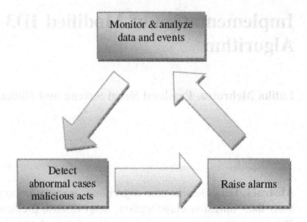

The Data mining-based intrusion detection systems (IDSs) have established precision in finding intrusions; at the same time provide simplified detection of new or unknown type of intruders. These IDS performs better and durably even in mutable environment. A major problem faced by them is the intensive computation required in the model generation phase.

Classification based on data collection strategies for any intrusion detection systems

(a) Signature based Intrusion Detection System
(b) Statistics based Intrusion Detection System

The intrusion detection systems [1–3] are based on either signature based techniques or the statistical based techniques. The signature based techniques make use of the training data or the signature to detect & prevent the intrusion. Therefore the signature based techniques are not good enough to detect the novel intrusion attacks. The statistical based techniques have an advantage over the signature based techniques that they can also detect the novel attacks. One most common method for classification is decision tree based classification. Main disadvantage of the network based intrusion detection [4] they work well only for local threats. The host based intrusion detection [4] runs on the local host.

2 Related Works

In [5] Jake Ryan et al. has worked on the concept of the neural network. The neural network performs learning on the basis of the test data and then it performs predictions. It can classify that the behavior of the node as normal or abnormal. Denning D.E. et al. [6, 7] has presented a sequential rule based model for the prediction of abnormal behavior. Sequential rules are based on the sequential data base. The training data is stored in the sequence in which may occur in the sequence data set & then the sequence rule mining algorithm is applied on the training data set to identify the patterns of the normal behavior and the abnormal behavior. The system developed in [8] has more accuracy in identifying whether the records are normal or attack one. Dewan M. et al. [9] proposed an improved version of the self-adaptive Bayesian algorithm (ISABA). It is based on the concept of the Bayesian network but accuracy rate is below expectation. S. Sathyabama et al. [10] Proposed a clustering based method. In this method the similarity based records are stored in the clusters. Also the dissimilar records are called the outliers. For the outliers the alarm is raised & the record is checked for the abnormal behavior. Amir Azimi Alasti et al. [11] proposed a self-organizing map based method for the intrusion detection. The map has been used successfully to classify the data records. In this method the false positive alerts have been reduced up to a good extent. Alan Bivens et al. [12] has proposed a self-organizing map based classification technique. The false negative has been reduced in this model. The authors of [13, 14] proposed the ensemble approach. This approach is a fusion of many existing algorithm. The experimental results have shown that it has outperformed many existing techniques. It has also outperformed support vector machine. This paper [15] presents a method which is based on the concept of the dimension reduction. The dimension reduction is achieved by the feature extraction technique of the data mining. Aly Ei-Senary et al. [16] has proposed a fusion of the apriori and the kuok algorithm. This proposed model also uses the concept of the fuzzy set i.e. partial membership function.

3   Proposed Work

Objective

- The objective is to classify the information of a flow available as normal or attack by an updated decision tree based classifier.
- The accuracy of the proposed decision tree based classifier will be better as compared to that of the existing classification techniques.

Proposed Methodology
Input:

1. Training Data
2. Testing Data

Output:

1. Classification Tree

Procedure:

1. If the input data set is empty then return a single node with value "unsuccessful".
2. If every record of the input data set contains similar value for the target element then create a node containing that value and return.
3. Otherwise if all the records of the training data set contains no then create a no node and stop.
4. Calculate the modified information gain of all the attributes using the following modified gain calculation formulae:
   The gain (d, a) is mutual information of sample data set d on an element a is defined as follows:

$$\text{Modified gain}(d, a) = (\text{entropy}(d) - s((|da|/|d|) * \text{entropy}(da))) * v$$

   Where:
   v is the weight associated with each element according to the importance in result da $\subseteq$ d and a $\in$ a.

$$|da| = \forall da$$

   select an element (discussed in element selection) with the highest mutual information and construct a decision node.
5. Repeat step 4 for each element.

4 Result Analysis

The data set used in the experimental study is kdd 99 data set. we have used following attribute for the classification.

- Packets sent
- Protocol
- Source Port
- Destination Port
- Target

Training Data Set:

The Training data set is as follows: after implementing the updated version of existing ID3 it is being found that ID3 is performing better in updated version. We implemented the existing and new ID3 on JAVA platform. The experimental study has proven that the accuracy of the proposed ID3 is better as compared to the existing one (Table 1 and Fig. 2).

Table 1 The training data set

Packets	Protocol	Sport	Dport	Target
2138	tcp	34	33	Ddos
12	tcp	2	3	Probe
20	tcp	3	12	Probe
230	tcp	2	120	Ddos
2138	tcp	34	33	Ddos
2138	tcp	34	33	Ddos
2138	tcp	34	33	Ddos
12	tcp	2	3	Probe
12	tcp	2	3	Probe
12	tcp	2	3	Probe
20	tcp	3	12	Probe
20	tcp	3	12	Probe
20	tcp	3	12	Probe
20	tcp	3	12	Probe
2138	tcp	34	33	Ddos
12	tcp	2	3	Probe
20	tcp	3	12	Probe
230	tcp	2	120	Ddos
2138	tcp	34	33	Ddos
2138	tcp	34	33	Ddos

Fig. 2 Result comparison

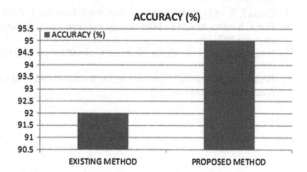

ACCURACY (%)

5   Conclusion

In this paper, a large number of papers are analyzed. On the basis of that analysis the problem is defined. It is also containing background and related work. Which describes why intrusion detection is needed? An enhanced data classification technique is also proposed. This version calculates information gain in a different way by giving more weightage to more important attribute instead of an attribute which is having more different values. The accuracy of proposed classifier is better.

References

1. Breiman L, Friedman JH, Olshen RA, Stone CJ Classification and regression trees. Wadsworth international
2. Quinlan JR (1986) Induction of decision trees. Machine learning, pp 81–106
3. Quinlan JR (1987) Simplifying decision trees. Int J Man-Mach Stud 27:221–234
4. Vijendra Singh (2011) Efficient clustering for high dimensional data: subspace based clustering and density based clustering. Inf Technol J 10(6):1092–1105
5. Langley P (1993) Induction of recursive bayesian classifiers. In: Brazdil PB (ed) Machine Learning: ECML-93. Springer, Berlin, New York/Tokyo, pp 153–164
6. Witten I, Frank E (2005) Data mining: practical machine learning tools and techniques, 2nd edn. Morgan Kaufmann, San Francisco, ch. 3, 4, pp 45–100
7. Yang Y, Webb G (2003) On why discretization works for Naive-Bayes classifiers. Lecture Notes in Computer Science, pp 440–452
8. Zantema H, Bodlaender HL (2000) Finding small equivalent decision trees is hard. Int J Found Comput Sci 11(2):343–354
9. Ming H, Wenying N, Xu L (2009) An improved decision tree classification algorithm based on ID3 and the application in score analysis. Software technology institute, Dalian Jiao Tong University, Dalian, China, June 2009
10. Chai R, Wang M (2010) A more efficient classification scheme for ID3. School of Electronic & Information Engineering, Liaoning University of Technology, Huludao, China, Version1, pp 329–345
11. Yuxun L, Niuniu X (2010) Improved ID3 algorithm. College of Information Science & Engineering, Henan University of Technology, Zhengzhou, China, pp 465–573
12. Jin C (2009) "Luo De-lin and Mu Fen-xiang" An improved ID3 decision tree algorithm. School of Information Science & Technology, Xiamen University, Xiamen, China, pp 127–134
13. Gama J, Brazdil P (1999) Linear tree. Intell Data Anal 3(1):1–22
14. Han J, Kamber M (2006) Data mining: concepts and techniques 2nd edn. Morgan Kaufmann, ch-3, pp 102–130
15. Singh RJ (2013) A survey of modern classification techniques. Int J Sci Res Dev 1(2): 209–211
16. Breast cancer statistics from centers for disease control and prevention. http://www.cdc.gov/cancer/breast/statistics/

A Data Classification Model: For Effective Classification of Intrusion in an Intrusion Detection System Based on Decision Tree Learning Algorithm

Latika Mehrotra, Prashant Sahai Saxena and Nitika Vats Doohan

Abstract Data classification is the heart favorite topic of many researchers. It has a huge array of real-world applications. Although many algorithms and tools are available for creating decision tree based classification, still improvements are required in many aspects. ID3 is a very popular decision tree-based data classification algorithm. A novel model is presented here with decision tree concepts for the data classification. Model that is suggested in this paper is based on the updated ID3 method. It uses a modified gain to select the attribute. This modified gain gives more weightage to most important attribute. The result analysis has shown that the accuracy of proposed model is better.

Keywords Data classification · Decision tree learning · KD3 data set
Modified gain · Data mining · Intrusion detection

1 Introduction

Data mining as we all know is popular and acclaimed technique for finding patterns and displaying remarkable associations between items in massive databases. The aim of data mining is to find out built in rules extracted from databases by applying different techniques. Based on the conception of robust rules, [1] introduced association rules for locating regularities between merchandise in large-scale dealing knowledge noted by point of sale (POS) systems in supermarket's.

L. Mehrotra (✉) · P.S. Saxena · N.V. Doohan
Computer Science Engineering Department, Jaipur National University, Jaipur, India
e-mail: latika19mehrotra@gmail.com

P.S. Saxena
e-mail: sahai.prashant@gmail.com

N.V. Doohan
e-mail: nitika.doohan@gmail.com

© Springer Nature Singapore Pte Ltd. 2018
D.K. Mishra et al. (eds.), *Information and Communication Technology
for Sustainable Development*, Lecture Notes in Networks and Systems 9,
https://doi.org/10.1007/978-981-10-3932-4_7

Pattern discovered in context of sales and marketing in a supermarket with the help of data mining techniques can give these kinds of results. If a customer is purchasing bread, he might buy butter as well. This kind of knowledge is often considered as the foundation for decisions making related to marketing activities. These rules or patterns of purchase may help in promotional evaluation or product placements. Apart from market basket-based analysis, association rules are used these days in several application areas, Web usage mining and intrusion detection that is network security. Association rule learning unlike sequence mining does not take into account order of thing, neither inside nor outside a particular transaction (Fig. 1).

Data Mining Tasks:

Data mining is used for extracting patterns as per user need from a transaction database. In general, there are four ways to find out these patterns:

- Classification
- Clustering
- Association rule mining
- Sequential pattern mining (Fig. 2).

In Jake Ryan et al. [2] have worked on the concept of the neural network. The neural network performs learning on the basis of the test data, and then, it performs predictions. It can classify the behavior of the node as normal or abnormal. Denning et al. [3, 4] have presented a sequential rule-based model for the prediction of abnormal behavior. Sequential rules are based on the sequential database. The training data are stored in the sequence in which they occur in the sequence data set, and then, the sequence rule mining algorithm is applied on the training data set to identify the patterns of the normal behavior and the abnormal behavior. The system developed in [5] has more accuracy in identifying whether the records are normal or

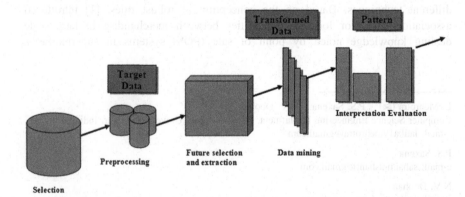

Fig. 1 General concept of data mining

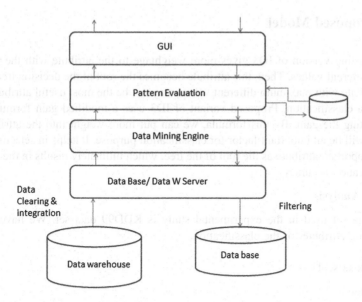

Fig. 2 KDD knowledge discovery in databases

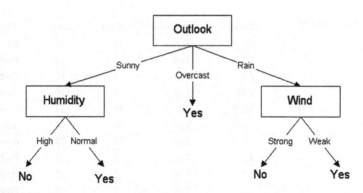

Fig. 3 A sample decision tree—partial view

attack one. Dewan et al. [6] proposed an improved version of the self-adaptive Bayesian algorithm (ISABA). It is based on the concept of the Bayesian network, but accuracy rate is below expectation.

Decision Tree and Classification: Decision trees are a common data classification process. Decision trees split the information for the data classification. Classification is the process of finding the different classes of the input data set on the basis of the common characteristics that the input data possess [7] (Fig. 3).

2 Proposed Model

The existing version of ID3 gives more weightage to the attribute with the maximum different values. Then, this attribute becomes the root of the decision tree. But an attribute with maximum different values may not be the most useful attribute for the data classification. Proposed variant of ID3 uses a modified gain formula for calculating the gain. By this formula, we can put more weight into the attributes, which will be an important factor for classification purpose. It helps in selecting the most important attribute as the root of the tree, which ultimately results in the better classification accuracy.

Result Analysis

The data set used in the experimental study is KDD99 data set. We have used following attribute for the classification.

- Packets sent
- Protocol
- Source port

Table 1 Input data set

Packets	Protocol	Sport	dport	Target
2138	Tcp	34	33	ddos
12	Tcp	2	3	Probe
20	Tcp	3	12	Probe
230	Tcp	2	120	ddos
2138	Tcp	34	33	ddos
2138	Tcp	34	33	ddos
2138	Tcp	34	33	ddos
12	Tcp	2	3	Probe
12	Tcp	2	3	Probe
12	Tcp	2	3	Probe
20	Tcp	3	12	Probe
20	Tcp	3	12	Probe
20	Tcp	3	12	Probe
20	Tcp	3	12	Probe
2138	Tcp	34	33	ddos
12	Tcp	2	3	Probe
20	Tcp	3	12	Probe
230	Tcp	2	120	ddos
2138	Tcp	34	33	ddos
2138	Tcp	34	33	ddos

Fig. 4 Result comparison

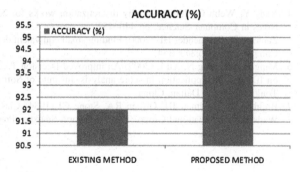

- Destination port
- Target

Training Data Set:

The Training data set is as follows (Table 1):

We implemented the existing ID3 and the modified ID3 on JAVA platform. The experimental study has proved that the accuracy of the proposed ID3 is better as compared to the existing one (Fig. 4).

3 Conclusion

In modern era of technology, the need of data and computation is increasing continuously. Each and every hand is mounted with the new generation gadgets and smart devices, which is increasing the data exponentially. Data classification makes the things easier. It categorizes the voluminous data. Also it increases the accuracy of the result. This paper has presented a data classification model for intrusion detection system by decision tree learning. This proposed model calculates information gain in a different way by giving more weightage to more important attribute instead of an attribute which is having more different values. The accuracy of proposed classifier is better.

References

1. Vijendra S (2011) Efficient clustering for high dimensional data: subspace based clustering and density based clustering. Inf Technol J 10(6):1092–1105
2. Langley P (1993) Induction of recursive bayesian classifiers. In: Brazdil PB (ed) Machine learning: ECML, vol 93. Springer, Berlin, pp 153–164
3. Witten I, Frank E (2005) Data mining: practical machine learning tools and techniques, 2nd edn. Morgan Kaufmann, San Francisco, ch. 3, 4, pp 45–100

4. Yang Y, Webb G (2003) On why discretization works for Naive-Bayes classifiers. Lecture Notes in Computer Science, pp 440–452
5. Zantema H, Bodlaender HL (2000) Finding small equivalent decision trees is hard. Int J Found Comput Sci 11(2):343–354
6. Ming H, Wenying N, Xu L (2009) An improved Decision Tree classification algorithm based on ID3 and the application in score analysis. Institute of Software Technologies, Dalian Jiao Tong University, Dalian, China
7. Breiman L, Friedman JH, Olshen RA, Stone CJ (1984) Classification and regression trees. Wadsworth International Group. The Wadsworth Statistics/Probability Series, Belmont, CA

Reliability-Aware Green Scheduling in Cloud Computing

Nidhi Rehani and Ritu Garg

Abstract In cloud computing scenario, workflow scheduling algorithms require multiple conflicting goals to be optimized. Optimal makespan, reduced energy consumption and reliability of execution are the most important goals to be optimized. In this paper, we propose a multi-objective workflow scheduling algorithm in cloud computing—ERAWS, which optimizes three conflicting criteria: makespan, reliability of task execution and energy consumption. We validate and analyze the performance of our algorithm by using the CloudSim toolkit. We use randomly generated task graphs and task graphs for Gaussian elimination and fast Fourier transformation to represent workflow applications. The simulation results show that ERAWS algorithm gains significantly in terms of makespan and energy consumption, in real-world scenarios where reliability and energy consumption are important issues.

Keywords Cloud computing · Reliability · Multi-objective workflow scheduling · Energy efficiency · Monte Carlo simulation · Green computing

1 Introduction

Cloud computing, the new ICT paradigm which offers infrastructure, platform and software as a utility service [1], has great significance for performing high-performance computing (HPC). The capacity of vast resources available in cloud can be utilized efficiently by using an appropriate scheduling algorithm for

N. Rehani (✉) · R. Garg
National Institute of Technology, Kurukshetra, Haryana, India
e-mail: nidhirehani@gmail.com

R. Garg
e-mail: ritu.59@gmail.com

© Springer Nature Singapore Pte Ltd. 2018
D.K. Mishra et al. (eds.), *Information and Communication Technology for Sustainable Development*, Lecture Notes in Networks and Systems 9, https://doi.org/10.1007/978-981-10-3932-4_8

service execution, such that the desired performance is achieved. It is also important to consider the huge amount of energy consumed by the vast data centers which leads to an increase in the data center operation cost, increase in the carbon-dioxide emissions to the environment, decrease in the reliability and lifetime of system components. Dynamic voltage and frequency scaling (DVFS) technique [2], which allows the dynamic adjustment of voltage and frequency of computing elements, can be appropriately used to reduce the energy consumption by servers. Reliability of service execution is another important issue in cloud. The failure of computing and networking resources available and the corresponding services executing on them can result in degradation of performance. Such failures become an important issue in cloud due to the heterogeneity of service requests, virtualization and the on-demand allocation of servers. However, work on providing reliable and energy-efficient schedule with performance constraints is still to be explored.

Scientific workflows are widely used to represent large applications with HPC requirements. The workflow scheduling problem on heterogeneous distributed systems is considered to be NP-Hard [3] in nature.

Garraghan et al. [4] analyzed the Google Cloud Trace Log to determine that failure characteristics of cloud servers follow Weibull distribution, whereas large variance is found for repair characteristics. The existing literature focuses on the use of Poisson distribution to estimate resource failure [5, 6]. In our work, we use *Weibull distribution with Monte Carlo Simulation* (MCS) [7] to ensure reliable task execution.

To incorporate energy efficiency, we use the DVFS technique to dynamically scale the voltage and frequency of the CPUs allocated to various virtual machines.

Thus, we propose energy-efficient and reliability-aware workflow scheduling (ERAWS) algorithm that optimizes the makespan (schedule length), reliability and energy consumption for workflow execution in cloud. In order to analyze the performance of our proposed ERAWS algorithm, we develop a cloud environment using CloudSim toolkit [8]. We use randomly generated task graphs and task graphs for real-world numerical problems like Gaussian elimination (GE) [9] and fast Fourier transformation (FFT) [10] to represent workflow applications.

2 Formalism

The cloud *user* submits the request as a workflow application to the *Cloud Coordinator (CC)* which passes it to the *Scheduler*. The *Scheduler* is responsible for producing an optimal workflow schedule. It works in coordination with *Reliability Predictor*, which provides the failure characteristics for virtual machines and the *DVFS Manager,* which provides the energy usage characteristics for the resources. The schedule generated by the *Scheduler* is used by *CC* to allocate tasks to various virtual machines, with the help of *DVFS Manager* to appropriately scale the voltage and frequency levels for the machines. The set of all virtual machines in the cloud is represented as $v_i \in V$, $(1 \le i \le m)$, where m is the total number of machines. Each virtual machine can adjust its frequency in levels, expressed in millions of instructions per second (MIPS) [11]. These frequency levels for $v_j \in V$ can be represented as $F_j = \{f_{j,1}, f_{j,2}, \ldots f_{j,l}\}$, where l is the number of frequency levels, $f_{j,b} < f_{j,c}$, *if* $b < c$.

The power consumed by a processor is static and dynamic power, dominated by the dynamic power consumption [12] which is directly proportional to the square of supply voltage (V_{dd}) and frequency (f). Since frequency is directly proportional to supply voltage $(f \propto V_{dd})$ [11], we calculate power consumption of a virtual machine as:

$$P = C(f)^3 \tag{1}$$

where C is the coefficient of proportionality. A decrease in the frequency of operation for the virtual machine results in an increase in the execution time of the corresponding task executing on it.

The workflow application is represented as a directed acyclic graph (DAG) $W = \langle T, E \rangle$ where $t_i \in T$, $(1 \le i \le n)$ represents the set of n precedence-constrained tasks and $E = e_{i,j}$, $(1 \le i \le n, 1 \le j \le n, i \ne j)$ represents the dependencies among the tasks. The set $pre(t_i)$ represents the immediate predecessors of a task t_i, and the set $suc(t_i)$ represents the immediate successors of the task. The computation requirement of a task $(w(t_i))$, expressed in millions of instructions (MI), can be used to calculate its *execution time (T_e)* on virtual machine v_j as:

$$T_e(t_i, v_j) = w(t_i)/f_{op}, f_{op} \in F_j \tag{2}$$

where f_{op} represents the frequency at which v_j operates during the execution of the task, expressed in millions of instructions per second (MIPS). The *communication time (T_c)* between two dependent tasks can be calculated as:

$$T_c(t_i, t_j) = w(e_{i,j})/bw(v_i, v_j) \tag{3}$$

where $bw(v_i, v_j)$ represents the bandwidth, expressed in Mbps, and $w(e_{i,j})$ represents the amount of data to be transferred from the predecessor task. The makespan of the workflow application is the time required for the execution of all the tasks in the set T on the virtual machines available to the user. It can be calculated as:

$$makespan = AFT(t_{exit}, v_j) \tag{4}$$

where *AFT* is the actual finish time of *exit task* on virtual machine v_j allocated to it.

3 Problem Definition

Monte Carlo Simulation (MCS) is a computation-intensive, statistical method that can be used to model the behavior of a complex system [7]. We use MCS method with *Weibull distributed failures* for realistic system reliability modeling, i.e., to determine the probable time-to-failure (T_F) and time-to-repair (T_R) for various virtual machines. To perform MCS in cloud environment, we realize the system a large number of times using a stream of random numbers. This stream is used to perform sampling by inverse transformation method, using the statistical properties of servers, in order to determine the probable T_Fs and T_Rs using the equation:

$$T_{F/R} = \beta(-\ln(1-r))^{1/\alpha} \tag{5}$$

where α (shape) and β (scale) are the Weibull distribution parameters and r is the random number generated. The cloud system is simulated a large number of times. Using the values of probable T_Fs and T_Rs, each virtual machine can be assumed to be in the *available* or *failed* state during its future operation.

The energy consumption for executing a task on a virtual machine can be obtained using the power model defined in the previous section as:

$$E(t_i, v_j) = C_j(f_{op})^3 T_e(t_i, v_j), \quad f_{op} \in F_j \tag{6}$$

where f_{op} represents the frequency with which v_j operates during the execution of the task. Thus, the total energy consumed by the virtual machine for execution of all the tasks allocated to it can be obtained as:

$$E_{v_j}^{Execution} = \sum_{t_i's \to v_j} C_j(f_{op})^3 T_e(t_i, v_j), \quad f_{op} \in F_j \tag{7}$$

When the virtual machine fails, it does not consume any energy. The virtual machine is assumed to operate at minimum frequency during idle time (T_{idle}) and communication time (T_{comm}). We assume that the communication time does not increase due to a decrease in the operating frequency of the virtual machine. Thus, the energy consumption during idle or communication time can be obtained as:

$$E_{v_j}^{Idle/Comm} = C_j(f_{j,1})^3 T_{Idle/Comm} \tag{8}$$

The total energy consumption by a virtual machine during the execution of the workflow application can be obtained as the sum of energy consumed by it during the execution time, idle time, communication time and failure time.

The total energy consumption for executing the workflow application is obtained as:

$$E = \sum_{v_j \in V} \left(E_{v_j}^{Execution} + E_{v_j}^{Comm} + E_{v_j}^{Idle} + E_{v_j}^{Failed} \right) \tag{9}$$

Thus, our aim is to produce a schedule for the workflow application that executes the tasks reliably, using an optimal makespan and reduced energy consumption. To achieve this objective, we propose the ERAWS algorithm.

4 Proposed ERAWS Algorithm

In this section, we present the energy-efficient and reliability-aware workflow scheduling (ERAWS) algorithm, derived from HEFT algorithm [13]. We calculate *mean execution time* and *mean communication time* for each task, considering maximum operating frequency, and use it to assign a rank value to each task as:

$$rank(t_i) = \overline{T_e(t_i)} + \max_{t_j \in suc_i} \left(T_c(t_i, t_j) + rank(t_j) \right) \tag{10}$$

where $\overline{T_e}$ is the *mean execution time* of the task. We then sort the tasks into a *task-rank list* based on these rank values. This list specifies the priority with which each task should be allocated a virtual machine. Selecting each task from the *task-rank list* in order, the scheduler chooses the appropriate virtual machine for the task while taking its reliability into consideration, using information provided by the *Reliability Predictor*. For this, we find out the *Expected Start Time (XST)* for each task as in Eq. (11), where $ready(v_j)$ represents the time at which v_j is ready for execution of the concerned task after finishing the previously allocated tasks, $available(k)_j$ and $failed(k)_j$ represent the kth availability and the kth failed state for the virtual machine at the concerned time, and $T_R(k)_j$ represents the repair time of the virtual machine for the kth *failed* state. We assume the virtual machine to be initially available at time 0.

$$XST(t_i, v_j) = \begin{cases} \max \left\{ ready(v_j), \max_{t_p \in pre_i} \left(\begin{matrix} XFT(t_p) + T_c(t_i, t_p), \ if \ t_p \notin v_j \\ XFT(t_p), \ if \ t_p \in v_j \end{matrix} \right) \right\}, & if \ v_j \in available(k)_j \\ T_R(k)_j, & if \ v_j \in failed(k)_j \end{cases} \tag{11}$$

XFT is the *Expected Finish Time (XFT)* for each task which is calculated as:

$$XFT(t_i, v_j) = \begin{cases} XST(t_i, v_j) + T_e(t_i, v_j(f_{j,l})), & \text{if } XST + T_e + \sum_{t_k \in suc_i} T_c(i,k) \in available(k)_j, f_{j,l} \in F_j \\ T_R(h)_j + T_e(t_i, v_j(f_{j,l})), & \text{if } h > k \ \& \ T_R(h)_j + T_e + \sum_{t_k \in suc_i} T_c(i,k) \in available(h)_j, f_{j,l} \in F_j \end{cases}$$

$$(12)$$

where h represents the index of the earliest availability state for v_j for which the machine is assumed to be available during the complete execution period. T_e represents the execution time for task t_i on v_j when v_j operates at maximum frequency ($f_{j,l} \in F_j$). We select the virtual machine which gives the minimum *XFT* for the task. While calculating the *XFT* for a task, we ensure that the virtual machine remains available until all the data transfer requirements of its successor tasks are complete. Using *XST* and *XFT* for the tasks, we calculate the deadline for workflow execution as the minimum *XFT* for the *exit task*. We then find out the critical path for the workflow using *Earliest Start Time (EST)* and *Latest Finish Time (LFT)*. *EST* for each task is:

$$EST(t_i) = \max_{t_k \in pre_i} \left\{ XFT(t_k, v_j) + T_c(t_k, t_i) \right\} \tag{13}$$

The *EST* for the *entry task* is assumed to be 0. *LFT* for each task is calculated as:

$$LFT(t_i) = \max_{t_k \in suc_i} \left\{ T_c(t_i, t_k) + XST(t_k, v_j) \right\} \tag{14}$$

The *LFT* for the *exit task* is equal to the *deadline* for the workflow application. To determine whether a task belongs to the critical path, we calculate its *slack* value as:

$$slack(t_i) = LFT(t_i) - EST(t_i) \tag{15}$$

If the *slack* for a task is equal to the difference between its *XFT* and *XST*, the task belongs to the critical path. Otherwise, the task is non-critical. For a critical task, we assign the *Actual Finish Time (AFT)* for the task as the minimum *XFT* and the *Actual Start Time (AST)* as the corresponding *XST*. For non-critical tasks in the workflow application, we make use of the *slack* available for their execution by assigning the non-critical task to a virtual machine which works on a reduced frequency level, such that it consumes minimum energy for the task execution, while completing the task before the *LFT* so that it does not hinder the execution of its successor tasks. The *Scheduler* passes this schedule S to the *CC* which works along with the *DVFS Manager* to execute the tasks on the concerned virtual machines. The results obtained are sent to the *user*. Algorithm 1 presents the ERAWS algorithm in detail.

Algorithm 1: Energy and Reliability-Aware Workflow Scheduling (ERAWS) algorithm

---

1. Compute *mean execution time* for each task using maximum operating frequency for each virtual machine in set V.
2. Compute *mean communication time* for each dependency among tasks.
3. Compute *rank(t_i)* for each task. //using Eq. (10)
4. Sort all tasks according to *rank* into a *task-rank list* in non-increasing order.
5. for each task t_i in *task-rank list*
6.   Calculate *XST* and *XFT* for the task on each virtual machine.
                     // using Eq. (11) and Eq. (12) resp.
7.   Select the virtual machine with the minimum *XFT* for the task.
8. Assign workflow *deadline* as minimun *XFT* for *exit task*.
9. for each task t_i in *task-rank list*
10.   Calculate *EST* for the task t_i. //using Eq. (13)
11.   Calculate *LFT* for the task t_i. //using Eq. (14)
12.   Calculate *slack* for the task t_i. //using Eq. (15)
13.   If (*slack* = $XFT(t_i, v_j) - XST(t_i, v_j)$)
14.   *Critical-path* ← t_i
15.   Assign $AST(t_i) = XST(t_i, v_j)$.
16.   Assign $AFT(t_i) = XFT(t_i, v_j(f_{j,1})), f_{j,l} \in F_j$
17.   Add task t_i assigned to virtual machine v_j with operating frequency $f_{j,l}$, *AST* and *AFT* to schedule S.
18.   else
19.   Task t_i is *non-critical*.
20.   for each virtual machine $v_j \in V$
21.   for each frequency level $f_{j,k} \in F_j$ in order
22.   if($T_e(t_i, v_j(f_{j,k})) < (LFT(t_i) - EST(t_i))$) &&($v_j \in$ *Idle* for time slot T_e between *EST* and *LFT* calculated using *XST* and *XFT*.)
23.   find out Execution Energy ($E(t_i, v_j)$)for task t_i on virtual machine v_j at operating frequency $f_{j,k}$. //using Eq. (6)
24.   exit for loop
25.   Select virtual machine v_j with operating frequency $f_{j,k}$ which gives minimum Execution Energy for the task and assign *AST* and *AFT*.
26.   Add task t_i assigned to virtual machine v_j with operating frequency $f_{j,k}$, *AST* and *AFT* to schedule S.
27. Return S.

---

5 Performance Evaluation

In this section, we analyze the performance of our proposed algorithm with the most popular workflow scheduling algorithm, HEFT [13], to verify its effectiveness.

We use the CloudSim toolkit [8] to model the cloud computing environment. We evaluate the performance of our algorithm over workflow structures with task size ranging from 40,00,000 MI to 1,00,00,000 MI. The maximum operating frequency for each virtual machine, mapped onto MIPS ratings, ranges from 1500 MIPS to 3000 MIPS. Each virtual machine has four different frequency levels of operation.

We choose the minimum operating frequency as 40% of the maximum operating frequency for the machine. The network bandwidth available for each virtual machine ranges from 50 Mbps to 100 Mbps.

Since the ERAWS algorithm proposes an energy-efficient and reliability-aware workflow scheduling algorithm, we use makespan and energy consumption as metrics to analyze the performance of the algorithm against HEFT.

For randomly generated task graphs, we analyzed the performance by first varying the size of the task graph. The comparative results, as shown in Fig. 1, show that our algorithm achieves significant gain over HEFT in terms of makespan and energy consumption. The performance of our algorithm improves in terms of makespan as we increase the number of tasks in the workflow application. This is because as the number of tasks increases, the probability of occurrence of failures also increases. Since ERAWS executes tasks reliably, its makespan gain increases with increase in the size of workflow application. The energy savings for ERAWS range from 10 to 25% over HEFT as the size of the application increases.

We also analyze the performance by varying the number of virtual machines. We conclude that as the number of virtual machines increases, as shown in Fig. 2, makespan for both ERAWS and HEFT decreases. ERAWS achieves significant makespan gain over HEFT. However, the makespan gain decreases as the number of virtual machines available increases. This is because as the number of virtual machines increases, the probability of occurrence of a failure decreases.

We performed similar simulation analysis for GE and FFT applications and conclude that the results for GE and FFT applications show similar performance as for randomly generated workflow applications.

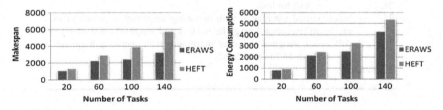

Fig. 1 Effect of varying the number of tasks on makespan and energy consumption for randomly generated workflow applications

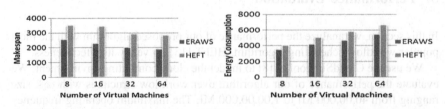

Fig. 2 Effect of varying the number of virtual machines on makespan and energy consumption for randomly generated workflow applications

6 Conclusion

In this paper, we proposed an energy-efficient and reliability-aware workflow scheduling algorithm, ERAWS, which takes three conflicting criteria into account: makespan, reliability of task execution and energy consumption. For reliable scheduling, we used Monte Carlo Simulation with Weibull distributed failures to determine the availability for each virtual machine. For energy-efficient scheduling, we used dynamic voltage and frequency scaling (DVFS) technique to reduce the energy consumption by processors. We simulated a cloud environment using CloudSim toolkit to analyze the performance of ERAWS algorithm. The simulation results show that the proposed algorithm gains significantly in terms of makespan and energy consumption, in real-world scenarios where reliability of task execution and energy consumption are important issues.

References

1. Sadiku MN, Musa SM, Momoh OD (2014) Cloud computing: opportunities and challenges. IEEE Potentials 33:34–36
2. Magkils G, Semeraro G, Albonesi DH, Dropsho SG, Dwarkadas S, Scott ML (2003) Dynamic frequency and voltage scaling for multiple-clock-domain microprocessor. IEEE Micro 23:62–68
3. Garey MR, Johnson DS (2002) Computers and intractability, vol 29. W.H. Freeman
4. Garraghan P, Townend P, Xu J (2014) An empirical failure-analysis of a large-scale cloud computing environment. In: 15th IEEE international symposium on high-assurance systems engineering (HASE). IEEE, pp 113–120
5. Tang X, Li K, Qiu M, Sha EHM (2012) A hierarchical reliability-driven scheduling algorithm in grid systems. J Par Dist Comp 72:525–535
6. Guo S, Huang HZ, Wang Z, Xie M (2011) Grid service reliability modeling and optimal task scheduling considering fault recovery. IEEE Trans Reliab 60:263–274
7. Zio E (2013) The Monte Carlo simulation method for system reliability and risk analysis. Springer, London
8. Calheiros RN, Ranjan R, Beloglazov A, De Rose CA, Buyya R (2011) CloudSim: a toolkit for modeling and simulation of cloud computing environments and evaluation of resource provisioning algorithms. Softw: Pract Exp 41:23–50
9. Cosnard M, Marrakchi M, Robert Y, Trystram D (1988) Parallel Gaussian elimination on an MIMD computer. Parallel Comput 6:275–296
10. Chung YC, Ranka S (1992) Applications and performance analysis of a compile-time optimization approach for list scheduling algorithms on distributed memory multiprocessors. In: Supercomputing'92, Proceedings. IEEE, pp 512–521
11. Kim KH, Beloglazov A, Buyya R (2011) Power-aware provisioning of virtual machines for real-time cloud services. Concurr Comput: Pract Exp 23:1491–1505
12. Minas L, Ellison B (2009) Energy efficiency for information technology: how to reduce power consumption in servers and data centers. Intel Press
13. Topcuoglu H, Hariri S, Wu MY (2002) Performance-effective and low-complexity task scheduling for heterogeneous computing. IEEE Trans Parallel Distrib Syst 13:260–274

Proposed Use of Information Dispersal Algorithm in User Profiling

Bhushan Atote, Saniya Zahoor, Mangesh Bedekar and Suja Panicker

Abstract For recommending the best result to the user as per his requirement, User Profiling plays an important role. In user profiling, the profiles are created from the past data of same user. Maintaining the security and privacy of this data becomes a big challenge for researchers. Here, we are proposing the algorithm for privacy and security purpose of different profiles, with the integration of Information Dispersal Algorithm. The use of vast data of profiles by the user from any location at any time would be achieved by the use of the private cloud. As the profiles of different devices are maintained on the central cloud server, the recommendation for user for particular device can be executed easily.

Keywords User profiling · Privacy · Security · Cloud · Mobile

1 Introduction

As the data on Internet is growing day-by-day, it is required to get the best possible result for users' search query. The task of finding relevant information from the history data and the activities of the user can be done through User Profiling.

Search engines are available for the same, but the problem of ever increasing data reduces the efficiency of the search results performed by a user. In User Profiling, the profiles that are created by the history of users play an important role.

B. Atote (✉) · M. Bedekar · S. Panicker
Computer Department, MAEER's MIT, Kothrud, Pune, India
e-mail: bhushanatote2408@gmail.com

M. Bedekar
e-mail: mangesh.bedekar@mitpune.edu.in

S. Panicker
e-mail: suja.panicker@mitpune.edu.in

S. Zahoor
IT Department, NIT Srinagar, Jammu & Kashmir, India
e-mail: saniya.zahoor@yahoo.com

© Springer Nature Singapore Pte Ltd. 2018
D.K. Mishra et al. (eds.), *Information and Communication Technology
for Sustainable Development*, Lecture Notes in Networks and Systems 9,
https://doi.org/10.1007/978-981-10-3932-4_9

So, the confidentiality, integrity, and availability can be provided through the cloud storage. Storing all the information on cloud gives us the best result.

Big data techniques provide excellent tools and techniques for more accurate User Profiling. However, there is an issue of privacy in User Profiling. In [1], we have shown the layered structure of User Profiling. Now, in this chapter, we are showing the different parameters that come under these layers and how we can manage the complete data with the use of private cloud. We will discuss more and explain the process in detail in the following sections.

This chapter is organized into six sections. In Sect. 2, we have given the motivation for doing this work. Section 3 will show the detailed literature survey of different concepts on privacy and security issues in User Profiling and how we can maintain the privacy of data in cloud. Problem finding and its solution is given in Sect. 4, with preceding of Proposed System and Mathematical Model in Sect. 5, which explain the architecture in detail with the working of our algorithm. Section 6 will conclude the theme of our chapter.

2 Motivation

User profiling is the process of creating profiles of user from his past data; this data may be his personal information, academic records, geographical data (locations), or other activities. By using these profiles, we can recommend the best suitable result for the search query by user. So, the security and privacy of these profiles is the main issue in User Profiling.

As the data collected from user day-by-day or time-to-time, the size is obviously more. So, there is storing and accessing problem for the given data (profiles). A model for privacy concerned in big data for user profiling has proposed in [1] and discussion about security and privacy issues in cloud computing is given in [2].

3 Literature Survey

The big data techniques and privacy issues in user profiling with case study of EEXCESS project is discussed in [1]. A brief introduction to personalization trends, social-, and location-based personalization is given in [3], but there is no solution for data leakage and un-authorized use of profiles, those are created from the huge amount of data for same. The cloud computing security with the issues in data security and privacy protection has discussed in [2]. Privacy protection for individual's data in cloud is not specified with the context of dynamic provisioning and real-time usage of cloud computing [2]. The client-side personalization approach which uses the keyword profiles on large data set for search engine is explained. It is limited for the advertising search engine as it uses advertisement keyword for profiles [4].

The SVDFS for public cloud with the secured virtual file system, which gives user to maintain the confidentiality of data on public cloud, is discussed in [5]. [6] shows the clustering and classification techniques using K-means and ANN for mobile content personalization. JigDFS provides the encryption of data that are stored as a segment in the form of slice, and also, it provides the privacy through plausible deniability. It is only limited to the peer-to-peer distributed file system, for the user privacy [7].

A well-known approach of k-anonymity in privacy preservation of published data, it focuses on the separation of information into sensitive attributes and quasi-identifiers, is introduced in this chapter. The location perturbation engine is complex and it requires more time map the quasi-identifier with the respective attributes [8]. A path perturbation algorithm which can maximize users' location privacy gives a quality of service constraint. It concentrates on a class of applications that continuously collect location samples from a large group of users has presented in this chapter. The reliance on a general optimization algorithm would improve computational efficiency of the path perturbation algorithm has to be removed. This algorithm cannot be used in applications of real-time data [9].

The authors investigate disclosure control algorithms that hide users' positions in sensitive areas and withhold path information that indicates which areas they have visited. This is not able to collect outdoor movement traces through GPS, and perhaps indoor traces through wireless LAN positioning mechanisms. This will also require developing and validating tools for automatically determining distinct areas, such as buildings, in a larger space, and for partitioning these areas as necessary for the k-area algorithm [10]. The method to hide the device's visited locations from third-party services and how to track device with the same services to get device location is discussed. For this, they have used Adeona which uses OpenDHT as third-party service. For tracking with Adeona, it is required to install it in the respective device; sometime, it is not compatible with different devices, which is the main drawback of this system [11].

The ontology-based model for tracking the user activities on mobile devices this is improvement for checking the internet services of devices is proposed in this chapter. It is limited to the Internet services of devices only [12]. An on-line distributed data repository e-Vault, which stores data across a network securely for maintaining integrity though servers having malfunction, has designed in this chapter. Cryptographic hash functions are used to get the information if the data is corrupted later [13].

4 Problem Definition

In [14], we have shown that there are privacy concerns that need to be taken care of. We consider the User Profile to be composed of three layers—Public Layer (Outer Layer), Private Layer (Intermediate Layer), and Personal Layer (Inner Layer) as shown in Fig. 1.

Fig. 1 Layered structure of user profile

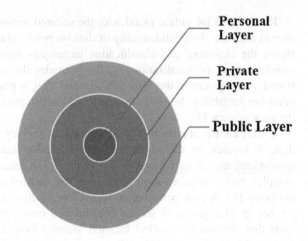

Personal Layer

Private Layer

Public Layer

Public Layer (Outer Layer):

(1) Name
(2) Gender
(3) Country

Private Layer (Intermediate Layer):

(1) Mobile No.
(2) Address
(3) Blood Group
(4) Education
(5) Family connected
(6) Friend circle
(7) Locations
(8) Devices connected

Personal Layer (Inner Layer):

(1) Actions (With date and time logs)
(2) Web history (Updated)
(3) Bookmarks
(4) Cookies
(5) Up-time of particular device
(6) Up-time of screen
(7) Up-time of keyboard
(8) Up-time of mouse
(9) Activities of Applications used

For real-time user profiling, the updating profiles and maintaining that profiles or further the part profiles (PP) as concerned with our proposed system is an important task. Here, for storing the part profiles of multiple devices, we can store the different

Fig. 2 Synchronization of
different devices

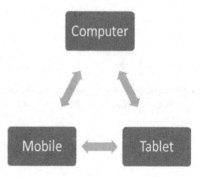

profiles on different devices. The synchronization of data (PP) is shown in the figure below, as the PP of computer usage can be stored on tablet and the PP of tablet usage can be stored on mobile. So, for suggesting the required result based on different profiles, we need to get that different profiles from the respective devices for that it is required that the other device also be in running condition (Fig. 2).

So, to overcome this problem, we can save the different PP of multiple devices at a central location. As we need to access these files, we used here a private cloud as a central location for our system. The different strategies and privacy problems in data repository are shown in above literature section. Working of private cloud is shown in the proposed system.

5 Proposed System and Mathematical Model

5.1 Proposed System

For security purpose of user profiles, here we are using Secured Virtual Diffused File System (SVDFS) by using private cloud. A user may have multiple devices, for each device, we collect different part profiles, which are mapped with the given device. Let, the user using the multiple devices like PC (Personal Computer), Mobile and Tablet for which we will be having part profiles of PC as PC_1, PC_2,....., PC_n for Mobile as M_1, M_2,....., M_n likewise for Tablet as T_1, T_2,...., T_n. Each PP is being stored in cloud, slice stores S1, S2, S3,...., Sn would be used here, and each slice store will have a PP in the form of private, public, and personal layer.

The mapping of PP stored in slice store, and different devices are stored in registry server; for retrieving the profile of particular device, only respective slice stores are active to give the best profile for the user (Fig. 3).

For getting the location of devices, we can track the IP address of PC and cell tower for GSM-enabled devices (mobile and tablet). The usage of multiple devices is stored in the form given (Table 1).

The date and time are the last used data for that particular device. The PP are created on the respective devices only and will be forwarded to the server if the

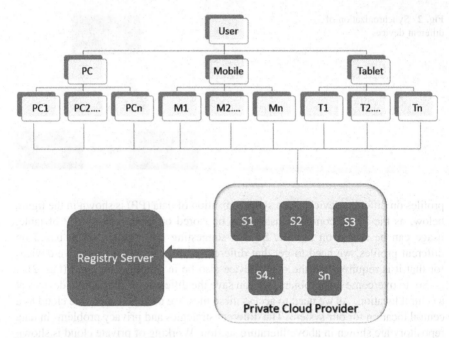

Fig. 3 Proposed system architecture

Table 1 Usage of multiple devices

Date	Time	Device	Device-Id
03/04/2016	5:58:14 IST	PC	192.168.0.12
03/04/2016	6:18:42 IST	Mobile	Mayur colony
03/04/2016	22:16:09 IST	Tablet	Mayur colony

network is not available the PP will remain as it is on the device itself and whenever it gets the internet connection enabled the PP will get send with the last entries only with respective date and time.

Figure 4 shows the flow of extended algorithm that we have explained in this chapter under the section of Mathematical Model.

5.2 Mathematical Model

The system is defined by the tuple, 'S', such that

1) $S = \{D, I, O, Ic, Oc, F\}$

   D = Multiple devices used by user,
   I   = Input to the system, a set of user activities on different devices.

O = Output of the system.

Ic = Input conditions, i.e., updated part profiles.

Oc = Output conditions, the changes in User Profiles over time as the behavior of the user changes.

F = Functions which map I to O, through Ic inputs to Oc outputs.

(2) Let D be the set of devices used by a user

$$D = \{D1, D2, .., Dn\}$$

(3) $I = \{(Pu, Pr, Pe) \mid Pu (N, G, C), Pr (Mo, A, Bg, Edu, Fc, Fr, Lo),$
$Pe (Ac, Wh, B, Ck, S, K, M, Ap)\}$

Where

Pu	= Public Layer,
Pr	= Private Layer,
Pe	= Personal Layer,
N	= Name,
G	= Gender,
C	= Country,
Mo	= Mobile No.,
A	= Age,
Bg	= Blood Group,
Edu	= Education,
Fc	= Family connected,
Fr	= Friend circle,
Lo	= Location,
Ac	= Action,
Wh	= Web history,
B	= Bookmarks,
Ck	= Cookies,
S	= Up-time of Screen,
K	= Use of keyboard,
M	= Use of mouse,
Ap	= Applications used

Now,

(4) Oc = {PPro, MPro, TPro}
Where

PPro = Updated profiles for PC,
MPro = Updated profiles for Mobile,
TPro = Updated profiles for Tablet.

(5) $F = \{(D, S) \mid D\ (D1, D2, D3)\ S\ (S1, S2, \ldots., Sn)\}$
     Where

     S = Set of data slices.

     $\{D1\} \rightarrow \{PP1, PP2\} \rightarrow \{S1, S2, S3, S4, S5\} \rightarrow \{PP1, PP2\} \rightarrow \{D1\}$

     M $= a_i.\ S_{(k\ -\ 1)}m + 1 + \ldots.. + a_{im}.\ b_{km},\ a = $ vector for each slice
     N $= a_{i1}.\ c_1\ k + \ldots\ldots + a_{im}.\ c_m k$

The above M will give the splitting of file and N shows reconstruction of file i.e., nothing but the slicing of file in S1, S2, and Sn for recovering the file (profile).

Fig. 4 Flowchart of extended algorithm

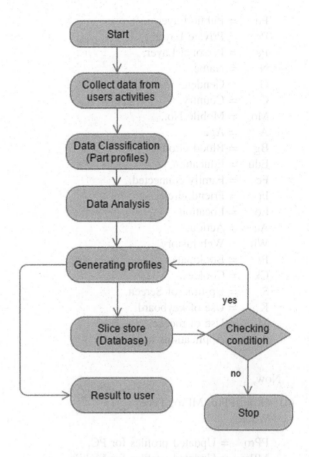

Likewise, mapping will be done for each device; as for single device we will be having number of part profiles, these part profiles will further divided into slices for storing on cloud slice store which has the mapping with the same device.

6 Conclusion

The profiles of user created on different devices would be forwarded to the central server for maintaining the privacy and security of profiles. By using IDA, indirectly we are minimizing the encryption of data. With private cloud we can overcome the problem of maintaining immense of data, which would be accessed by the user's devices. For accessing the server, it is required to be connected with the Internet, every time the usage of user would be recorded as a new part profile for the respective device and uploaded to the server.

References

1. Atote B, Zahoor S, Dangra B, Bedekar M (2016) Personalization in user profiling privacy security. In: IEEE sponsored an international conference on internet of things and applications (IOTA 2016) at MIT, Pune
2. Toch E, Wang Y, Cranor LF (2012) Personalization and privacy: a survey of privacy risks and remedies in personalization-based systems. J User Model User-Adap Inter 22(1–2):203–220
3. Hasan O, Habegger B, Brunie L, Bennani N, Damiani E (2013) A discussion of privacy challenges in user profiling with big data techniques: the EEXCESS use case. In: IEEE 2nd international congress on big data, Santa Clara Marriott, CA, USA
4. Narayanan A, Shmatikov V (2008) Robust de-anonymization of large sparse datasets (how to break anonymity of the Netix Prize Dataset). In: IEEE symposium on security and privacy. Oakland, pp 111–125
5. Hull G, Lipford HR, Latulipe C (2011) Contextual gaps: privacy issues on facebook. Ethics Inf Technol 13(4):289–302
6. Chen D, Zhao H (2012) Data security and privacy protection issues in cloud computing, In: International conference on computer science and electronics engineering, Hangzhou, pp 647–651
7. Mar KK (2011) Secured virtual diffused file system for the cloud. In: 6th international conference on internet technology and secured transactions, Abu Dhabi, pp 116–121
8. Bian J, Seker R (2009) JigDFS: a secure distributed file system. In: IEEE symposium on computational intelligence in cyber security, Nashville, TN, pp 76–82
9. Sasajima M, Kitamura Y, Naganuma T Toward task ontology-based modeling for mobile phone users activity. In: PID-29
10. Iyengar A, Cahn R, Garay JA, Jutla C (1998) Design and implementation of a secure distributed data repository, In: Proceedings of the 14th IFIP international information security conference, pp. 123–135(1998)
11. M. Bilenko, M. Richardson (2011) Predictive client-side proles for personalized advertising. In: KDD'11, proceedings of the 17th ACM SIGKDD international conference on knowledge discovery and data mining, San Diego, California, pp 413–421

12. Cassel LN, Wolz U (2001) Client side personalization. In: Proceedings of the joint DELOSNSF workshop on personalization and recommender systems in digital libraries. Dublin City University, Dublin
13. Ceri S, Dolog P, Matera M, Nejdl W (2004) Model-driven design of web applications with client-side adaptation. In: International conference on web engineering, ICWE04, vol 3140. Springer, Munich, pp 201–214
14. Kay J (2006) Scrutable adaptation: because we can and must. In: Adaptive hypermedia and adaptive web-based systems. Springer, Berlin, pp 11–19

Recent Research in Wireless Sensor Networks: A Trend Analysis

Prerak S. Shah, Neel N. Patel, Dhrumil M. Patel, Devanshi P. Patel and Rutvij H. Jhaveri

Abstract Wireless sensor networks (WSNs) have emerged as one of the huge research areas in this digital world since last few years. In this paper, we review total of 150 research papers from January 2015 to December 2015 in order to enlighten the researchers and educators about baseline of the current trends in the field. This study is a graphical and systematic review of various research works carried out in WSNs. These findings show that the research in WSNs received more attention over the past few years. For the primary research method of studying, highly cited research papers from top-rated publishers were included. Although a vast number of research objectives have been floated in this field, we analyzed that there are two main topics which are trending in 2015 as discussed in the paper. This analysis would provide insights for researchers, students, publishers, and experts to study current research trends in WSNs.

Keywords Wireless sensor networks · Trend analysis · Graphical interpretation

P.S. Shah (✉) · N.N. Patel · D.M. Patel · D.P. Patel · R.H. Jhaveri
Department of Computer Engineering, Shri S'ad Vidya Mandal Institute
of Technology, Bharuch, India
e-mail: prerak.14co50@gmail.com

N.N. Patel
e-mail: neelpatel9793@gmail.com

D.M. Patel
e-mail: dhrumilcse14@gmail.com

D.P. Patel
e-mail: devanshipatel42@gmail.com

R.H. Jhaveri
e-mail: rhj_svmit@yahoo.com

© Springer Nature Singapore Pte Ltd. 2018
D.K. Mishra et al. (eds.), *Information and Communication Technology
for Sustainable Development*, Lecture Notes in Networks and Systems 9,
https://doi.org/10.1007/978-981-10-3932-4_10

1 Introduction

Wireless sensor networks (WSNs), sometimes called wireless sensor and actuator networks (WSANs), are wireless network consisting of different devices using sensors to monitor conditions such as area, health care, environmental, air pollution, water quality, industrial, and structural health monitoring [1].

Trend analysis is an aspect of technical analysis to predict movement of particular research area in future based on past data. Since WSNs are a vast field, our objective behind this trend analysis is to find in which particular field 'what' and 'how much' of work has been done. In this paper, we propose a trend analysis on WSNs showing the current trend of year 2015. We have gone through 150 research papers of different well-known journals and reputed publishers. During trend analysis, we found that recent improvement of WSNs results in new technology and new generation of equipment that added new applications making it more accurate [2].

Observation of research papers leads us to energy and security. Energy in WSNs considered as a most important resource, because it determines life time of sensor node [3]. Most of the work during year 2015 has been done in improving energy consumption and security of WSNs. Talking about security, improving security will increase accuracy, prevent data loss, and improve privacy; researchers studied that the security in wireless sensor networks is tremendously poor. Our analysis shows that most of researchers and publishers have kept security as their primary objective.

Organization of paper is as follows: in Sect. 2, we include our graphical representation of this trend analysis. Followed by, Sect. 3 which comprises the final conclusion of paper.

2 Graph Interpretation

2.1 Simulation Versus Real-world Implementation

In this article, we have distributed every research papers according to their implementation. The works are implemented either on a simulator or in real world.

As shown in Fig. 1, evaluation of the data that we extracted guides us to the result that around 93% of works in 2015 were implemented on certain simulator followed by rest of 7% of works which are implemented on a real-world scenario.

2.2 Percentage of Papers Published by Countries

WSNs are an emerging topic these days. Different authors from different countries are involving in research of WSNs. We have analyzed 150 journal and conference

Fig. 1 Simulation versus
real-world implementation

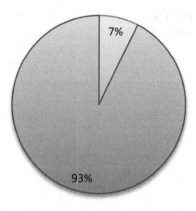

☐ Real World ☐ Simulation

Fig. 2 Percentage of papers
published by countries

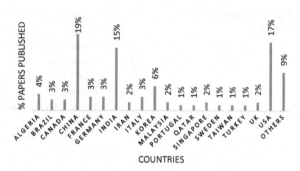

papers from different publishers, i.e., IEEE, Springer, Elsevier, Wiley, Taylor & Francis from January to December 2015.

As shown in Fig. 2, in our dataset authors from USA (17%), China (19%), India (15%) are having maximum contribution as well as other countries are also involving in the comprehensive work of WSNs.

2.3 Percentage of Publishers

For proper and accurate analysis, it is necessary to use high-quality literatures as resource.

As shown in Fig. 3, IEEE is the most well-recognized publisher covering around 29% of chart followed by Springer (25%) and others. "Others" part include every other remaining publisher which were not included individually.

Fig. 3 Percentage of
publishers

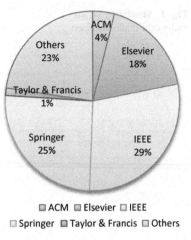

☐ ACM ☐ Elsevier ☐ IEEE

☐ Springer ☐ Taylor & Francis ☐ Others

Fig. 4 Citations per
publishers

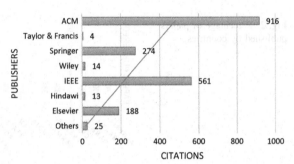

2.4   Citations Per Publisher

This analysis of representing the citation for every publisher shows the quality of comprehensive work carried out by well-recognized publisher. The higher the citation the better quality of paper is published.

As shown in Fig. 4, ACM has the maximum number of citations (916) in 2015, followed by IEEE (561), Springer (274), Elsevier (188), and others. The "Others" include rest of the remaining publishers which are undefined in the graph individually.

2.5   Conference Papers Versus Journals Versus Books

Conference papers are the articles that are written with the goal of being accepted to a conference where researchers can present their research to community, whereas journal papers refer to an article that is published in an issue of the journal.

As shown in Fig. 5, the percentage of journal papers from our dataset is 82%. This interprets that most of the work is done to be published in journals in 2015 as per dataset.

2.6 Frequency of Number of Authors

Its next to impossible for individual researcher to explore the whole topic by himself. So efforts are made to work in co-authorized manner to cover whole topic accurately.

As shown in Fig. 6, maximum research papers (51) were published possessing coauthorship of three authors and only one paper possessing a total number of 13 authors.

2.7 Objective Wise Percentage of Citations

Citations are viewed as a numerical value that acknowledges the quality of work in given research paper for the publishers, educators, and the students.

Fig. 5 Conference papers versus journal versus book count

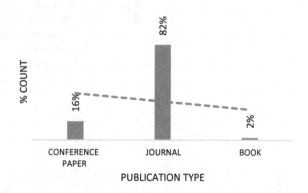

Fig. 6 Frequency of number of authors

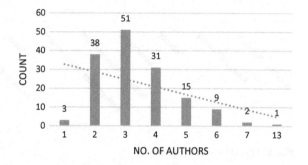

As shown in Fig. 7, network management scores the maximum citation percentage (34%), followed by monitoring at 19%, improving network performance at 15%, improving security (in WSNs) at 14%, energy efficiency at 6%, survey through wireless sensor networks at 5%, artificial intelligence (A.I), and routing at 2%, while tracking is cited at 1%.

2.8 Objective Wise Frequency

Objective wise frequency represents the most trending objective of 2015 followed by other objective, respectively.

As shown in Fig. 8, energy efficiency (22%) and security in WSNs (17%) were the most trending objective in 2015 and were followed by network management

Fig. 7 Objective wise percentage of citations

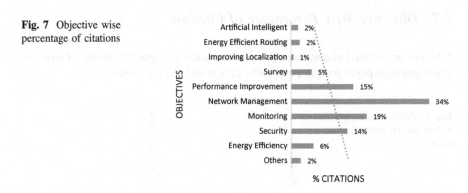

Fig. 8 Objective wise frequency

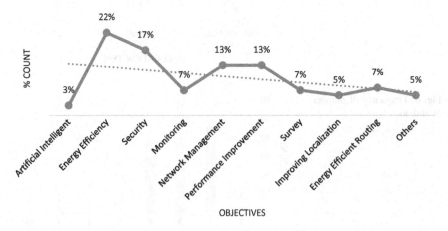

and network performance both at 13%, while monitoring, survey, and routing all are at 7%. Moving down, we got tracking at 5% and artificial intelligence at 3%.

2.9 Paper Efficiency/Ranking

For the efficiency analysis, we first need to standardize the efficiency measure for all papers of our dataset published in recent years (2015). The following equation is proposed to normalize the efficiency scores for all research articles of our dataset [5]:

$$E = C/(15 - M)$$

where **E**—efficiency of the paper, **C**—citation of the paper, **M**—published months
   This following derived equation helps us find efficiency of the individual paper. The citation of the paper is divided by the value obtained by excluding published months from total number of months till the paper published (till month of march of year 2016). Total 12 months of year 2015 and month of January, February, and March that leads us to a total of 15 months. According to our data, the results are categorized into two fragments as follows:

1. High-efficiency papers
2. Low-efficiency papers

Note: On the top of that, the remaining papers left to categorize have efficient less than one (<1). Thus, they do not fall under any of the above mentioned category (Tables 1 and 2).

Table 1 High-efficiency papers

Title	M	C	E
Application of smart antenna technologies in simultaneous wireless information and power transfer	12	28	9.3
Wireless powered communication: opportunities and challenges	4	61	5.6
On managing quality of experience of multiple video streams in wireless networks	10	23	4.6
On the delay performance in a large-scale WSN: measurement, analysis, and implications	11	16	4.0
A secure temporal-credential-based mutual authentication and key agreement scheme with pseudo-identity for WSNs	2	34	2.6
Environmental parameters monitoring in precision agriculture using WSNs	4	22	2.0
Energy management in WSNs: a survey	6	18	2.0
Deploying WSNs with fault tolerance for structural health monitoring	2	24	1.9

Table 2 Low-efficiency papers

Title	M	C	E
Evaluating energy cost of route diversity for security in wireless sensor networks	12	5	1.7
A 10 mW bluetooth low-energy transceiver with on-chip matching	12	5	1.7
A review on stochastic approach for dynamic power management in WSNs	12	5	1.7
Secure data aggregation technique for WSNs in the presence of collusion attacks	1	23	1.6
Data gathering with compressive sensing in WSNs: a random walk-based approach	7	13	1.6
A distributed algorithm for energy efficient and fault-tolerant routing in WSNs	8	11	1.6
Saliency-directed prioritization of visual data in wireless surveillance networks	8	11	1.6
Enabling high-level application development for the Internet of things	1	22	1.6
Mobile data gathering with load-balanced clustering and dual data uploading in WSNs	4	17	1.6
Efficient intelligent energy routing protocol in WSNs	8	9	1.3
Security in the integration of low-power Wireless Sensor Networks with the Internet	8	9	1.3
Energy-efficient probabilistic area coverage in WSNs	1	17	1.2
BOD-LEACH: Broadcasting over duty-cycled radio using LEACH clustering for delay/power efficient dissimilation in WSNs	9	7	1.2
Opportunistic routing algorithm for relay node selection in WSNs	2	13	1.0

3 Conclusion

The research conducted a systematic analysis of various fields in wireless sensor networks. It provides interpretation and implication of most recent findings which conclude the following results for researchers, authors, and educators.

We draw conclusions that: Initially, the most trending topics of 2015 were energy efficiency in WSNs and security management in WSNs. Secondly, the country which plays a huge role in research in WSNs is India, USA, and China followed by other countries across the globe. In 2015, IEEE has the maximum research paper published compared to other well-known publishers concluded from our dataset. Here, ACM publisher has attracted maximum citations among every other well-recognized publisher. Concluding from the graphs, the maximum citations were attracted in the field of network management. In addition, a good amount of work has been carried out in improving energy consumption and conservation. Furthermore, we conclude that maximum amount of research works are limited to simulation, while only 7% out of all research works have carried out real-world implementations.

Looking forward through the amount of work carried out for improving various performance metrics in WSNs, it shows a good potential for further research in this area.

References

1. Bokare M (2012) Wireless sensor network: a promising approach for distributed sensing tasks. Excel J Eng Technol Manag Sci I(1). ISSN 2249-9032
2. Wu et al (2012) Review trends from mobile learning studies: a meta analysis. Comput Educ 59:817–827
3. Yudo et al (2015) Effective initial route construction for mobile relay on wireless sensor network. Artif Life Robot 20:49–55. doi:10.1007/s10015-014-0194-5
4. Shahadat U et al (2011) Trend and efficiency analysis of co-authorship network. Scientometrics 90(2):687–699

Profit Analysis of a Computing Machine with Priority and s/w Rejuvenation

Ashish Kumar, Monika Saini and Devesh Kumar Srivastava

Abstract The main concentration of the present study is to carry out the profit analysis of a computing machine using the concepts of priority to software rejuvenation over hardware component's preventive maintenance and hardware repair. For this purpose, a stochastic model is designed which comprises of two identical units. Cold standby redundancy technique is used in the development of the model. A single repair facility remains available with the system to do all repair activities, and he takes the system for preventive maintenance after a specific period of time. All repair and maintenance are perfect. The random variables are all Weibull-distributed. Using semi-Markov processes, Laplace transformation, and regenerative point technique recurrence relations and numerical results are obtained for various measures of system effectiveness.

Keywords Weibull failure and repair laws · Priority · Computing machine Preventive maintenance · Software rejuvenation

1 Introduction

With the advance development in information technology, the use of computing gadgets for commercial purpose is rapidly increasing. In computing gadgets (machines or devices), hardware and software components work simultaneously. The main concentration of the present work is to design a reliability model for the performance measures of computer systems and to give such suggestions that improve the availability and profit generated by the system. In the present reliability

A. Kumar (✉) · M. Saini
Department of Mathematics and Statistics, Manipal University Jaipur,
Jaipur, Rajasthan, India
e-mail: ashishbarak2020@gmail.com

D.K. Srivastava
Department of Information Technology, Manipal University Jaipur,
Jaipur, Rajasthan, India

© Springer Nature Singapore Pte Ltd. 2018
D.K. Mishra et al. (eds.), *Information and Communication Technology
for Sustainable Development*, Lecture Notes in Networks and Systems 9,
https://doi.org/10.1007/978-981-10-3932-4_11

model, two computer systems are used: one as an operative unit and other as cold standby with one server which deals with all software and hardware failures. The computer systems suffer from software failures more in comparison with hardware failures. Most of the software developers are trying to develop more and more advanced version of the software for the execution of complex computational calculations. Due to the complexity of software, it is always preferred to make a study about software faults.

The software faults are classified into various categories such as Heisenbugs, Bohrbugs, and software ageing faults. These faults are identified according to their repetitive nature and functionality. Software ageing also reduces the performance of the system and hangs the system. To overcome the software faults and hardware failures for enhancing the performance and reliability of computing gadgets' preventive maintenance of the system, hardware repair and software rejuvenation policies are recommended. Software rejuvenation is one proactive technique which stops all the application of computer system and restarts it again in a healthy state and stops the failures before their actual occurrence. Software rejuvenation policy is classified into three categories on the basis of stoppage time, an effect on applications, and restart process. In the present study, the effect of third-level software rejuvenation is studied in which all the applications going on the computer system is stopped. It takes a lot of time, and all the applications of computer systems are very much affected by this software rejuvenation. In the present investigation, a computer system with priority to software rejuvenation over hardware failure and preventive maintenance have been studied. The rest of the manuscript is arranged in the following sections. In the second section, a relevant literature review is appended. The necessary assumptions, notations, and possible states of the model are given in third section. In the fourth section, transition probabilities, various recurrence relations, and their Laplace transformation are given. Numerical results are given in the fifth section. In sixth section, final conclusion is given.

1.1 Literature Review

A computing machine is a combination of lot of software and hardware components. As the number of components increased, the complexity of design of the system also increased. The complexity of system effects the reliability of the system. Computing machines are most commonly used in information technology sector, mobile industry, and much more. Asif et al. [1] analyzed the performance of a cluster system in which many computer systems are connected. Zhou and Ippoliti [15] studied the resource allocation of server clusters. Jain et al. [3] developed some reliability models to study the performance of hardware and software system with standby hardware unit and multi-component circular consecutive k-out-of-n:f repairable system using the concepts of switching failures, common cause failure, and two types of repair. Availability analysis of a deteriorated multi-component system is discussed by Hajeeh [2] with inspection, human error, and common cause failure.

Recently, Freedman and Tran [6], Welke et al. [7], Lai et al. [8], and Kumar and Malik [5] studied various stochastic models for computing devices such as computer system, integrated h/w and s/w system with independent h/w and s/w components. Many researchers such as Kumar et al. [11], Kumar and Malik [10], and Malik and Kumar [9] developed some stochastic models using preventive maintenance and priority for computer systems and establish that preventive maintenance plays a very important in reliability enhance of computer systems. Software rejuvenation is an important technique for caring of software components by stopping its operation and controlling the internal faults. Some researcher such as Wang et al. [14], Koutras and Platis [12], and Okamura and Dohi [13] described some reliability models for a redundant system with different software rejuvenation policies. Recently, Jain and Preeti [4] analyzed the availability of cluster system with software rejuvenation.

1.2 Model Description

With the excessive use of computing devices, the necessity of preventive maintenance of h/w components, software rejuvenation, and h/w repair is felt for maintaining the performance and cost-effectiveness of the system. In the present investigation, we considered software rejuvenation, h/w repair, and h/w preventive maintenance for a cold standby unit stochastic model of computing devices. There are two units in the system used in cold standby redundancy. The rejuvenation and preventive maintenance are carried out upon failures of software and hardware components. When both units failed, then system is in shutdown state. For the construction of stochastic model, following assumptions are considered:

- The system is composed with the help of cold standby redundancy having two units. And in both units, failure time of s/w and h/w components, repair time and PM time of h/w components, s/w rejuvenation time of s/w components are Weibull-distributed with common shape parameter η and scale parameters λ, β, and θ.
- For all repair facilities, a single repairman is available.
- Cold standby unit becomes operative immediately after every type of failure.
- Upon failure of both units, priority to software rejuvenation is given over hardware repair and h/w preventive maintenance.
- After a maximum operation time, h/w components undergo preventive maintenance.

For the model description, the stochastic model for integrated h/w and s/w system is defined state by state as follows:

State-0 It is the full working state of the system. After repair and preventive maintenance of h/w components and s/w rejuvenation of failed software components, the system comes back to this state if one unit is available in operation. Here, one unit is operative and other is in cold standby.

State-1 Operative state where one unit in operation and other unit under software rejuvenation.

State-2 Operative state with one operative and other under repair due to hardware failure.

State-3 Operative state with one operative and in other h/w components are under preventive maintenance.

State-4 Failed state in which failed hardware components of first unit under repair from preceding state and h/w components of second unit are waiting for preventive maintenance.

State-5 Shut down state, both units are failed due to failure of hardware components in which one unit continuously under h/w repair and other waiting for repair.

State-6 Shut down state, one unit under software rejuvenation and other waiting for h/w repair. It is priority state.

State-7 Shut down state, both units are failed due to failure of software in which one unit continuously under s/w rejuvenation and other waiting for rejuvenation.

State-8 Shut down state. One unit continuously under s/w rejuvenation and other waiting for PM of hardware components.

State-9 Shut down state. One unit continuously under s/w rejuvenation and other waiting for repair of failed hardware components.

State-10 Shut down and priority state. One unit under s/w rejuvenation and other waiting for PM of hardware components.

State-11 Shut down state. One unit continuously under PM of h/w components and other waiting for repair of failed hardware components.

State-12 Shut down state. One unit continuously under PM of h/w components and other waiting for PM of hardware components.

2 Performance Indices

The probability density function of all random variables is as follows:

$$f(t) = \theta \eta t^{\eta-1} e^{-\theta t^\eta}, f_1(t) = \theta_1 \eta t^{\eta-1} e^{-\theta_1 t^\eta}, h(t) = \lambda \eta t^{\eta-1} e^{-\lambda t^\eta},$$
$$h_1(t) = \lambda_1 \eta t^{\eta-1} e^{-\lambda_1 t^\eta}, S(t) = \beta \eta t^{\eta-1} e^{-\beta t^\eta} \ S_1(t) = \beta_1 \eta t^{\eta-1} e^{-\beta_1 t^\eta} \quad where \ t \geq 0$$
$$and \ \beta, \beta_1, \eta, \theta, \theta_1, \lambda, \lambda_1 > 0$$

2.1 Transition Probabilities

By using simple probabilistic concepts and probability distributions appended above, the transition probabilities are obtained as follows:

$$p_{01} = \int [\text{Probability that system suffers by software failure during time } (t, t + dt)]dt$$

$$= \int_0^\infty S(t)\overline{H(t)F(t)}dt$$

$$= \beta\eta \int t^{\eta-1}e^{-(\lambda+\beta+\theta)t^\eta}dt, \quad p_{02} = \lambda\eta \int t^{\eta-1}e^{-(\lambda+\beta+\theta)t^\eta}dt,$$

$$p_{03} = \theta\eta \int t^{\eta-1}e^{-(\lambda+\beta+\theta)t^\eta}dt, \quad p_{10} = \eta\beta_1 \int t^{\eta-1}e^{-(\beta_1+\beta+\lambda+\theta)t^\eta}dt,$$

$$p_{18} = \theta\eta \int t^{\eta-1}e^{-(\beta_1+\beta+\lambda+\theta)t^\eta}dt, \quad p_{17} = \beta\eta \int t^{\eta-1}e^{-(\beta_1+\beta+\lambda+\theta)t^\eta}dt,$$

$$p_{19} = \lambda\eta \int t^{\eta-1}e^{-(\beta_1+\beta+\lambda+\theta)t^\eta}dt, \quad p_{20} = \lambda_1\eta \int t^{\eta-1}e^{-(\lambda_1+\lambda+\beta+\theta)t^\eta}dt, \tag{1}$$

$$p_{24} = \theta\eta \int t^{\eta-1}e^{-(\lambda_1+\lambda+\beta+\theta)t^\eta}dt, \quad p_{25} = \lambda\eta \int t^{\eta-1}e^{-(\lambda_1+\lambda+\beta+\theta)t^\eta}dt,$$

$$p_{26} = \beta\eta \int t^{\eta-1}e^{-(\lambda_1+\lambda+\beta+\theta)t^\eta}dt, \quad p_{30} = \theta_1\eta \int t^{\eta-1}e^{-(\theta_1+\lambda+\beta+\theta)t^\eta}dt,$$

$$p_{3.12} = \theta\eta \int t^{\eta-1}e^{-(\theta_1+\lambda+\beta+\theta)t^\eta}dt, \quad p_{3.10} = \eta\beta \int t^{\eta-1}e^{-(\theta_1+\lambda+\beta+\theta)t^\eta}dt,$$

$$p_{3.11} = \lambda\eta \int t^{\eta-1}e^{-(\theta_1+\lambda+\beta+\theta)t^\eta}dt,$$

$$p_{43} = p_{52} = p_{62} = p_{71} = p_{83} = p_{92} = p_{10.3} = p_{11.2} = p_{12.3} = 1 \ \& \ p_{24} = p_{23.4},$$

$$p_{25} = p_{22.5}, p_{11.7} = p_{17}, p_{18} = p_{23.4} = p_{19} = p_{12.9}, p_{3.11} = p_{32.11}, p_{33.12} = p_{3.12}$$

2.2 Mean Sojourn Times

The mean Sojourn times at various states are given as follows:

$$\Psi_0 = \int_0^\infty e^{-(\beta+\lambda+\theta)t^\eta}dt = \frac{\Gamma(1+\frac{1}{\eta})}{(\beta+\lambda+\theta)^{1/\eta}}, \Psi_1 = \frac{\Gamma(1+\frac{1}{\eta})}{(\beta_1+\beta+\lambda+\theta)^{1/\eta}}, \Psi_2 = \frac{\Gamma(1+\frac{1}{\eta})}{(\lambda_1+\beta+\lambda+\theta)^{1/\eta}},$$

$$\Psi_3 = \frac{\Gamma(1+\frac{1}{\eta})}{(\theta_1+\beta+\lambda+\theta)^{1/\eta}}, \quad \Psi_6 = \frac{\Gamma(1+\frac{1}{\eta})}{(\beta_1)^{1/\eta}}, \quad \Psi_{10} = \frac{\Gamma(1+\frac{1}{\eta})}{(\beta_1)^{1/\eta}}$$

$$\tag{2}$$

2.3 Mean Time to System Failure

Let $Z_i(t)$ denote the cumulative density function of first passage time between the regenerative states of the system model by considering failed state as an absorbing state. By using probabilistic arguments, the following recurrence relations are obtained:

$$Z_i(t) = \sum_j Q_{i,j}(t) \circledR Z_j(t) + \sum_k Q_{i,k}(t) \tag{3}$$

After applying Laplace Stieltjes transformation on (3), we obtained the following system of equation:

$$Z_i^{**}(s) = \sum_j Q_{i,j}(s) * Z_j^{**}(s) + \sum_k Q_{i,k}(s) \tag{4}$$

Evaluate the value of $Z_0^{**}(s)$ from equation. The mean time to system is given by following relation

$$\lim_{s \to o} \frac{1 - Z_i^{**}(s)}{s} = \frac{N_1}{D_1}$$

where

$$N_1 = \psi_0 \text{ and } D_1 = 1 - p_{01}p_{10} - p_{02}p_{20} - p_{03}p_{30} \tag{5}$$

2.4 Availability Analysis

By using simple probabilistic arguments and regenerative processes, the recurrence relations of steady state availability $V_i(t)$ are given by

$$V_i(t) = M_i(t) + \sum_j q_{i,j}^{(n)} \copyright V_j(t) \tag{6}$$

where $V_i(t)$ denotes the probability of transition between regenerative states in transitions. $S_i(t)$ be the probability to remain available at a regenerative state without failure. The Laplace transformation of above recurrence relations (6) is as follows: $V_i(s) = M_i(s) + \sum_j q_{i,j}^{(n)} * V_j(s)$.

The steady state availability is given by

$$A(\infty) = \lim_{s \to 0} s V_0^*(s) = \frac{N_1(0)}{D_1'(0)} \tag{7}$$

where

$$D_1(s) = \begin{vmatrix} 1 & -q_{01}^*(s) & -q_{02}^*(s) & -q_{03}^*(s) & 0 & 0 \\ -q_{10}^*(s) & 1-q_{11.7}^*(s) & -q_{12.9}^*(s) & -q_{13.8}^*(s) & 0 & 0 \\ -q_{20}^*(s) & 0 & 1-q_{22.5}^*(s) & -q_{23.4}^*(s) & -q_{26}^*(s) & 0 \\ -q_{30}^*(s) & 0 & -q_{32.11}^*(s) & 1-q_{33.12}^*(s) & 0 & -q_{3.10}^*(s) \\ 0 & 0 & -q_{62}^*(s) & 0 & 1 & 0 \\ 0 & 0 & 0 & -q_{10.3}^*(s) & 0 & 1 \end{vmatrix}$$

and

$$N_1(s) = \begin{vmatrix} S_0^*(s) & -q_{01}^*(s) & -q_{02}^*(s) & -q_{03}^*(s) & 0 & 0 \\ S_1^*(s) & 1-q_{11.7}^*(s) & -q_{12.9}^*(s) & -q_{13.8}^*(s) & 0 & 0 \\ S_2^*(s) & 0 & 1-q_{22.5}^*(s) & -q_{23.4}^*(s) & -q_{26}^*(s) & 0 \\ S_3^*(s) & 0 & -q_{32.11}^*(s) & 1-q_{33.12}^*(s) & 0 & -q_{3.10}^*(s) \\ 0 & 0 & -q_{62}^*(s) & 0 & 1 & 0 \\ 0 & 0 & 0 & -q_{10.3}^*(s) & 0 & 1 \end{vmatrix}$$

2.5　Busy Period Analysis for Repairman

By using simple probabilistic arguments and regenerative processes, the recurrence relations of busy period of the repairman $B_i(t)$ are given by

$$B_i(t) = K_i(t) + \sum_j q_{i,j}^{(n)}(t) \circledcirc B_j(t) \tag{8}$$

where $B_i(t)$ denotes the probability of transition between regenerative states in transitions for busy period of repairman. $M_i(t)$ be the probability to remain busy of repairman at a regenerative state. The LST of above recurrence relations (8) is as follows:

$$B_i^{**}(s) = K_i(s) + \sum_j q_{i,j}^{(n)}(s) * B_j^{**}(s).$$

The busy period of the repairman is given by

$$B(\infty) = \lim_{s \to 0} s B_0^*(s) = \frac{N_2(0)}{D_1'(0)} \tag{9}$$

where

$$N_2(s) = \begin{vmatrix} 0 & -q_{01}^*(s) & -q_{02}^*(s) & -q_{03}^*(s) & 0 & 0 \\ W_1^*(s) & 1-q_{11.7}^*(s) & -q_{12.9}^*(s) & -q_{13.8}^*(s) & 0 & 0 \\ W_2^*(s) & 0 & 1-q_{22.5}^*(s) & -q_{23.4}^*(s) & -q_{26}^*(s) & 0 \\ W_3^*(s) & 0 & -q_{32.11}^*(s) & 1-q_{33.12}^*(s) & 0 & -q_{3.10}^*(s) \\ W_6^*(s) & 0 & -q_{62}^*(s) & 0 & 1 & 0 \\ W_{10}^*(s) & 0 & 0 & -q_{10.3}^*(s) & 0 & 1 \end{vmatrix}$$

and $D_1(s)$ is obtained already.

2.6 Profit Analysis

The profit generated by the integrated h/w and s/w system for a particular set of values of all parameters in long run can be obtained as follows:

$$P = K_0 A - K_1 B \tag{10}$$

where K_0 and K_1 indicate the gross income generated by the computing gadget and expenditure on repairman for performing repair activities per unit time, A represents availability, and B denotes the busy period of server.

3 Numerical Study

In the present investigation, numerical values for availability, mean time to system failure, and profit function of the integrated h/w and s/w system are depicted for a particular case by assuming that shape parameter (η) for all random variables is equal to one.

Let the parameter assume the following set of values:

(i) $\eta = 1, \theta = 2, \theta_1 = 8, \beta = 0.0009, \beta_1 = 1.2, \lambda_1 = 1.9$
(ii) $\eta = 1, \theta = 2, \theta_1 = 8, \beta = 0.009, \beta_1 = 1.2, \lambda_1 = 1.9$
(iii) $\eta = 1, \theta = 5, \theta_1 = 8, \beta = 0.0009, \beta_1 = 1.2, \lambda_1 = 1.9$
(iv) $\eta = 1, \theta = 2, \theta_1 = 8, \beta = 0.0009, \beta_1 = 2.1, \lambda_1 = 1.9$
(v) $\eta = 1, \theta = 2, \theta_1 = 8, \beta = 0.0009, \beta_1 = 1.2, \lambda_1 = 2.9$
(vi) $\eta = 1, \theta = 2, \theta_1 = 15, \beta = 0.0009, \beta_1 = 1.2, \lambda_1 = 1.9.$

Table 1 Software rejuvenation and other repair activities on MTSF with respect to hardware failure rate

Λ	$\eta = 1$, $\beta = 0.0009$, $\theta_1 = 8$, $\theta = 2$, $\beta_1 = 1.2$, $\lambda_1 = 1.9$	$\eta = 1$, $\beta = 0.009$, $\theta_1 = 8$, $\beta_1 = 1.2$, $\lambda_1 = 1.9$, $\theta = 2$	$\eta = 1$, $\beta = 0.0009$, $\theta_1 = 8$, $\beta_1 = 1.2$, $\lambda_1 = 1.9$, $\theta = 5$	$\eta = 1$, $\beta = 0.0009$, $\theta_1 = 8$, $\beta_1 = 2.1$, $\lambda_1 = 1.9$, $\theta = 2$	$\eta = 1$, $\beta = 0.0009$, $\theta_1 = 8$, $\beta_1 = 1.2$, $\lambda_1 = 2.9$, $\theta = 2$	$\eta = 1$, $\beta = 0.0009$, $\theta_1 = 15$, $\beta_1 = 1.2$, $\lambda_1 = 1.9$, $\theta = 2$
0.01	2.4543	2.4164	0.5172	2.4550	2.4606	4.1300
0.02	2.4142	2.3775	0.5146	2.4150	2.4265	4.0268
0.03	2.3753	2.3397	0.5121	2.3760	2.3932	3.9283
0.04	2.3375	2.3029	0.5096	2.3382	2.3607	3.8342
0.05	2.3008	2.2672	0.5071	2.3015	2.3289	3.7441
0.06	2.2651	2.2325	0.5046	2.2658	2.2979	3.6579
0.07	2.2304	2.1987	0.5022	2.2311	2.2676	3.5753
0.08	2.1967	2.1658	0.4997	2.1973	2.2380	3.4960
0.09	2.1639	2.1338	0.4973	2.1644	2.2090	3.4199

Table 2 Software rejuvenation and other repair activities versus steady state availability with respect to hardware failure rate

λ	$\eta = 1$, $\beta = 0.0009$, $\theta_1 = 8$, $\theta = 2$, $\beta_1 = 1.2$, $\lambda_1 = 1.9$	$\eta = 1$, $\beta = 0.009$, $\theta_1 = 8$, $\beta_1 = 1.2$, $\lambda_1 = 1.9$, $\theta = 2$	$\eta = 1$, $\beta = 0.0009$, $\theta_1 = 8$, $\beta_1 = 1.2$, $\lambda_1 = 1.9$, $\theta = 5$	$\eta = 1$, $\beta = 0.0009$, $\theta_1 = 8$, $\beta_1 = 2.1$, $\lambda_1 = 1.9$, $\theta = 2$	$\eta = 1$, $\beta = 0.0009$, $\theta_1 = 8$, $\beta_1 = 1.2$, $\lambda_1 = 2.9$, $\theta = 2$	$\eta = 1$, $\beta = 0.0009$, $\theta_1 = 15$, $\beta_1 = 1.2$, $\lambda_1 = 1.9$, $\theta = 2$
0.01	0.9494	0.9465	0.8032	0.9497	0.9505	0.9816
0.02	0.9468	0.9439	0.8006	0.9470	0.9490	0.9789
0.03	0.9442	0.9412	0.7979	0.9444	0.9476	0.9762
0.04	0.9416	0.9386	0.7953	0.9418	0.9461	0.9735
0.05	0.9389	0.9360	0.7927	0.9392	0.9446	0.9709
0.06	0.9363	0.9334	0.7901	0.9366	0.9432	0.9682
0.07	0.9337	0.9308	0.7875	0.9340	0.9417	0.9655
0.08	0.9311	0.9282	0.7850	0.9314	0.9402	0.9629
0.09	0.9285	0.9256	0.7824	0.9288	0.9387	0.9602

Putting all these values in Eqs. (5), (7), and (10) and obtained the values for mean time to system failures, availability and profit function and respectively shown in Tables 1, 2, and 3. From Tables 1, 2, and 3, we find that the mean time to system failure, availability, and profit decrease with the increase of hardware failure (λ), software failure rate (β), and maximum operation time (θ) while the values of these measures increase with the increase of preventive maintenance rate (θ_1),

Table 3 Software rejuvenation and other repair activities versus profit function w.r.t. hardware failure rate

Λ	$\eta = 1,$ $\beta = 0.0009,$ $\theta_1 = 8,$ $\theta = 2,$ $\beta_1 = 1.2,$ $\lambda_1 = 1.9$	$\eta = 1,$ $\beta = 0.009,$ $\theta_1 = 8,$ $\beta_1 = 1.2,$ $\lambda_1 = 1.9,$ $\theta = 2$	$\eta = 1,$ $\beta = 0.0009,$ $\theta_1 = 8,$ $\beta_1 = 1.2,$ $\lambda_1 = 1.9,$ $\theta = 5$	$\eta = 1,$ $\beta = 0.0009,$ $\theta_1 = 8,$ $\beta_1 = 2.1,$ $\lambda_1 = 1.9,$ $\theta = 2$	$\eta = 1,$ $\beta = 0.0009,$ $\theta_1 = 8,$ $\beta_1 = 1.2,$ $\lambda_1 = 2.9,$ $\theta = 2$	$\eta = 1,$ $\beta = 0.0009,$ $\theta_1 = 15,$ $\beta_1 = 1.2,$ $\lambda_1 = 1.9,$ $\theta = 2$
0.01	4651.0	4635.6	3861.0	4652.2	4656.5	4848.9
0.02	4637.0	4621.6	3847.8	4638.2	4648.5	4834.4
0.03	4623.0	4607.6	3834.7	4624.3	4640.4	4820.0
0.04	4609.1	4593.7	3821.7	4610.4	4632.4	4805.6
0.05	4595.2	4579.8	3808.7	4596.5	4624.3	4791.2
0.06	4581.4	4565.9	3795.8	4582.6	4616.3	4776.9
0.07	4567.6	4552.1	3783.0	4568.8	4608.2	4762.5
0.08	4553.8	4538.3	3770.3	4555.0	4600.2	4748.2
0.09	4540.0	4524.5	3757.6	4541.2	4592.1	4734.0

software rejuvenation rate (β_1), and hardware repair rate (λ_1). So, from this analysis, we conclude that an integrated h/w and s/w system with priority to software rejuvenation over hardware repair and hardware preventive maintenance can be made more reliable and profitable by increasing software rejuvenation rate.

References

1. Asif M, Majumdar S, Kopec G (2007) Load sharing in call server clusters. Comput Commun 30(16):3027–3045
2. Hajeeh MA (2011) Availability of deteriorated system with inspection subject to common-cause failure and human error. Int J Oper Res 12(2):207–222
3. Jain M, Agrawal SC, Preeti (2010) Availability analysis of software-hardware system with common cause shock failure, spare and switching failure. Int J Int Acad Phys Sci 14(1):1–13
4. Jain M, Preeti (2015) Availability analysis of software rejuvenation in active/standby cluster system. Int J Ind Syst Eng 19(1):75–93
5. Kumar A, Malik SC (2012) Reliability modeling of a computer system with priority to s/w replacement over h/w replacement subject to MOT and MRT. Int J Pure Appl Math 80 (5):693–709
6. Friedman MA, Tran P (1992) Reliability techniques for combined hardware/software systems. In: Proceedings of annual reliability and maintainability symposium, pp 209–293
7. Welke SR, Johnson BW, Aylar JH (1995) Reliability modeling of hardware/software systems. IEEE Trans Reliab 44(3):413–418
8. Lai CD, Xie M, Poh KL, Dai YS, Yang P (2002) A model for availability analysis of distributed software/hardware systems. Inf Softw Technol 44:343–350
9. Malik SC, Kumar A (2011) Profit analysis of a computer system with priority to software replacement over hardware repair subject to maximum operation and repair times. Int J Eng Sci Technol 3(10):7452–7468

10. Kumar A, Malik SC (2012) Stochastic modeling of a computer system with priority to pm over s/w replacement subject to maximum operation and repair times. Int J Comput Appl 43 (3):27–34
11. Kumar A, Malik SC, Barak MS (2012) Reliability modeling of a computer system with independent h/w and s/w failures subject to maximum operation and repair times. Int J Math Achiev 3(7):2622–2630
12. Koutras VP, Platis AN (2010) Semi-Markov performance modelling of a redundant system with partial, full and failed rejuvenation. Int J Critic Comput-Based Syst 1(1–3):59–85
13. Okamura H, Dohi T (2011) Application of reinforcement learning to software rejuvenation. Tenth international symposium on autonomous decentralized system, pp 647–652
14. Wang D, Xie W, Trivedi KS (2007) Performability analysis of clustered systems with rejuvenation under varying workload. Perform Eval 64(9):247–265
15. Zhou X, Ippoliti D (2008) Resource allocation optimization for quantitative service differentiation on server clusters. J Parallel Distrib Comput 68(9):1250–1262

V2V-DDS Approach to Provide Sheltered Communication Over ALTERATION ATTACK to Control Against Vehicular Traffic

P. Prittopaul, M. Usha, M.V.S. Santhosh, R. Sharath and E. Kughan

Abstract The objective of this paper is to focus on secure communication over VANET which is a more questioning issue nowadays. Vehicular adhoc networks mainly focus on road safety which is used to provide a way to prevent road accidents, to pass information about road traffic etc. In vehicular to Infrastructure (V2I) scenario, an RSU unit is used to provide security by giving certification in order to identify authorized users and also uses digital signature approach to provide confidentiality and authentication for secure communication. But in case of vehicular to vehicular (V2V) scenario, no such security is provided between the communicating parties and so there might arise a situation in which the nodes act maliciously over other nodes.

Keywords Vehicular-to-vehicular · Vanet · Digital signature · Authentication Malicious node

P. Prittopaul (✉) · M. Usha · M.V.S. Santhosh · R. Sharath · E. Kughan
Velammal Engineering College, Chennai, India
e-mail: p.prittopaul@gmail.com

M. Usha
e-mail: umahalingam@gmail.com

M.V.S. Santhosh
e-mail: santhoshmvs38@gmail.com

R. Sharath
e-mail: rsharath17@gmail.com

E. Kughan
e-mail: ekughan@gmail.com

© Springer Nature Singapore Pte Ltd. 2018   109
D.K. Mishra et al. (eds.), *Information and Communication Technology for Sustainable Development*, Lecture Notes in Networks and Systems 9, https://doi.org/10.1007/978-981-10-3932-4_12

1 Introduction

Vehicular Ad Hoc Networks (VANETs) is a cutting-edge technology that provides Wi-Fi services to all the vehicles and on-the-road communication facilities. There are two main types of vehicular communication. Namely, Vehicle-to-Vehicle (V2V) and Vehicle-to-Roadside (VRC) or Vehicle-to-Infrastructure (V2I) Communications. VANETs can be utilized for a broad range of safety and non-safety applications. Namely,

- Additional services such as vehicle safety, automated toll payment, traffic administration, improved navigation.
- Location-based services such as finding the neighboring fuel station, restaurant or travel lodge.
- Infotainment applications such as providing right to use the Internet.

When it comes to vehicle safety, it is known that 60% of accidents can be avoided if the drivers get a notice of the possible danger 0.50 s before it. VANETs are really useful in giving such threat warnings to the drivers, and thereby avoiding major accidents. And in cases of traffic jams, if the police controlling the traffic get some information about the density of the traffic in different lanes, then they can easily clear the traffic. VANETs can also be used to give directions to the drivers about anything like a fuel station or some other place they want to go. One of the major benefits of VANETs is that they provide internet access to drivers, the minimum requirement for a driver to use the above given features is to have internet access and that is provided by VANETs. In V2I communication, two vehicles communicate through a special infrastructure called RSU (Road Side Unit) that is placed on the either sides of roads at equal distances. They have a phenomenon called Digital Signature instilled in them. The communication takes place as follows

- The vehicle, v1 sends a request to the nearest RSU requesting communication with another vehicle, v2. The RSU then sends parameters called Certified Authentication (CA), Trust (T), and a unique ID to v1. v1 sends a data packet to v2.
- v2 asks for verification to the RSU. This in turn verifies and connects v1 and v2 so that their communication starts.

To cut the cost of infrastructure, vehicles began to communicate with each other without the help of RSU units known as V2V communication. Though the cost was reduced, a new set of problems surfaced. In this paper, we are giving a solution on two major problems. First, when two vehicles communicate directly without any external infrastructure, a third party can hack into the system and alter the messages. To avoid this, we have come up with a phenomenon called Direct Digital Signature. Second, when vehicles communicate with each other, it is difficult to know whether a vehicle is still present within the range of communication. This leads to loss of data packets. To overcome this problem, we calculate the trust value of how long

t

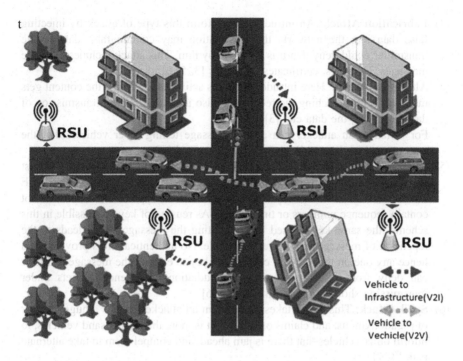

Vehicle to
Infrastructure(V2I)

Vehicle to
Vechicle(V2V)

2　Security Issues of Vehicular Ad Hoc Networks

VANETs, in spite of having so many advantages, have few security concerns and challenges [1] as well. Namely,

A. Attacks and Threats

(1) **Denial of Service attack**: This type of attack appears when the intruder takes control of a vehicle's resources, and hence deny the information from reaching to the specified receiver over the communication channel used by the Vehicular Network, which leads in preventing the most priority information to be reached at the destination on time. For instance, a malicious node can also create fake data to get transmitted in the network which creates a mesh up in order to make the resources to deny of data packets reaching the receiver on a highway [2–5].

(2) **Message Suppression Attack**: This type of attack is performed by an attacker by simply dropping packets selectively from the network. These packets may contain important information [3]. The target of such an invader would be to thwart registration and cover authorities from learning about collisions to avoid delivering collision reports to roadside access points [6].

(3) **Fabrication Attack**: An intruder can perform this type of attack by injecting false data into the network, the information may contain false data or the transmitter could deny that it is not done by him. This attack includes fabricate messages, warnings, certificates, identities [3, 5, 6].

(4) **Alteration Attack**: Here intruder performs active attack where the content gets modified before reaching the receiver. It also includes delay of transmission of data, replaying the data etc. [3].
For instance, an attacker can alter a message telling other vehicles that the current road is clear while the road is congested [6].

(5) **Replay Attack**: This type of attack is performed by the intruder by just replaying the old transmitted messages just to divert the receiver [3]. IEEE 802.11 does not have security measures against such type of threats. It does not contain sequence numbers or timestamps. As reusing of keys is possible in this scheme, the same key is used for encrypting the message which leads to the repetition of messages on the recipient side. No authentication is provided and hence any one on the network can use the key to encrypt the message and later can deny from it. This causes non-repudiation act. Hence packets under transmission should contain timestamps [6].

(6) **Sybil Attack**: This attack takes place when an attacker creates a huge number of pseudonymous, and claims or acts like it is more than a thousand vehicles, to inform other vehicles that there is jam ahead, and compel them to take alternate path [3, 7].

2.1 Location Aided Routing

Design of routing protocols is a crucial problem in ad hoc networks [8, 9], one such type of routing protocol is the table driven protocol where each vehicle has a table. In that table, it contains information about its neighbouring vehicles, their distances and possible routes. This is a tedious process as the table needs to be updated every time a vehicle goes out of range. This table also requires constant battery supply. Another type of routing protocol is the on-demand routing protocol where the message is sent by a vehicle dynamically.

LAR1 in VANET

The Location-Aided routing protocol1 [10–12] (LAR1) is a reactive on-demand source routing protocol that uses the location information of the moving nodes. Such information about mobility is obtained by using Global Positioning System (GPS) [13, 14]. In LAR, location information about moving nodes flood a route request packet in a forwarding zone called request zone instead to entire network [15]. LAR1 protocol uses GPS to communicate between any two nodes. V2V communication in VANETs uses this LAR1 protocol to provide communication between vehicles.

For instance, if an accident occurs, the following steps take place:

- The vehicle, v1 which is in the closest proximity of the occurred accident will send the information of accident to the vehicles which are its immediate neighbors.
- The neighbors who receive the information will in turn send the information to their immediate neighbors.
- This continues until there are no more vehicles in the domain.
- The vehicles in the same range/network as that of v1 will not receive any information of the accident, as it would have already got that information.
- The information is sent to each vehicle with a unique sequence number to avoid collision.

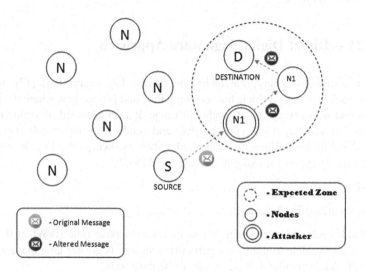

3 Detection of Malicious Nodes

In vanet, it is important to identify the malicious nodes [16], as they can tamper or alter the message that is being sent over the network. Hence, malicious nodes are detected if any of the following scenarios take place

- **Lost Acknowledgement**
  - Every vehicle sends an acknowledgement to the sender vehicle.
  - If an acknowledgement is not received, the sender retransmits the message again to the receiver.

- If the acknowledgement is not received for the third time, the receiver becomes a malicious node.

- **Energy Level**

  - The energy levels of each vehicle are published to every other vehicle in the domain.
  - If a particular vehicle does not send its energy level till a particular time-limit, then that vehicle becomes a malicious node.

- **Inactive Vehicles**

  - All the vehicles in a domain must be involved in at least one communication.
  - If a vehicle remains inactive for a long duration of time, then, that becomes a malicious node.

4 V2V—Direct Digital Signature Approach

It is an asymmetric encryption technique (public key encryption) [17]. In this method, each and every vehicle has its own public and private key, where the public key is broadcast to every vehicle within its range. If a vehicle needs to communicate with another vehicle, it encrypts the data and sends it to the vehicle using that vehicles' public key. When the vehicle identifies its own public key, it reads the message by decrypting it using its private key [18–20].

TRUST

Trust is identified by 3 parameters:

Threshold: each and every vehicle has its own response time (t).When the time taken for that vehicle to transmit a particular message is greater than that response time, then that particular vehicle is said to be malicious.

Frequent broadcast of messages: Whenever a message is transmitted to a particular node, it must be able to either respond to it or transmit it to any vehicle. This process must be done on a continuous basis. Whenever there is no reply, that vehicle is removed out of that particular range. It comes back inside the range when it begins to transmit again.

Energy level: Sometimes, when a vehicle is having low battery power, it ceases to forward/send/reply to messages. Thereby becoming a malicious node.

4.1 Pre-requisites for Direct Digital Signature Algorithm

- **group head**: each domain (A collection of vehicles within a finite range) has a leader which establishes connection between the sender and the receiver.

- **sender**: A vehicle that wants to send a message is called as a sender.
- **receiver**: A vehicle to which a message is sent, is called as a receiver.
- **public key**: Each vehicle should have a public key and this key should be readable to every other vehicle in the domain.
- **private key**: Each vehicle should have a private key that is readable and writable only to that vehicle. (i.e.), the private key of a vehicle is accessible only to that vehicle.

4.2 Procedure

- A sender vehicle, v1 sends a request to the group head requesting a connection with a receiver vehicle, v2.
- The group head communicates with the receiver, v2 and waits for an acknowledgement from v2.
- If the acknowledgement arrives, then a connection is established between the sender, v1 and the receiver, v2.
- A message consists of two parts, namely, A sign and the body of the message.
- v1 encrypts the message with its private key, put a sign which is readable only to v2, and sends it to v2 through its public key which is accessible to v2.
- Only v2 will be able to decrypt the message because the sign of v1 is readable only to v2. Thereby, avoiding a third party to hack into the message and manipulate it.
- v2 decrypts the message using the public key of v1.
- v2 sends a reply to v1 using the same process that is described above.

4.3 Selection of the Group Head

The group head is selected by the following steps,

- The vehicle with the highest energy level becomes the group head.
- If two vehicles have the same energy level and they happen to be the highest. Then, the selection of the group head is done by voting.

4.4 Detection of Malicious Node

- **Members**

The group head identifies the malicious members of a domain based on timeouts, energy levels and involvement in the communication.

- **Group Head**

If the energy level of a group node decreases, it becomes slow in establishing a communication between the vehicles. Then, the group is said to be a malicious node and the vehicle which is having the second best energy level will become the new group head.

4.5 Comparative Analysis

As it can be seen from the graph 1, the through put though is comparatively steady for the proposed system unlike the existing one that shows a rise in the graph initially and then maintains a steady state after a minor drop. The steadiness in our proposed system is because it uses Asymmetric encryption technique that takes constant time for authentication providing procedure for every node. However it more than makes up in the security features it provides.

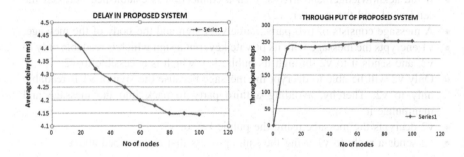

5 Conclusion

As movement of people from place to place on a regular basis increases, they look for safety and security in their travel. Seeing the technology takes leap after leap, they expect a lot from these to protect them from any kinds of danger including the ones they meet on roads. VANET, though is an excellent invention to serve this purpose, still needs a long way to go before they can earn the trust of the people. Along the process comes or proposal wherein we use a trust based model with the highly regarded Digital Signature Standard approach in a VANET network. Because of the usage of Location Aided Routing Protocol, the latency caused due to distance calculation has been avoided. Adopting Trust based model in node authentication prevents participation of ineligible node and maintains continuity in message passing through the network. Since DDS mechanism is session oriented and Asymmetric, it is secure from external interface or hacking and messages are safely transmitted between the nodes. The trust based model along with the DSS approach has proven to improve the communication between two VANET nodes in terms of Authentication and confidentiality.

References

1. Ghassan S, Al-Salihy WAH, Sures R (2010) Security analysis of vehicular ad hoc networks (VANET). In: 2010 second international conference on network applications, protocols and services, national advanced IPv6 center. Universiti Sains Malaysia Penang, Malaysia
2. Raya M, Papadimitratos P, Hubaux JP (2006) Securing vehicular communications. IEEE Wirel Commun 13
3. Parno B, Perrig A (2005) Challenges in securing vehicular networks. In: Proceedings of HotNets-IV
4. Aad I, Hubaux JP, Knightly EW (2008) Impact of denial of service attacks on ad hoc networks. IEEE/ACM Trans Network 16
5. Raya M, Pierre Hubaux J (2005) The security of VANETs. In: Proceedings of the 2nd ACM international workshop on vehicular adhoc networks
6. Security & Privacy for DSRC-based Automotive Collision Reporting (2010) In: Second international conference on network applications, protocols and services
7. Douceur J (2003) The Sybil attack. In: First international workshop on peer-to-peer systems, 1st edn. Springer, USA
8. Corson S, Batsell S, Macker J (1996) Architectural considerations for mobile mesh networking (Internet draft RFC, version 2). In: Mobile ad-hoc network (MANET) working group, IETF
9. Ramanathan S, Steenstrup M (1996) A survey of routing techniques for mobile communication networks. Mobile Netw Appl 89–104
10. Ko Y-B, Vaidya NH (1997) Using location information to improver outing in ad hoc networks. Technical report 97–013, Texas A&M University
11. Ko Y-B, Vaidya NH (1998) Location-aided routing in mobile ad hoc networks. Technical report 98-012, Texas A&M University
12. Ko Y-B, Vaidya NH (1998) Location-aided routing (LAR) in mobile ad hoc networks. In: Proceedings of MOBICOM'98 (1998)
13. Dommety G, Jain R (1996) Potential networking applications of global positioning systems (GPS). Technical report TR-24, The Ohio State University
14. Parkinson BW, Gilbert SW (1983) NAVSTAR: global positioning system—ten years later. Proc IEEE 71(10):1177–1186
15. Ko Y-B, Vaidya NH (2000) Location-aided routing (LAR) in mobile ad hoc networks in wireless networks vol 6, pp 307–321
16. Golle P, Greene D, Staddon J (2004) Detecting and correcting malicious data in vanets. In: VANET'04: proceedings of the 1st ACM international workshop on vehicular ad hoc networks. ACM, New York, NY, USA, pp 29–37
17. IEEE, IEEE 1609.2-Standard for Wireless Access in Vehicular Environments (WAVE)—Security Services for Applications and Management Messages, available from ITS Standards Program
18. Djenouri D, Khelladi L (2005) A survey of security issues in mobile ad hoc and sensor networks. IEEE Commun Surv Tutor 7(4):2–28
19. Stallings W (2003) Cryptography and network security, principles and practices, 3rd edn. Prentice Hall, pp 67–68 and 317–375
20. Forouzan BA (2008) Cryptography and network security. McGraw Hill, Boston

21. Kaur P, Arora N (2015) A comprehensive study of cryptography and digital signature. Int J Comput Sci Eng Technol (IJCSET)
22. Shiva Rama Krishna D (2015) Providing security to confidential information using digital signature. Int J Adv Res Comput Sci Manag Stud
23. Rehman S, Arif Khan M, Zia TA, Zheng L (2013) Vehicular ad-hoc networks (VANETs): an overview and challenges. J Wirel Network Commun
24. Zeadally S, Hunt R, Chen Y-S, Irwin A, Hassan A (2010) Vehicular ad hoc networks (VANETS): status, results and challenges. Springer Science + Business Media, LLC
25. Agrawal A, Garg A, Chaudhiri N, Gupta S, Pandey D, Roy T (2013) Security on vehicular ad hoc networks (VANET): a review paper. Int J Emerg Technol Adv Eng
26. Husain A, Kumar B, Doegar A (2010) A study of location aided routing (LAR) protocol for vehicular adhoc networks in highway scenario. Int J Eng Inform Technol (IJEIT)

Enhancing Performance of Data Centers Using Location-Aware Live VM Migration

Narander Kumar and Swati Saxena

Abstract Virtualization in cloud data centers ensures many benefits to cloud providers and customers, such as availability, efficiency, transparency, reduced costs, reliability, easier backups. An inefficient virtualization, however, can not only degrade the performance of a cloud data center but can also prove to be energy inefficient. To bring virtualization to our benefit, an efficient virtual machine migration is required where a demanding virtual machine is transferred from an overloaded host machine to another for the purpose of load balancing. This paper combines the benefits of existing virtual machine migration techniques in non-clustered data centers and presents a new improved live migration technique that takes place in clustered data centers, thereby taking full advantage of virtual machines' identities in clusters. Simulation experiments also show that our proposed migration technique is energy efficient by 28% as compared to migration in non-clustered environment, while maintaining low round-trip time, low latency and high throughput.

Keywords Cloud data centers · Virtualization · Migration · Clusters Efficiency

1 Introduction

The heart of cloud computing system lies in virtualization. This technique multiplies a single physical machine into several virtual machine (VM) instances. These VMs are placed on physical machines, using an appropriate VM Placement technique, and as and when the need arises, VM is transferred to a different physical machine

N. Kumar (✉) · S. Saxena
Department of Computer Science, Babasaheb Bhimrao
Ambedkar University (A Central University), Lucknow, India
e-mail: nk_iet@yahoo.co.in

S. Saxena
e-mail: swatesaxena@gmail.com

© Springer Nature Singapore Pte Ltd. 2018   119
D.K. Mishra et al. (eds.), *Information and Communication Technology for Sustainable Development*, Lecture Notes in Networks and Systems 9,
https://doi.org/10.1007/978-981-10-3932-4_13

causing VM migration. Reasons to initiate a VM migration can be an overloaded physical machine, increased energy consumption or a routine periodic migration to ensure new and effective placements of existing VMs.

Desired requirements of an effective VM migration are transparency, negligible service degradation, fast migration with minimum migration volume and security. Essential features of VM migration are as follows:

Destination host location of migration plays a crucial role as it adds to profits if destination machine is close by the source machine. This location-bound feature will be utilized by our proposed migration technique to reap maximum benefits. Choice of a VM to be migrated can be based on many factors, such as maximum CPU utilization, largest memory image and/or minimum migration time [1]. A virtual machine must preferably retain its IP address even after migration so as to avoid network redirection mechanisms. VM migration involves transfer of three types of virtual machine files, namely memory files (*.vmem), local disk files (*.vmdk) and network interface files (*.vmx) [4].

Most data centers today incorporate pull phase with stop-n-copy. Our proposed migration technique makes use of live migration and combines push and pull phases with stop-n-copy. The organization of this paper is as follows: Section 2 presents the review of related work on VM migration, Sect. 3 explains our proposed technique, Sect. 4 describes the simulation environment. Section 5 presents the experimental results, merits and limitations of the proposed technique, and Sect. 5 concludes the paper with a brief discussion on future works.

2   Related Work

Resources consumed by virtual machines, time taken for migration and energy consumption are some of the important factors which affect the live migration process [2, 3]. The effects of virtual machine migration on XEN machines are discussed in [4], and minimization of overheads incurred during live migration is suggested. Also a performance-based comparison of existing migration approaches, i.e., post-copy and pre-copy, is presented in [5] along with their advantages and disadvantages. An application-level solution, VMFlockMS, is described in [6] which discusses inter-data center virtual machine migrations in groups. A strategy to initiate migration at the right time is proposed in [1] to make it more efficient, profitable and reliable. VM migration introduced in [7] uses checkpoint/recovery and trace/replay mechanisms rather than transferring full VM image, and this approach reduces the migration volume overhead. A comparison of pre- and post-copy approaches of VM migration is also given in [8], which suggests a method to increase the migration speed by lowering down the migration volume. Replication is incorporated in cloud data centers as a tool to reduce migration latency and improve energy efficiency in [9]. Based on identical resource demands, virtual machines are grouped together and group migration is considered making use of the datacenter's topology in [10].

Power usage and energy consumption of servers are studied in [11–13] where VM migrations are initiated with the aim to reduce cost and carbon footprints of the cloud data center. Static and uneven dynamic user demands in a data center and planning of energy conservation based on these demands are discussed in [14]. In order to reduce SLA violations, [15] introduces dynamic grouping of servers and virtual machines for placements and migrations. Unnecessary VM migrations are reduced in [16] by initially prioritizing them based on their resource capacities and usages. Importance is laid on VM placements that can eventually reduce the need for further VM migrations, thereby making data centers more efficient and stable [17]. A detailed review of existing migration technologies and their comparison is given in [18–20], and stress is laid on the need of a strong, efficient migration scheme that strengthens trust among cloud users. Consolidation of physical machines based on RAM and CPU capacities is presented in [21] to save energy consumption, and a VM migration methodology is introduced which is based on local information. VM migrations based on renewable energies using various resource distribution techniques are considered in [21].

A study of heterogeneous workloads and implementation of live VM migration techniques by consolidating these workloads is presented in [22] to save energy consumption in cloud data centers. Implementation of a live migration technique using Amazon EC2 clone, Eucalyptus, is given in [23] with the aim to reduce energy consumption in the cloud computing environment. Also in [12], migration policies are suggested to move a server from a local data center to a more cost-effective location in a secure way. The use of virtual machines running same application or operating system instances and their migration among wide area networks is given in [19] to reduce migration workload. Most of the reviews discussed above attempt to make migration as effective as possible; however, they either stress on energy reduction or on improving migration statistics. A combined approach is lacking where not only VM migration is improved but also 'green computing' is applied. Our proposed work combines these two goals of energy reduction and performance enhancement by proposing a location-aware live migration mechanism which restricts migration of identical VMs within a cluster, thereby providing low latency, low round-trip time, low energy consumption but better throughput as compared with traditional approaches of VM migration.

3 Proposed Technique

The proposed location-aware live VM migration requires physical machines (PMs) to be arranged in clusters. There can be various ways to group PMs in a cluster; we are considering 'application type' as a benchmark for clustering, so that all the PMs servicing same or similar applications are grouped under one single cluster. Next we choose a virtual machine (VM) for migration and call it as 'victim' VM; it is usually the most demanding VM on an overloaded PM. Thereafter, a destination PM is selected to host the migrated VM. It accesses the profile file and execution log of the victim VM and synchronizes its functioning at the new destination host.

Finally, the current input–output requests from cloud users for the victim VM are directed toward this new host. Now, we present our proposed live VM migration technique, in detail

1. **Data center topology**: The first step in the proposed VM migration technique is to choose the data center topology. We have considered a fat-tree topology for a cloud data center for reasons like use of low-cost switches, scalability and contention prevention [5]. Figure 1 shows the proposed data center topology.

2. **Clustered environment**: Clusters are formed depending on the type of applications processed by physical machines [18]. Virtual machines, belonging to the same or similar applications, have identical memory images, so they can be hosted on PMs which form one cluster [21]. Further, we are restricting VM migrations within its own cluster, so as to take full advantage of identical memory images. This reduces migration volume as well as migration time. *The decision to restrict VM migration inside its cluster makes it a location-aware migration.* Advantages of using location-aware migration are: IP address of a migrating VM need not change after migration, network disk need not be transferred during migration, and very less amount of memory image needs to be migrated, as the destination PM in the cluster also has an identical image of the migrating virtual machine. This migration approach can also be seen as location-restricted virtual machine migration.

3. **Selection of source host, destination host and victim VM**: This step selects a source host which is overloaded and needs load shedding, a destination host which will host the migrated virtual machine and a 'victim' VM which is migrated from a source host to a destination host for load balancing.

   a. **Selections of an overloaded source host**: Physical machine in a cluster is continuously monitored for any sign of overload. This includes a sudden spike in a VM's traffic suggesting an increase in its I/O activity in case of a

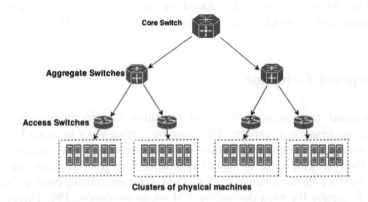

Fig. 1 Proposed data center topology

heavier load application or a spike in CPU or RAM usage. We are monitoring the utilization of four resources, namely CPU, disk, memory and bandwidth resources of each physical machine. Suppose a single PM is hosting n active VMs, each with its maximum resource capacity/demand as given in Table 1.

Let the total capacity of a PM w.r.t. the four resources be A (CPU), B (memory), C (disk) and D (bandwidth). These n VMs will be hosted by a PM iff Eq. 1 holds true

$$\sum_{i=1}^{n} a_i < A \,\&\&\, \sum_{i=1}^{n} b_i < B \,\&\&\, \sum_{i=1}^{n} c_i < C \,\&\&\, \sum_{i=1}^{n} d_i < D \qquad (1)$$

An overloaded PM will exhibit a characteristic as shown in Eq. 2

$$\sum_{i=1}^{n} a_i \cong A \,\|\, \sum_{i=1}^{n} b_i \cong B \,\|\, \sum_{i=1}^{n} c_i \cong C \,\|\, \sum_{i=1}^{n} d_i \cong D \qquad (2)$$

Also, total load on a PM_i caused by active VMs, at any instant, will be calculated as

$$Load(PM_i) = \Theta_{CPU} + \Theta_{mem} + \Theta_{disk} + \Theta_{net} \qquad (3)$$

In Eq. 3, $\Theta_{CPU},$ $\Theta_{mem},$ Θ_{disk} and Θ_{net} are the total utilization ratios of CPU, memory, disk and network bandwidth, respectively, by all active virtual machines on any PM_i and their individual values range from 0 to 1. For example, if all the active VMs are using its host memory up to their maximum demand (i.e., $b_1, b_2, \ldots, b_n$) then value of Θ_{mem} is 1, whereas in case, a particular resource is not in use by any of the active virtual machines, then its utilization ratio is 0. Moreover, load calculated in Eq. 3 ranges from 0 to 4. Accordingly, a threshold value of 3.5 is chosen to declare a physical machine as overloaded, i.e., if the load on a physical machine is equal to or exceeds 3.5, then it is considered 'overloaded' otherwise not.

b. **Selection of 'victim' VM for migration**: A crucial task in VM migration is to choose which VM(s) to migrate. In this paper, we are considering the migration of a single virtual machine terming it as a 'victim' VM. An active virtual machine, whose load on its overloaded host is the maximum, is selected for migration. To choose a victim VM, we calculate the load incurred by each VM on a PM as:

Table 1 Example of VMs demands on a single server

Max demand →	CPU	Memory	Disk	B/w
VM_1	a_1	b_1	c_1	d_1
VM_2	a_2	b_2	c_2	d_2
:	:	:	:	:
VM_n	a_n	b_n	c_n	d_n

$$Load\ incurred\ by\ a\ VM\ on\ a\ PM = \frac{Resources\ consumed\ by\ a\ VM}{Resource\ demanded\ by\ a\ VM} \quad (4)$$

c. **Selection of a destination host**: To choose a destination host, we maintain an 'availability' graph of physical machines in the cluster whose resource availabilities satisfy the resource requirements of the migrating VM. These available physical machines constitute the nodes of the graph, and the edges represent the distance between these physical machines. This graph will include the source host as well. Consider the graph shown in Fig. 2.

In Fig. 2, PM3 is the source host containing VM11 which needs to be migrated. All the other nodes represent physical machines which satisfy the resource availability criteria of the migrating VM, i.e., all these PMs (PM1, PM2, PM4, PM5 and PM6) can host the migrating VM. In order to select the best possible destination host, we consider the distance from source host to all other available hosts. As is shown in the graph, PM3 is directly linked with PM2 and PM4 with distance 7 and 2, respectively. With the aim to reduce the migration time, we select the minimum distance of the destination host from source host PM3, and hence, PM4 is chosen as the destination physical machine. Note that all these PMs belong to a single cluster.

The destination host machine retrieves the profile file of victim virtual machine from network-attached storage (NAS) device. Profile files (.nvram files) of virtual machines contain the data blocks which are required at boot time and application startup time. They assist in reconstructing the full image of the victim VM at the destination host machine. Next, the execution log of the victim VM, up to the latest checkpoint, is transferred from source host machine to the destination host machine. This helps in synchronizing the victim VM at the source host with the newly started destination VM. So far, the source VM is servicing the input–output requests of the cloud customer.

4. After the completion of step 5, the new migrated VM at the destination is in a position to service the input–output requests of cloud customers. At this juncture, source VM is stopped and all input–output requests are directed to the new VM at a new host. If the requested data are not available with the new migrated VM, then it is requested from the old VM at the source machine on a higher priority basis, reflecting the pull phase of traditional migration. The diagrammatic view of the proposed procedure is shown in Fig. 3.

Fig. 2 'Availability' graph example

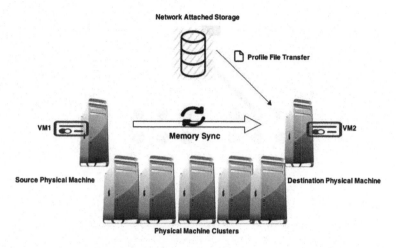

Fig. 3 Proposed live VM migration technique

4 Evaluation with Simulation

We have simulated the proposed technique of virtual machine migration in clusters using ns3 and compared it with migration in a non-clustered environment. Two scenarios are generated. Figure 4 shows the proposed live virtual machine migration in a clustered data center consisting of 12 physical machines, while Fig. 5 shows non-clustered data center with multiple physical machines. For simplicity, Fig. 5 shows only source and destination host machines. The data rate considered in non-clustered data center is 1 Gbps with a delay of 2 ms, whereas in clusters it is taken as 100 Mbps with 6560 ns delay. VM image size is same in both the scenarios, taken as 1024 bytes. Both source and destination physical machines are assumed to be containing single virtual machine.

Performance results: Figure 6 shows the comparison of the RTT during migrations in a clustered environment and in a non-clustered environment. As the proposed migration is taking place within cluster, its round-trip time is lower than the non-clustered migration, thereby making proposed migration faster. Through our proposed migration technique in clusters, we have tried to reduce the power usage of servers considerably, as shown in Figs. 7 and 8.

Energy consumption is calculated with respect to time in our experiment. Intra-cluster migration takes lesser time than inter-cluster migration; hence, energy consumed is less within cluster. Faster migration in clusters also enables a source host to switch off early as compared to delayed VM migrations in non-clustered data centers, i.e., *EnergyConsumed* $\propto$ *Time a server is switched on.*

Thus, the above performance claims show that our proposed location-aware migration reaps better benefits as compared to migration in non-clustered datacenters.

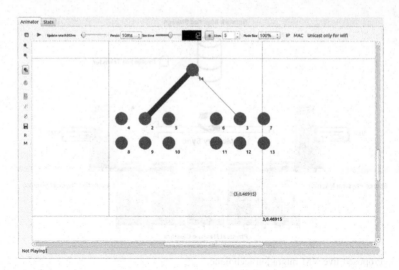

Fig. 4 Clustered data center topology

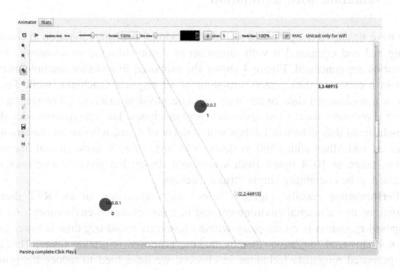

Fig. 5 Non-clustered data center topology

Fig. 6 RTT comparison

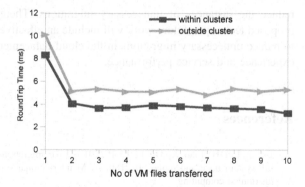

Fig. 7 Latency comparison

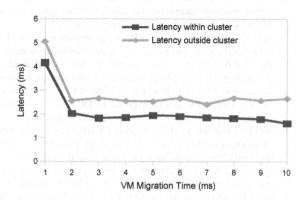

Fig. 8 Energy consumption

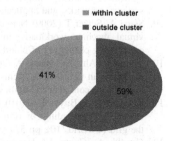

5 Conclusions and Future Work

This paper proposes a live VM migration scheme which is restricted to group of server hosts so that the migration process is fast, convenient, efficient and saves energy of a server. Our simulation scenario compares the proposed technique with other migration process that take place in non-clustered topology, and the results retrieved show that restricting VM migrations within clusters prove to be beneficial in saving cost and energy while maintaining the service-level agreements. Our proposed techniques handle migration successfully; however, it does not try to reduce the instances of unnecessary migrations. Therefore, as extension to the proposed technique, future work will include an effective VM placement technique to reduce unnecessary migrations in the cloud datacenters, thereby improving user experience and service performance.

References

1. Jin H, Liu H, Liao X, Hu L, Li P (2010) Live migration of virtual machine based on full-system trace and replay. In: 18th ACM international symposium on high performance distributed computing
2. Timothy W, Prashant S, Arun V, Mazin Y (2007) Black-box and gray-box strategies for virtual machine migration. In: 4th USENIX symposium on networked systems design USENIX association & implementation (NSDI 07)
3. Anja S, Waltenegus D (2013) Does live migration of virtual machines cost energy? In: 27th IEEE international conference on advanced information networking and applications (AINA)
4. Diego P-B (2013) A brief tutorial on live virtual machine migration from a security perspective. In: Proceedings of the 2013 international workshop on security in cloud
5. Samer A-K, Dinesh S, Prasenjit S, Matei R (2011) VMFlock: virtual machine co-migration for the cloud. In: Proceedings in HPDC 2011 proceedings of the 20th international symposium on High performance distributed computing
6. Wei D, Fangming L, Hai J, Xiaofei L (2012) Lifetime or energy: consolidating servers with reliability control in virtualized cloud datacenters. In: 2012 IEEE 4th international conference on cloud computing technology and science (CloudCom)
7. Hines MR, Kartik G (2009) Post-copy based live virtual machine migration using adaptive pre-paging and dynamic self-ballooning. VEE'09, March 11–13, Washington, DC, USA
8. Boru D, Kliazovich D, Granelli F, Bouvry P, Zomaya AY (2013) Energy-efficient data replication in cloud computing datacenters. Globecom 2013 workshop—cloud computing systems, networks, and applications
9. Stage A, Setzer T (2009) Network-aware migration control and scheduling of differentiated virtual machine workloads. In: Proceedings of the 2009 ICSE workshop on software engineering challenges of cloud computing
10. Verma A, Ahuja P, Neogi A (2008) pMapper: power and migration cost aware application placement in virtualized systems. In: Proceedings of ACM/IFIP/USENIX ninth international middleware conference (Middleware '08), pp 243–264
11. Beloglazov A, Buyya R (2010) Energy efficient allocation of virtual machines in cloud data centers. In: Proceedings of 10th IEEE/ACM international symposium cluster computing and the grid (CCGrid '10), pp 577–578
12. Gandhi A, Gupta V, Harchol-Balter M, Kozuch MA (2010) Optimality analysis of energy-performance trade-off for server farm management performance evaluation 67 (11):1155–1171
13. Graubner P, Schmidt M, Freisleben B (2013) Energy-efficient virtual machine consolidation. Professional 15(2):28–34
14. Bobroff N, Kochut A, Beaty K (2007) Dynamic placement of virtual machines for managing SLA violations. In: Proceedings of 10th IFIP/IEEE international symposium on integrated network management (IM '07)
15. Ferreto T, Netto M, Calheiros R, De RC (2011) Server consolidation with migration control for virtualized data centers. Future Gener Comput Syst 27(8):1027–1034
16. Do AU, Chen J, Wang C, Lee, YC, Zomaya AY, Zhou BB (2011) Profiling applications for virtual machine placement in clouds. In: 2011 IEEE 4th international conference on cloud computing, pp 660–667

17. Jamshidi P, Ahmad A, Pahl C (2013) Cloud migration research: a systematic review. IEEE Trans Cloud Comput 1(2):142–157
18. Mishra M, Das A, Kulkarni P, Sahoo A (2012) Dynamic resource management using virtual machine migrations. IEEE Commun Mag 50(9):34–40
19. Deng W, Liu F, Jin H, Liao X, Liu H (2014) Reliability-aware server consolidation for balancing energy-lifetime tradeoff in virtualized cloud datacenters. Int J Commun Syst 27:623–642
20. Mastroianni C, Meo M, Papuzzo G (2013) Probabilistic consolidation of virtual machines in self-organizing cloud data centers. IEEE Trans Cloud Comput 1(2):215–228
21. Hongyou L, Jiangyong W, Jian P, Junfeng, W, Tang L (2013) Energy-aware scheduling scheme using workload-aware consolidation technique in cloud data centres. Communications 10(12):114–124
22. Huang D, Yi L, Song F, Yang D, Zhang H (2013) A secure cost-effective migration of enterprise applications to the cloud. Int J Commun Syst
23. Riteau P, Morin C, Priol T (2013) Shrinker: efficient live migration of virtual clusters over wide area networks. Concurr Comput: Pract Exp 25:541–555

Progressive Visual Analytics in Big Data Using MapReduce FPM

Amit Kumar and Prabhat Ranjan

Abstract Visual analytics uses interactive visualizations in order to incorporate user's knowledge and cognitive capability into data analytics processes. The progressive visual analytic paradigm simplifies the analytic process when it comes to large datasets. It uses the interactive sequential pattern mining algorithm which reports patterns as it finds them. But, the sequential pattern mining algorithms like SPAM, SPADE and PrefixSpan are suited for a single-node environment only. It is also constrained by the available size of memory and computational power while handling a very large quantity of data. So to overcome these challenges, the proposed MapReduce frequent pattern mining (MR-FPM) algorithm distributes data across various nodes in the Hadoop cluster, finds the candidate itemsets and counts their support using the MapReduce paradigm. The patterns with supportless than the user-defined *minsup* are discarded. Experimental results show that MR-FPM continuously outperforms SPAM when the *minsup* is decreased.

Keywords MapReduce FPM · Progressive visual analytics · Sequential pattern mining (SPAM) algorithm

1 Introduction

In this era of big data, the role of visual analytics [1] becomes very crucial from analytics perspective. Visual analytics integrates automated analytics alongside visualizations that are interactive in nature, for a simple viewing, perception and making logical decisions based from huge and complex datasets. It is an iterative

A. Kumar (✉) · P. Ranjan
Central University of South Bihar, Patna, India
e-mail: amit_kumar@live.in

P. Ranjan
e-mail: prabhatranjan@cub.ac.in

© Springer Nature Singapore Pte Ltd. 2018
D.K. Mishra et al. (eds.), *Information and Communication Technology for Sustainable Development*, Lecture Notes in Networks and Systems 9,
https://doi.org/10.1007/978-981-10-3932-4_14

process which includes gathering information, data preprocessing, knowledge representation, interaction and decision making. Various approaches to handle big data in visual analytics have been formulated, but the progressive visual analytics [2] stands out among them all. This paradigm enables us to utilize the time spent in waiting for completion of analytics in evaluation of incomplete results. Concluding these results, the input parameters may be adjusted or modified by the users and even the dataset may be changed, if needed. Any misdirected analytics which may produce useless results can be terminated. However, with the unprecedented rise in the data size even this paradigm has been challenged. The sequential pattern mining and other similar algorithms are capable of mining patterns in a single node only due to their design constraints. Therefore, they cannot take advantage of parallel processing, the need of the time is to incorporate parallel processing so as to handle big data. The proposed method of MR-FPM distributes data on various nodes using the MapReduce paradigm of Hadoop to generate candidate patterns and find the frequent ones among them. It can find out the candidate patterns of interest, discard obsolete itemsets and significantly reduce the time it takes to process very large datasets. The paper is organized in the following manner. Section 2 highlights the background and discusses the literature related to it. Sections 3 and 4 explain the preliminary and proposed approach, respectively. Section 5 elaborates the implementation part and analyses the experimental results. At last, Sect. 6 concludes the paper and explains the future scope.

2   Background and Related Work

A. Visual Analytics: Visual analytics [1] is not merely visualization of data. It is better defined as intrinsic methodology that combines visualization, data analytics and human–computer interaction. Closely relating to visualization, visual analytics integrates approaches from fields like information analytics, geospatial analytics and scientific analytics [3]. However, human capabilities like interacting, information perception, cognitive measures, presentation, conjunction and propagation have very a significant role in the information exchange and decision making process (Fig. 1).

Above figure represents the various stages in visual analytics and their flow from one stage to another. The visual analytics process is marked by relations between data, the data visualizations, the underlying data models and its user's with the intent of knowledge discovery. The visual analytics process is concerned with conglomerating the analysis automation methods with visual representations capable of interactivity.

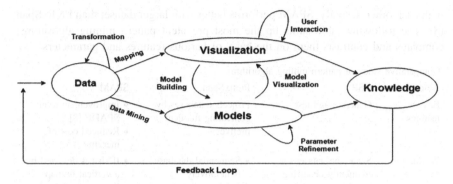

Fig. 1 Visual analytics process

B. Progressive Visual Analytics: As proposed by Stopler et al. [2] a progressive visual analytic system is capable of producing semantically meaningful incomplete outputs amidst execution. The results obtained in this manner can thus be integrated into interactivity supported visualizations allowing users to instantly dig deep into those results, explore the new one's as early as they can be computed, and conduct further incremental analytics without having to wait for prior on-going analytic to terminate. Therefore, this approach has several great positives for visual analytics. For instance, the time wasted in waiting for the completion of analytics can henceforth be utilized for exploring the incomplete outputs. These can help in adjustment of the input parameters, changing the datasets and on a simple note terminate the wrongly processed analytics which may produce trash outputs.

C. The SPAM Algorithm: The sequential pattern mining algorithm [4] identifies and reports the frequently occurring candidate sequences and subsequences in a data sequence. In order to avoid multiple scanning in Apriori-based approaches and for enhancing the efficiency of mining, Ayres et al. came up with the SPAM algorithm [4]. It proceeds by scanning the parent database one time and transforms it into a vertical bitmap table. It involves the construction of a lexicographic sequence tree based on the depth-first search traversal technique. However, several attempts have been made to make it more compatible to the visual analytic scenario. Like the breadth first traversal of SPAM algorithm as proposed by Stopler and perer, the incremental SPAM algorithm offers several advantages useful in progressive visual analytic scenario [2]. However, all of them are constrained by the inherent drawback of this class of algorithm [5] which restricts them from making use of parallel processing.

D. Comparative Study: The sequential pattern mining algorithms discussed before use various approaches to solve the pattern mining problem. A comprehensive performance study shows that PrefixSpan [6], in most cases outperforms algorithms like GSP and SPADE [7]. However, SPAM [4] and PrefixSpan closely compete with each other. Usually, SPAM performs better with larger dataset than PrefixSpan [8]. The following table sums up the most prevalent pattern mining algorithms, compares and contrasts them on the basis of various features and parameters.

Comparative study of pattern mining algorithms			
Algorithms	GSP	PrefixSpan	SPAM
Key features	Generate and test	Projects databases by generating frequent prefixes	• Improvement over SPADE [8] • Reduced cost of merging
Working	Scans the dataset for frequently occurring candidates. For candidates that need more memory, only the candidates that closely fit in are generated. If it is found frequent, flush to disk; otherwise discard	• Sequential database is iteratively fragmented into a number of smaller projected databases • Sequential patterns are generated in individual projected database by scanning the locally frequent patterns	• ID-list is depicted by a vertical bitmap • Dataset is of the form <CID, TID, Itemsets> where, CID: customer-id and TID: transaction-id
Memory	Does not use primary memory	Main memory algorithm	Dataset is saved in primary memory
Data structure	The itemsets generated are saved as Hash tree	Stored as Hash Tree	Vertical bitmap
Limitations	Multiple scanning of database	Generation of projected databases is costlier	<CID, TID, Itemset> should be in main memory

3 Problem Definition

As the dataset increases in size and the need to process large amounts of data at promising speeds becomes evident, even the existing sequential pattern mining algorithm turns out to be obsolete. However, the SPAM algorithm lags behind when the data size becomes huge due to the limitations imposed by the implicit drawback of the underlying algorithm. The SPAM [4] algorithm and other traditional pattern mining algorithms based on Apriori, pattern growth and projection techniques have been framed to scan and find frequent patterns on an individual node, and therefore they cannot take advantage of the parallel processing. It also suffers from other problems like memory size constraints as not all data can be kept in memory at once due to the limited memory size. Most of the other such algorithms are unable to effectively carry out the mining task when it comes to a large scale of transactional databases as they are not completely loaded into the memory [9]. The computation power boost which is required to process the tremendous amount of data at an affordable pace requires significant hardware upgrades which turn out to be very costly.

4 Proposed Approach

The proposed approach makes use of the Hadoop platform to design a MapReduce-based frequent pattern mining algorithm as explained in the algorithm below. This algorithm is composed of a couple of MapReduce tasks namely the candidate pattern finder and the support calculator task. When the data input is received from the sequences at any instant of time, the itemsets belonging to the various sequences are disbursed among the various individual machines in the Hadoop [10] cluster. Each machine in the cluster finds candidate patterns generated from individual sequences in the given time period. In the meantime, the candidate pattern finder gathers details about each sequence.

Algorithm 1: Map-Reduce FPM

Input: Itemset of sequences at any time t, t' = time period
1: **While**(New sequence arrives at time t)
2: CandidatePatternFinder;
3: SupportCalculator;
4: t=t+t'
5: output frequent sequential patterns;
6: **end while**
Output: Frequent patterns found at every time period.

The framework for the proposed MR-FPM is depicted in Fig. 2. The candidate pattern finder task is followed by the support calculator task. Both the Mapper and Reducer work at both the levels. Input data at time t are fed into the MapReduce engine of candidate pattern finder which outputs candidate itemsets. This is given as input to the support calculator task which counts the node-specific support and adds them up after merging; therefore, producing only those itemsets whose support is greater than the minsup defined for it. For the subsequent time periods, this

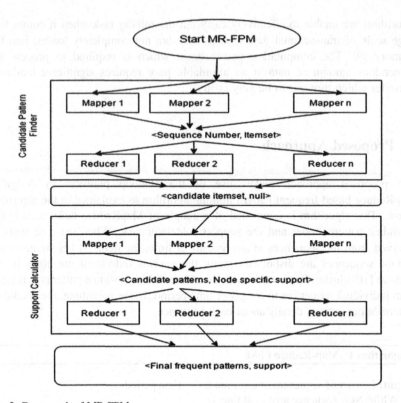

Fig. 2 Framework of MR-FPM

algorithm continues to perform these MapReduce tasks. In this manner, it reports the most updated frequent patterns in each time period. Since, our algorithm reports the frequent patterns progressively; it is named as MapReduce frequent pattern mining (MR-FPM).

Candidate Pattern Finder: The candidate pattern finder takes input data at time t and distributes it to various Mappers. Mappers help generate many pairs of <sequence number, itemsets> as evident from Algorithm 3. The pairs having same sequence numbers are placed with the same Reducer. The Reducer finds the candidate patterns from the input data and outputs <candidate sequential patterns, null>. Apart from this, the Reducer keeps track of the details of every sequence and discards useless data simultaneously. Candidate patterns at the current time are output for the computation at the next time period.

Algorithm 2: Candidate Pattern Finder

Input: Itemset of sequences at current time,
Candidate sequences from previous timestamp, var x: = read input data
var y := stores input data,
var z := stores candidate itemsets from last timestamp
var result := stores results

Mapper:
1: **if**(x) has timestamp) //data from previous time
2: **return:** <x.sequenceNumber, x.itemset, x.timestamp>
3: **else** //new data at current time
4: **return:** <x.sequenceNumber, x.itemset>

Reducer: (key,value)
1: **for**(all items in value){
2: **if**(value is a previous timestamp itemset) {
3: z := <value.timestamp, value.itemset>
4: } **else** //if value is current data
5: y := current data
6: } **for**(all items in y.itemset) {
7: **if**(y.itemset.time > t)
8: **return:** <key, newCandidateItemsets, timestamp>
9: **else**
10: **return:** <key, allCandidateItemsets, timestamp> }

Algorithm 3: Support Calculator

Input: var localsup := <itemset, localSupport> //local support count
var x:= reads input data, var total := 0 //adds total support
Mapper:
1: **for**(x.itemset in each local node)
2: support(itemset) := localsup;
3: **return:** <itemset, localSupport>
Reducer: (key,value)
1: **for**(each itemset in value)
2: total = total+value;
3: **if**(total>= minimum support)
4: **return:** < itemset,total >
Final Output: <Frequent patterns, support>

Support Calculator: In this phase, the Mapper in the support calculator algorithm takes input data from the previous phase and counts local frequencies of each candidate patterns on individual nodes as described in Algorithm 3. The Mapper here generates pairs of <candidate patterns, node-specific support> as outputs. After this, sets having the same candidate patterns have to be transferred to the same Reducer. The Reducers add up the individual supports of each pattern and generate the final frequent patterns for the given time period. Finally, the process reiterates for new data arriving at time t + t'.

5   Implementation

We conducted experiments in order to check the time of execution of MR-FPM in case of big data. We implemented MR-FPM on Hadoop 2.6.4 and jdk-8u66 in fully distributed mode. In this mode, three nodes were deployed in the Hadoop cluster with two datanodes and a namenode controlling them. The experiments have been run on a machine with 3.40 GHz Intel Core i7 CPU and 8 GB main memory. We calculated the efficiency of MR-FPM and various other sequential pattern mining algorithms with varied values of *minsup* on *kosark25 k* dataset which consists of 25,000 click-stream sequences from a Hungarian news portal from the SPMF open source library (Fig. 3).

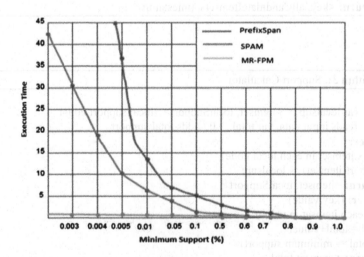

Fig. 3 Execution time of algorithms (kosark25 k)

Algorithms (Execution time)			
minsup	SPAM	PrefixSpan	MR-FPM
1.0	312 ms	94 ms	182 ms
0.9	281 ms	91 ms	182 ms
0.8	327 ms	174 ms	199 ms
0.7	281 ms	267 ms	255 ms
0.6	312 ms	1.282 s	247 ms
0.5	265 ms	3.255 s	526 ms
0.4	374 ms	3.459 s	1.209 s
0.3	437 ms	3.996 s	1.240 s
0.2	436 ms	4.109 s	1.244 s
0.1	421 ms	5.421 s	1.265 s
0.09	312 ms	5.564 s	1.291 s
0.08	359 ms	6.887 s	1.232 s
0.07	889 ms	6.655 s	1.288 s
0.06	2.399 s	6.991 s	1.297 s
0.05	3.422 s	7.328 s	1.290 s
0.04	3.537 s	8.227 s	1.417 s
0.03	4.101 s	10.975 s	1.419 s
0.02	5.733 s	12.021 s	1.759 s
0.01	6.763 s	14.763 s	1.511 s
0.009	6.996 s	18.010 s	1.623 s
0.008	7.792 s	22.923 s	1.621 s
0.007	7.633 s	28.434 s	1.943 s
0.006	8.044 s	33.767 s	1.786 s
0.005	10.734 s	37.734 s	1.822 s
0.004	18.448 s	51.448 s	2.151 s
0.003	43.580 s	77.332 s	2.900 s
0.002	131.221 s	192.454 s	6.322 s
0.001	449.119 s	698.432 s	8.884 s

To contrast between the execution of various algorithms like SPAM and PrefixSpan that are basically framed to run on a single workstation, we choose this dataset as, SPAM will not be handling very large datasets easily because of the memory limitations of single machine. It can be clearly deduced from the Figure and analysis table that higher mining overhead occurs when *minsup* is decreased. As we keep decreasing the value of user-defined minimum support, the time taken by various traditional algorithms in contrast to our proposed approach is drastically high implying that the traditional methods are far too obsolete to be applied in visual analytic scenario. It can also be observed that MR-FPM outperforms SPAM and PrefixSpan when the *minsup* is gradually decreased to very lower values.

6 Conclusion and Future Work

In the present paper, the challenging issue of scalability has been presented while handling humungous datasets which may not be completely loaded into memory in a progressive visual analytic scenario. We proposed a MapReduce-based pattern mining algorithm which efficiently generates and prunes candidate patterns in accordance with the user's choice of minimum support and implemented the MR-FPM on Hadoop ecosystem with large transactional sequences, elucidating that it can drastically lessen the execution time in case of big data. Besides, the current system is still constrained with the MapReduce frequent pattern mining algorithm. It could be made compatible with other more advanced algorithms.

References

1. Keim DA, Mansmann F, Schneidewind J, Thomas J, Ziegler H (2012) Visual analytics: scope and challenges. Springer
2. Stolper CD, Perer A, Gotz D (2014) Progressive visual analytics: user-driven visual exploration of in-progress analytics. IEEE Trans Vis Comput Graph 20(12):1653–1662
3. Keim D, Huamin Q, Ma K-L (2013) Big-data visualization. Comput Graph Appl IEEE 33 (4):20–21
4. Ayres J, Flannick J, Gehrke J, Yiu T (2008) Sequential pattern mining using a bitmap representation. In: Proceedings of the eighth ACM SIGKDD international conference on knowledge discovery and data mining. ACM, pp 429–435
5. Mabroukeh NR, Ezeife CI (2010) A taxonomy of sequential pattern mining algorithms. In: ACM computing surveys (CSUR'10), 2010
6. Pei J, Han J, Mortazavi-asl B, Pinto H, Chen Q, Dayal U, Hsu M-C (2001) PrefixSpan: mining sequential patterns efficiently by prefix-projected pattern growth. IEEE Comput Soc, 215
7. Zaki MJ (2001) SPADE: an efficient algorithm for mining frequent sequences. Mach Learn 42:31–60
8. Huang J-W, Tseng C-Y, Ou J-C, Chen M-S (2008) A general model for sequential pattern mining with a progressive database. In: IEEE transactions on knowledge and data engineering (TKDE'08), 2008
9. Dean J, Ghemawat S (2008) MapReduce: simplified data processing on large clusters. CACM 51(1):107–113
10. Apache Hadoop. http://hadoop.apache.org/

Energy Based Recent Trends in Delay Tolerant Networks

Nimish Ukey and Lalit Kulkarni

Abstract Energy plays very important role in any kind of network. As most of routing protocols and strategies are designed by considering the well-connected network but because of the mobility of nodes these routing protocols do not work effectively. For such kind of intermittent connectivity, Delay Tolerant Network is used. Amongst all the mobile nodes, many have a very limited amount of energy and to perform any kind of operation energy is required. So, it is necessary to reduce the consumption of energy. The effective and efficient use of energy increases the lifetime of node and network. Hence, there are some approaches which proposed different schemes to reduce the energy consumption of node and also increase the delivery probability. By using different aspects of the conservation of energy the energy efficiency has been increased. Amongst them, few recent approaches are explained in this papers which consider the energy prominently.

Keywords Energy-aware · Priority transmission · Robustness
Location-based · Store-carry-forward · Delay tolerant networks

1 Introduction

Many Routing protocols have been proposed to reduce the number of copies of the message and to increase the delivery probability ratio, but very few consider the energy constraints into a consideration. Now a day's many mobile devices like cell phone, laptop, PCs, tablets have the limited amount of energy and these devices consume much more energy for the communication purpose. So it is necessary to conserve energy in DTN. There are various things which are responsible for the

N. Ukey (✉) · L. Kulkarni
Department of Information Technology, Maharashtra Institute of Technology,
Pune, India
e-mail: nimishukey@gmail.com

L. Kulkarni
e-mail: lvkulkarni@gmail.com

© Springer Nature Singapore Pte Ltd. 2018
D.K. Mishra et al. (eds.), *Information and Communication Technology
for Sustainable Development*, Lecture Notes in Networks and Systems 9,
https://doi.org/10.1007/978-981-10-3932-4_15

depletion of energy for ex. node mobility, the size of a buffer and inefficient forwarding scheme etc. So it is necessary to utilize the appropriate routing strategy or routing protocol to reduce the consumption of energy.

The Delay Tolerant Network (DTNs) [1] is characterized as the intermittent network. To overcome the intermittent problem the DTN [2] has been used. The main reason behind the connectivity disruption is the mobility of nodes. The performance evaluation [3] in DTN shows that the most of the routing strategies are extensive energy consumptive and hence the energy consumption of node also increased. This is mainly caused due to the mobility of a node, message size, and forwarding strategy.

There are various routing strategies in DTN. Amongst them, some of the very famous routing strategies are: epidemic [4], Spray-and-Wait [5], and PROPHET [6, 7]. These strategies are very much famous, but they did not consider the energy constraints. DTN uses the store carry and forward mechanism to forward the packet from source to destination. This mechanism stores the message in every node which comes into the contact range, which increases the replication of message, thus results in the consumption of the resources which are very limited in amount. Hence, the routing protocols described in [8–11] are used to improve the utilization of resources effectively and also to increase the delivery ratio of the message.

In this paper, in Sect. 2 related work is explained. In Sect. 3 various recent trends based on energy in DTN are explained in details and at the end of this section, the comparative study has been done. In the final section, the conclusion is explained.

2 Related Work

In [12], node forwards the packet only when it has n neighbor nodes in its maximum transmission range to reduce the energy consumption. But it is difficult to decide the value of n. In [13], F. D. Rango, S. Amelio, and P. Fazio implement the protocol, which was an improvement over the [12]. Previously the value of the n has to be set statically but in [13] the value has been chosen dynamically. It chooses a value based on the current neighbor node and the energy level. In [14], the trade-off has been achieved in between the forwarding efficiency and the energy preservation.

The routing protocols generally classified into two categories: topology based and location based. The topology based protocol uses the link information to find the path. And the Location based protocol uses the location information of the node, destination and neighbor nodes. In [15–17], the topology based routing protocols are being studied. In [18–22], few location-based routing protocols has been designed in DTN and described in detail.

3 Energy Based Strategies

3.1 A Novel Energy Aware Priority Transmission Scheme Based on Context-Metric

This ensures to achieve the high probability to deliver the HIGH priority messages. It also introduces the new buffered model which consists of the Admission Unit and Transmission Control Unit. This Context Metric queuing method can also be called as CEAMS [23]. It also introduces the Energy Aware message transmission strategy.

In many cases the importance of the message is not measured while transmitting the data. Sometimes the importance of the message needs to be taken into consideration when the transmission of the message is crucial i.e. disaster relief application or military networks. With the depletion of energy the chance of the message delivery decreases. In queuing method the message gets dropped if the priority of the message is not considered. As the priority is not given, the message may be queued at the end of the list and the message get drop later on due to the message Time to live (TTL) or the unavailability of resources. If the message does not have the priority assigned, then it may suffer from the multiple replications due to the long delay. So here a context metric queuing method has been implemented to accomplish the high delivery ratio of the HIGH priority message and also to use the energy of node effectively.

Node Delivery Capability. The NDC states that the higher the Energy, Speed and the lesser the Distance to the destination, increase the probability of delivering. It identifies the probability that the message will reach to destination successfully or not.

Remaining Energy, Estimated Distance and Speed are taken into consideration and utilized while taking the queuing decision of message.

Energy Aware Transmission Scheme. The source node stores the information until and unless it comes in the contact of another node. When an intermediate node comes in the range of source node then the summary vector is exchanged between both the nodes i.e. index of messages. The nodes can request the message they don't have from another node. But this does not consider the priority of the message.

In the implemented system the authors initialized the three kinds of level priorities to the messages, i.e. LOW, MEDIUM, and HIGH. Two kinds of Buffers are implemented, i.e. connectivity buffer and persistent storage. To manage the message effectively, two kinds of units are implemented here Admission Control Unit and Transmission Control Unit.

Admission control Unit. This checks whether the buffer is full or not. If the buffer is full then it drops the LOW priority message. Based on the level of priority it decides whether to send a message to connectivity buffer or to persistent storage. If a message has the HIGH priority then it is sent to Persistent storage otherwise to connectivity buffer.

Transmission Control Unit. Based on the remaining energy it determines whether the intermediate node is capable of delivering the message or not. Depends on the certain criteria it takes a particular action. If the energy of an intermediate node is greater than 75%, then the intermediate node accepts all the messages. If it is in between 75% to 25%, then it accepts HIGH and MEDIUM messages only, and if remaining energy is less than 25% then it accepts only HIGH priority messages.

3.2 A Robust Energy Efficient Epidemic Routing Protocol

In this protocol mainly two things are taken into consideration while forwarding the message. It checks whether the energy of the intermediate node is greater than the current node. If it is greater then and only then it forwards the message to the intermediate node. And it also checks the buffer space is available or not, to store the data. The network is said to be robust [24] if node consumes very less energy and the lifetime of node increase.

This method takes these things into consideration:

- The node exchanges the summary vector, free buffer size and the current energy available with its neighbor node
- Then the node checks the remaining energy of a neighbor node. It checks whether the neighbor nodes energy is greater or not. It also checks messages which node has to copy to neighbor node and the free buffer space available or not, to store the message
- If the neighbor node has the more remaining energy and the free buffer space than the current node, then node copies the message to Transmit List
- If the current node has the remaining energy more than neighbor node and also has the free buffer space, then it will keep the message information to Receiving From List
- After the completion of comparison with all the nodes, the message is broadcasted and the neighbor who has the higher remaining energy and high free buffer space receives the message and it stores the copy of that message

In epidemic routing if the buffer space is not available, then it drops the old message from the buffer which results in the reduction of delivery probability. Hence, this robust energy efficient epidemic routing protocol methodology is implemented to increase the message delivery probability and also use the energy effectively.

3.3   3D Location-Based Energy Aware Routing Protocol

This routing uses the location information for real DTN to reduce the overhead of it. It also implements the energy aware routing scheme to increase the lifetime of the network. Most of the routing protocols are designed on the basis of the 2-Dimensional (2D) location based information. Only few routing strategies are discovered by considering the 3D location based information in Ad hoc networks, which does not work well in DTN. So, it is the first time the 3D location based routing protocol [25] is implemented for the DTN.

In many cases, all nodes in DTN are wireless which is distributed all over the 3D environment. Nodes have been considered as the sphere shaped and with radius r, which denotes the coverage range of that node. The coordinates of the node are denoted by x, y and z. This presented scheme uses the location information by using Global Positioning System (GPS) to reduce the external overhead of the network. The following assumptions are taken into consideration:

- All nodes know their current location information by using the 3D coordinates of nodes(x, y, z).
- All nodes also exchange the message information by using the "hello-reply" message.
- The limited amount of energy and buffer space.
- Transmission speed has to be pretty enough to forward the message when both nodes come into the contact range.

Expected Zone and Request Zone.

Expected zone. Before starting the sending of messages from source to destination the source node first finds the path from source to destination at time t0. And it is possible that the destination may move at the speed of v from its position. So source node estimates the expected zone of destination at time t1. Hence, the expected zone must be a sphere with the radius of v (t1-t0). If the source does not know the speed of destination, then it is difficult to estimate the expected zone.

Request zone. If the source does not find the path for the destination then it increases the coverage range to a maximum radius. After increment of the radius if a source node does not find the proper next hop for forwarding the message, then the source node broadcast the message to all the nodes which come under the request zone. It uses the certain condition strategy for broadcasting the message.

Routing schemes.

Neighbor discovery. Here the priority of the message is not considered. The message comes first is added to buffer prior. When a source node comes in the contact range of any better hop node, then it sends the message as per the buffer queuing. The source node firstly broadcasts the HELLO message to the entire node which comes into the transmission range u i.e. half of the maximum transmission range of nodes. The receiving node checks the neighbor list. If the source node is already in its neighbor list, then it drops the HELLO packet. Otherwise, accept the Hello packet and reply with an ACK packet. It is necessary to update the neighbor list in a frequent manner.

Transmission scheme. The node firstly checks for the recent paths to the destination. If any recent path exists, then it forwards the packet through that recent path. If there is not any recent path, then it will check for the next best hop. If a next best hop is also not available, then it will check the maximum transmission range of the node. If the maximum transmission range of the node is u then it is increased to r to find the neighbor. If it does not find the neighbor then source forward the message by conditional flooding to all nodes which come under the request zone. In the best case, the message is forwarded by choosing the best hop node with the help of implemented scheme. In the worst case, the condition flooding scheme has to be applied to flood the message to request zone, which is determined by the speed and location of the source and destination node.

3.4   Augmented DTN Based Energy Efficient Routing Protocol for Vehicular Ad Hoc Networks

Vehicular Ad hoc Network (VANET) is derived from the Mobile Ad hoc Network (MANET). Augmented DTN based Energy Efficient Routing Protocol for Vehicular Ad hoc Networks (ADTNEER) [26] is a combination of various strategies such as store stay spray, connection lifetime and link state information. It utilizes those different strategies to provide better performance than the VANET. It also uses the angular region strategy for selecting the best suitable next hop. VANET basically consist of two entities: vehicle and Road-side Infrastructure Units (PSU). The vehicle is the moving entity and PSU is the fixed entities which are connected to the internet. VANET node can communicate with other vehicles directly. Due the mobility of nodes, it is challengeable to perform end to end communication.

There are various factors which are responsible for the performance of the network, energy, mobility, etc. It is very difficult to achieve the energy efficiency and reliability for the transmission of data when the movement of the vehicle is fast in the network. Hence, for that purpose ADTNEER has been proposed to achieve better performance than the VANET. It uses various approaches like store stay spray, connection lifetime and link state information to take the appropriate decision for forwarding the data.

Following two assumptions are being considered:

- All nodes know their location information.
- The system can able to estimate the distance between two nodes, path between them and the time required.

Transmission range selection based on the angular region. The selection of the specific region for the transmission of data reduces the overhead of transmission. On the road, a vehicle travels in the one direction. Hence, by limiting the angle of transmission overhead is reduced. Firstly, source node finds its location and information about the destination and neighbors. Initially, the angle of 30 is initialized and can be increased up to 180. If within the transmission range the node is not found, then the transmission angle is increased by 15 on both the side with considering the source and destination node vector.

Link quality estimation. Link Quality (LQ) is calculated by the broadcasting the ping message to every node.

$$LQ(S, D) = \frac{No.\ of\ packet\ transmitted\ successfully \times No.\ of\ packet\ retransmitted}{Total\ no.\ of\ packets\ transmitted}$$

Connection Lifetime. The energy draining rate and the mobility of nodes are used to predict the connection lifetime. Each node has memory space which stores the information of remaining energy; energy spent on successful communication and received signal strength. This information also helps to predict the connection lifetime.

Store-stay-spray mechanism. In this proposed mechanism the message is copied to the node or forwarded to the node which assures to go closer the destination node. Each node always finds the better next hop for the further transmission and chooses the best path for the packet transmission. This helps to improve the efficiency. Table 1 shows the concise analysis which was discussed in detail in Sect. 3.

Table 1 Comparative Study

Objective of study	Strategies or schemes	Advantages	Limitations	Results
A novel energy aware priority transmission scheme based on context-metric [23]	Queuing Management	Improve message delivery in DTN	Only a limited amount of energy efficiency is achieved	It achieves a high delivery rate of message with high Priority and effective utilization of nodes energy in transmission
	Energy-aware transmission scheme		The message redundancy ratio is not reduced to a full extent. Still, it is possible to reduce the message redundancy ration with maintaining delivery rate	
	Message forwarding			
A robust energy efficient epidemic routing protocol [24]	By considering	No need to set any predefined value	Size of buffer, size of message, varying mobility speed and data rate is not considered in simulation setting of this protocol	Network lifetime is extended and delivery probability is also improved
	Nodes remaining energy	Number of nodes does not affect the energy consumption of the node		
	Free buffer space available			
3D Location-based energy aware routing protocol [25]	Location based (3D Location information i.e. GPS)	No need to establish and maintain global routing table	Congestion control and Priority of the message is not considered here	Introduced energy aware scheme to save power and extend the network lifetime
		Less control overhead	Does not work fine in large density of node	Reduce network overload
Augmented DTN based energy efficient routing protocol for vehicular ad hoc networks [26]	Geographical region Angular based	It is reactive protocol, i.e. it takes routing decision to forward the packet to the destination	The message redundancy ratio is not reduced to full extent still it is possible to reduce the message redundancy ratio with maximum delivery rate	ADTNEER given results shows better performance than VANET
	Store-stay-spray			
	Connection Lifetime prediction			
	Link state information			

4 Conclusion

All the strategies mentioned above take energy into consideration while designing the strategies. They achieved their goal to implement the strategy with increasing the energy efficiency. All the strategies have some limitations which are mentioned in the comparison table. So by reducing all these limitations the effective routing protocol can be implemented which will consume very less energy and the network lifetime increase. All above-mentioned strategies are recently proposed strategies. Each strategy has some specific specialty and a different result. But all utilizes the energy very effectively.

References

1. Fall K, Farrell S (2008) DTN: an architectural retrospective. IEEE J Sel Areas Commun 26 (5):828–836
2. Fall K (2003) A delay-tolerant network architecture for challenged internets. In: Proceedings of the 2003 conference on applications, technologies, architectures, and protocols for computer communications, SIGCOMM'03. ACM, New York, NY, USA, pp 27–34
3. Cabacas R, Nakamura H, Ra I (2013) Performance evaluation of routing protocols in delay tolerant networks in terms of energy consumption. In: Proceedings of the 2nd international conference on smart media and applications
4. Vahdat A, Becker D (2000) Epidemic routing for partially-connected ad hoc networks, CS-200006, Duke University, Technical Report, Apr 2000
5. Spyropoulos T, Psounis K, Raghavendra CS (2005) Spray and wait: an efficient routing scheme for intermittently connected mobile networks. In: Proceedings of the 2005 ACM SIGCOMM workshop on delay-tolerant networking, WDTN'05. ACM, New York, NY, USA, pp 252–259
6. Lindgren A, Doria A, Schelén O (2003) Probabilistic routing in intermittently connected networks. SIGMOBILE Mob Comput Commun Rev 7(3):19–20
7. Lindgren A, Doria A, Davies E, Grasic S (2012) Probabilistic routing protocol for intermittently connected networks. Internet Requests for Comments, RFC Editor, RFC 6693, Aug 2012
8. Daly E, Haahr M (2009) Social network analysis for information flow in disconnected delay-tolerant manets. IEEE Trans Mob Comput 8(5):606–621
9. Altman E, Azad A, Basar T, De Pellegrini F (2010) Optimal activation and transmission control in delay tolerant networks. In: 2010 proceedings of IEEE INFOCOM, pp 1–5, Mar 2010
10. de Oliveira ECR, de Albuquerque CVN (2009) NECTAR: a DTN routing protocol based on neighborhood contact history. In: Proceedings of the 2009 ACM symposium on applied computing, SAC'09. ACM, New York, NY, USA, pp 40–46
11. Nelson S, Bakht M, Kravets R (2009) Encounter-based routing in DTNs. In: INFOCOM 2009, IEEE, pp 846–854, Apr 2009
12. Lu X, Hui P (2010) An energy-efficient n-epidemic routing protocol for delay tolerant networks. In: 2010 IEEE fifth international conference on networking, architecture and storage (NAS), pp 341–347, July 2010
13. De Rango F, Amelio S, Fazio P (2013) Enhancements of epidemic routing in delay tolerant networks from an energy perspective. In: 2013 9th international wireless communications and mobile computing conference (IWCMC), pp 731–735, July 2013

14. Khouzani M, Eshghi S, Sarkar S, Shroff NB, Venkatesh SS (2012) Optimal energy-aware epidemic routing in DTNs. In: Proceedings of the thirteenth ACM international symposium on mobile ad hoc networking and computing, MobiHoc'12. ACM, New York, NY, USA, pp 175–182
15. Jain S, Fall K, Patra R (2004) Routing in a delay tolerant network, Portland, OR, United states, vol 34(4), pp 145–157
16. Zhang Z (2006) Routing in intermittently connected mobile ad hoc networks and delay tolerant networks: overview and challenges. IEEE Commun Surv Tutor 8(1):24–37
17. Zhang L, Zhou X-W, Wang J-P, Deng Y, Wu Q-W (2010) Routing protocols for delay and disruption tolerant networks. J Softw 21(10):2554–2572
18. Shen J, Moh S, Chung I (2010) A priority routing protocol based on location and moving direction in delay tolerant networks. IEICE Trans Inf Syst e93-d(10):2763–75
19. Yasmeen F, Urushidani S, Yamada S (2009) A probabilistic position based routing scheme for delay-tolerant networks, Piscataway, NJ, USA, pp 88–93
20. Luo G, Zhang J, Qin K, Sun H (2012) Location-aware social routing in delay tolerant networks. IEICE Trans Commun E95-B(5):1826–9
21. Lu X, Hui P, Towsley D, Pu J, Xiong Z (2010) Lopp: a location privacy protected anonymous routing protocol for disruption tolerant network. IEICE Trans Inf Syst E93-D(3):503–9
22. Tian Y, Li J (2010) Location-aware routing for delay tolerant networks, Beijing, China
23. Cabacas R, Ra IH (2014) A novel energy-aware priority transmission scheme based on context-metric queuing for delay tolerant networks. In: Proceedings of the IEEE information science, electronics and electrical engineering, vol 2, pp 1095–1099
24. Bista BB, Rawat DB (2015) A robust energy efficient epidemic routing protocol for delay tolerant networks. In: Proceedings of the IEEE international conference on data science and data intensive systems, pp 290–296
25. Tian C, Ci L, Cheng B, Li X (2014) A 3D location-based energy aware routing protocol in delay tolerant networks. In: Proceedings of the IEEE dependable, autonomic and secure computing, pp 485–490
26. Paramasivan B, Bhuvaneswari M, Pitchai KM (2015) Augmented DTN based energy efficient routing protocol for vehicular ad hoc networks. In: Proceedings of the IEEE SENSORS, pp 1–4

Smart Solar Panel: Wireless Sensor Network-Based Measurement and Monitoring of Performance Parameters of Solar Panel

Dhiraj Nitnaware

Abstract Solar photovoltaic (PV) technology generates electricity in any area, where there is need, by installation of solar PV modules. So we need an instrument that can measure important parameter like open-circuit voltage, short-circuit current, power, fill factor which helps in installation of solar PV modules. After installation of solar plant, it is required to monitoring all these parameters. To determine the performance of a panel or an array of panels, it is important to understand the characteristics of a particular panel under different light conditions. This can be obtained by finding the I–V characteristics of panels or arrays under different real-world environmental conditions. These parameters can vary due to the effect of panel shading, temperature, depending on solar irradiance intensity, cell type, etc. Finding the maximum power of the solar panel is important to detect faulty cells or panels and insuring that panels in arrays are matched to get maximum power. After installation of solar plant, it is also required monitoring of all the parameters of solar energy. Monitoring system parameters through wireless sensor network gives flexibility to the technician not to be in the actual area where solar panels are located. Also it allows the simultaneously performance monitoring of solar panels. In this paper, a prototype is developed and tested with application software to measure and monitor the important parameters of solar panel like voltage, current and power. This module is tested in 100-m range and observed the good communication performance.

Keywords Wireless sensor network · Solar panel · Voltage–current–power parameters · Zigbee module · Wireless monitoring and control application

D. Nitnaware (✉)
Electronics and Telecommunication Department,
Institute of Engineering and Technology, DAVV, Indore, Madhya Pradesh, India
e-mail: dhiirajnitnawwre@gmail.com

© Springer Nature Singapore Pte Ltd. 2018
D.K. Mishra et al. (eds.), *Information and Communication Technology for Sustainable Development*, Lecture Notes in Networks and Systems 9,
https://doi.org/10.1007/978-981-10-3932-4_16

1   Introduction

As the sun is the most powerful energy source, most forms of renewable energy come either directly or indirectly from the sun. The authors [1] have explained the working of solar power and its types. The types of solar technology, i.e., active and passive are also explained in great detail in the same. F. Salvadori et al. have explored the radio frequency (RF) communication for wireless technologies which finds its application including cordless phones, radar, ham radio, GPS, and radio and television broadcasts. RF are electromagnetic waves that propagate at the light speed and are invisible to human eye because RF waves are slower than visible light [2]. The Federal Communications Commission's (FCC) regulations responsibility for RF assignment with the National Telecommunications and Information Administration (NTIA) is responsible for regulating federal uses of the radio spectrum, which is given in [3].

In paper [4], the authors focus on an interface between a data acquisition device and three sensors using Linux operating system. The sensors will sense voltage, current and temperature of the SuPER system and display the respective values on a computer screen. Briskman et al. [5] have developed a photovoltaic cell curve tracer that is low cost and user-friendly. The instrument will be used in conjunction with an IBM compatible PC, or a Mac and a printer. The curve tracer will trace out I–V curve and compute the parameters such as V_{OC}, I_{SC}, FF, V_{MP}, I_{MP} and P_{MAX}. The authors of paper [6] have explored how to extract maximum power from the solar panel by using MPPT controller with P&O algorithm using the effective embedded wireless communication technology. This helps to overcome the existing problem in manual monitoring of solar PV modules. The paper also monitors the voltage, current, maximum power and solar light irradiance of PV modules.

The maximum power point (MPP) of an array which is coupled to power inverter is determined in paper [7]. I–V characteristics of the photovoltaic array can be generated by putting capacitor as a load on a DC-bus side of the inverter. Various parameters like V_{OC}, I_{SC}, V_{MP}, I_{MP} and P_{MAX} are determined based on charging–discharging of capacitor which helps to calculate the MPP. The authors of paper [8] have developed the initial device setup to measure the above parameters of solar panel.

2   Problem Domain

Every manufacturer tests their solar panel modules under system called standard test conditions (STC). These are a set of rules that all has to follow and which allow consumers and solar designers to compare panels. But their performance will be different in the real world. The effects such as insulation, shading, orientation, tilt angel, temperature and load resistance affect the panel performance. Solar cell is mostly affected by the temperature. Also the main problem at the solar plant is to measure the performance parameter of solar panel and to detect faulty cells or panels in the array. Thus, there is a requirement of manpower or technician to check for the faulty panels.

3 Proposed Solution

The proposed system not only helps to measure these parameters to detect faulty cells or panels (which insuring that panels in arrays are matched to get maximum power) but also reduced the manpower/technician not have to be in the actual area where solar panels are located to monitor the solar panel parameters. The total system is combination of different sections like voltage and current sensor, digital load, microcontroller and RF module embedded on a single PCB.

3.1 Block Diagram of the Proposed System

The block diagram of the proposed system is shown in Fig. 1 for one solar panel module. Four such modules were taken for the measurement and analysis. Block diagram of transmitter and receiver section is given in Figs. 2 and 3. The sensor node do not have any external power supply as the voltage is generated by the solar panel which is stored in battery. The microcontroller read open-circuit voltage and short-circuit current by controlling the pulse width of PWM send via RF module to base station.

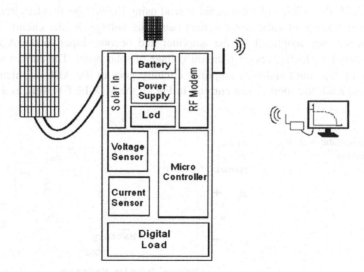

Fig. 1 System block diagram

Fig. 2 Transmitter section block diagram

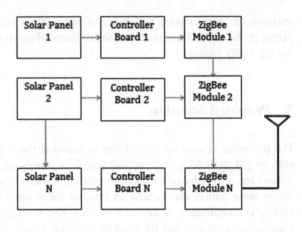

Fig. 3 Receiver section block diagram

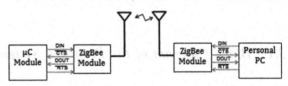

3.2 Schematic of the Proposed System

Figures 4, 5, 6 and 7 give the schematics of power supply, voltage measuring section, current measuring section and digital load section. The voltage that is generated by the solar panel is stored in rechargeable batteries which are regulated by voltage regulator LM 7805 as in Fig. 4.

In Fig. 5, the voltage of solar panel is read using the voltage divider circuit. In Fig. 5, the voltage of solar panel is read using the voltage divider circuit. These voltages are then amplified by the amplifier and become input to the ADC of microcontroller which gives open-circuit voltage of solar panel. The current sensor is made by the shunt and after amplifier circuitry goes to the ADC of microcontroller and reads the short-circuit current. In Fig. 7, the MOSFET is work as a load.

Fig. 4 Schematic of power supply section

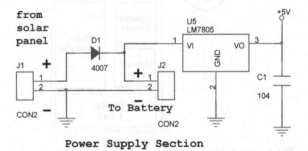

Fig. 5 Schematic of voltage measuring section

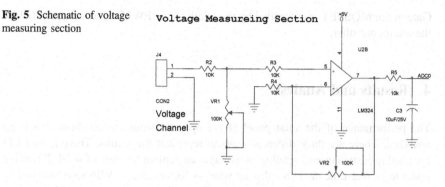

Fig. 6 Schematic of current measuring section

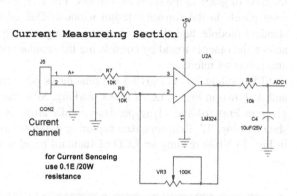

Fig. 7 Schematic of digital load section

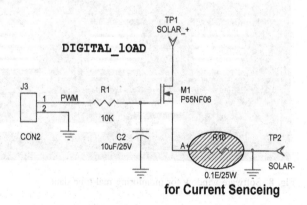

Gate of the MOSFET is controlled by the width of the PWM those are generated by the microcontroller.

4 Results and Analysis

The performance of the solar panel at the open environment condition has been analyzed. There are three different ways to represent the results. There is an LCD for display instantaneous reading of voltage, current and power of 4 DUT (device under test), real-time data on software window (developed in VB6 software) and in the form of graph in the software window. Figure 8 shows the real-time data of four solar panels in the software design window. One of the panels is considered as standard module, and all are connected in ring topology. Detection of faulty panel among the other is found by considering the standard panel data of voltage, current and power as reference.

SPV module 1 is not giving the reading as per standard within 10–20% range and is shown in Fig. 9, i.e., it is not working. SPV module 2 and 3 is faulty and is given in Figs. 10 and 11, respectively, while all three modules are faulty and are shown in Fig. 12. Real-time data capturing in the environmental condition is given in Fig. 13 while reading on LCD of standard panel is shown in Fig. 14.

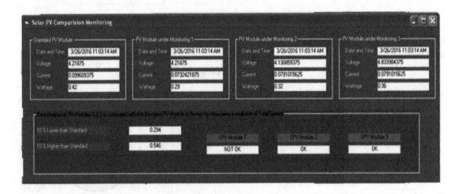

Fig. 8 Software window for monitoring real-time data

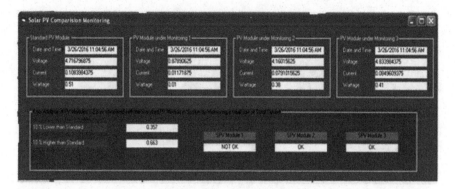

Fig. 9 Software window showing SPV module 1 faulty

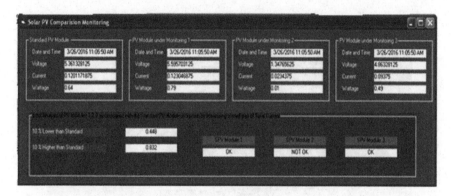

Fig. 10 Software window showing SPV module 2 faulty

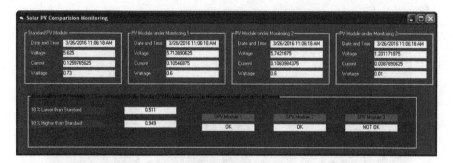

Fig. 11 Software window showing SPV module 3 faulty

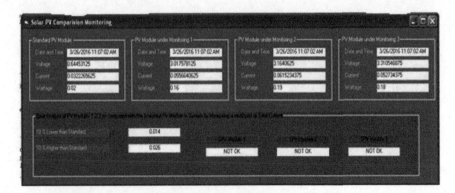

Fig. 12 Software window showing all SPV module faulty

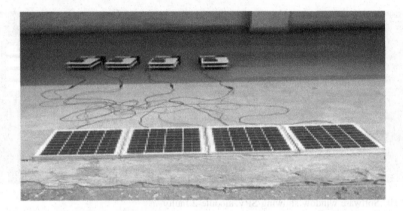

Fig. 13 Real-time setup for data capturing

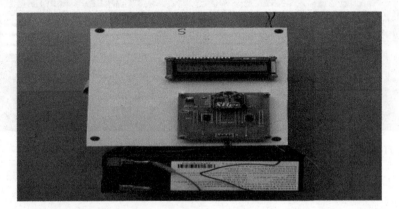

Fig. 14 Real-time data display on standard panel

5   Conclusion

The paper is aimed to provide solar photovoltaic characteristics by monitoring the important parameters like open-circuit voltage, short-circuit current and power of solar panel. These parameters are necessary to study the solar energy which helps in the installation of solar plant. We analyze the solar panel at various places and have taken the reading of solar parameters under all climate conditions. Thus, on the basis of above results, the conclusion can be drawn that a prototype system has been successfully tested. Wireless sensor data are transmitted and received at PC and displayed on screen. In this system, the battery is also charging with solar panel so there is no replacement of battery which provides longer backup for the system.

References

1. Solanki CS (2012) Solar photovoltaics fundamentals, technologies and applications, 2nd edn. PHI Learning Private Limited Publications, Delhi
2. Salvadori F, de Campos M, de Figueiredo R, Gehrke C, Rech C, Sausen PS (2007) Monitoring and diagnosis in industrial systems using wireless sensor networks: IEEE international symposium on intelligent signal processing, pp 40–45
3. Stallings W (2002) Wireless communication and networking. Prentice Hall, ISBN 0-13-040864-6
4. Vasquez GJ (2006) Data acquisition and sensor circuits for the SuPER project. California Polytechnic State University, San Luis Obispo
5. Briskman RN, Livingstone PE (1997) A low cost, user friendly photovoltaic cell curve tracer. Sol Energy Mater Sol Cells 6:187–199
6. Go O, Katsuya H, Shigeyasu N (2006) Development of a high-speed system measuring a maximum power point of PV modules: in photovoltaic energy conversion, conference record of the 2006 IEEE 4th world conference on, pp. 2262–2263
7. Bhavaraju V, Grand KE, Tuladhar A (2007) Method and apparatus for determining a maximum power point of photovoltaic cells: application. US, Ballard Power Systems Corporation, USA
8. Ranhotitogamage C, Mukhopadhyay SC, Garratt SN, Campbell WM (2011) Measurement and monitoring of performance parameters of distributed solar panels using wireless sensors network: IEEE international conference, New Zealand

Efficiency Evaluation of Recommender Systems: Study of Existing Problems and Possible Extensions

Mugdha Sharma and Laxmi Ahuja

Abstract Recommender systems can help people to choose the desired product from a choice of various products. But there are various issues with the existing recommender systems. Thus, new systems which can efficiently recommend the most appropriate item to users based on their preferences are in demand. As a step towards providing the users with such a system, we present a general overview of the recommender systems in this paper. This paper also proposes potential solutions to the different problems which are found in current recommendation methods. These extensions to the current recommendation systems can improve their capabilities and make them appropriate for a broader area of applications. These potential extensions include integration of contextual information into the recommendation method, an improvement in understanding of items and users, developing less intrusive recommendation approaches, utilization of multi-criteria ratings, and providing more flexible types of recommendations.

Keywords Contextual information · Extensions to recommender systems
Multi-criteria ratings · Multidimensionality · Recommender systems

1 Introduction

Today, there are enormous amount of information and products which people can find and buy online. Due to this, it is not convenient for the users to find the correct information they need or choose the appropriate product to buy. To overcome this problem, we had various personalization techniques, and one aspect of personalization is the recommendation systems [1]. These recommender systems learn from

M. Sharma (✉) · L. Ahuja
Amity University, Noida, Uttar Pradesh, India
e-mail: mugdha.sharma145@gmail.com

L. Ahuja
e-mail: lahuja@amity.edu

© Springer Nature Singapore Pte Ltd. 2018
D.K. Mishra et al. (eds.), *Information and Communication Technology for Sustainable Development*, Lecture Notes in Networks and Systems 9,
https://doi.org/10.1007/978-981-10-3932-4_17

the previous behaviour of user transactions records and recommend very few products which may be useful to the users.

Till now, a lot of work has been done to find new recommendation approaches for academia as well as industry. But, there is so much to explore in this field since there are various practical applications related to this domain that can help people to handle the ample amount of information we have online today and give useful recommendations, relevant content, and services to them.

Even after various advancements, the current recommendation approaches still needs to improve a lot to make it more useful which can be used to a wider range of practical applications. These enhancements include more advanced methods for representing user behaviour, utilization of multi-criteria ratings, integration of contextual information into recommendation method, improved and less intrusive recommendation approaches, and development of more flexible recommendation methods.

This paper illustrates various methods to extend the capabilities of recommender systems. This paper is structured as follows: Section 2 provides a comprehensive review of recommender systems. In next section, we discover various issues in the current recommendation methods and suggest various approaches to extending their capabilities. Conclusions of the research are mentioned in Sect. 4.

2 Related Work

Recommendation system gained popularity in the 1990s because of its most popular algorithm collaborative filtering [1]. It started as an independent research domain, where the focus was on those recommendation problems that were dependent on rating mechanism. So, initially the problem of recommendation is transformed into the problem of approximating ratings for those items that are still need to be rated. After that, we can approximate the ratings for those unrated items. Then finally, we can recommend the highest rated items to the user.

Let I be the set of potential items that can be recommended, such as books, movies, or restaurants and U be the set of users. The space I of probable items can be very big, ranging in hundreds of thousands or in fact millions of items in few applications, such as recommending CDs or books. Similarly, the user space can too be very big. Let X be a utility function that defines the usefulness of item i to user u,

$$X: U \times I \to \text{Rating}$$

Here, rating is an ordered set. And, for each user $u \in U$, we should select a particular item $i' \in I$ that maximizes the utility of the user.

Suppose a user "Sam" gave the ratings 7 out of 10 for a particular movie "Avengers". It depicts that the utility of that item, i.e., movie depends upon the ratings given by the user. Now, this utility can be considered in two different ways

—one is arbitrary function, and another is a profit function. If it is an arbitrary function, it can be mentioned by the user explicitly. Or in case of a profit-based utility function, utility is calculated by the application.

One of the most common problems in recommendation systems is that normally, utility X is not meant for the whole $U \times I$ space but only for a small subset of it. That means X requires to be extrapolated to the whole $U \times I$ space. Generally in recommender systems, utility is defined only on the items which have been rated by the users previously. Consider the case of a movie recommendation system, a user initially rates few subsets of movies that he/she has already seen. The estimation of new ratings for not-yet-rated items can be done through various methods such as approximation theory, various heuristics, and machine learning techniques.

Furthermore, recommendation system employs various techniques that include —content-based [2]: in this the system recommends items similar to the items that the users have liked in the past; collaborative filtering [3]: recommends items that similar taste users have liked in past; demographic [4]: recommends items based on demographics (language, country) of user; knowledge-based [5]: recommends item based on the knowledge that how a particular item meets user demand; hybrid recommender system [6, 7]: combination of one or more techniques described. Since recommendation systems have been proposed, they are being used in variety of applications such as for tourism [8], mobile [9], music [10], news [11], and social networking sites [12–14].

2.1 Content-Based Methods

In this recommendation technique, system recommends the items similar to the items that people have liked earlier in the past. The features linked with the compared items help in calculating the similarity of the items. For instance, in a movie recommendation system, the content-based recommender method understands the similarities among the movies a particular user has rated well in the past. Then, only those movies would be recommended to the user, which have a high extent of similarity to the user's taste and preferences [2].

As was discussed in [15, 16], content-based recommender systems have various problems that are mentioned below in Table 1.

2.2 Collaborative Methods

In this recommendation system, the items are recommended to a user based on the items which are earlier rated by other people. For this purpose, "peers" of a particular user are discovered by the system, who have same preferences and interests in movies. Then, the movies that are mainly liked by the "peers" would be recommended to the user.

Table 1 Problems in content-based recommender system

New user problem	• A particular number of items need to be rated by a user before the system can really recognize the preferences of the user and recommend the relevant items to the user • But when we have new users, then we do not have any prior ratings from those users • So this results in poor recommendations since the recommender system would not be able to get accurate and quality recommendations
Overspecialization	• This recommendation technique can suggest only those items that attain high ratings against a user's profile. That is why the user always gets the recommendations similar to the items that are already rated previously • For example, consider a case where people who never showed any interest in Mexican food will never get a recommendation even for the most famous Mexican eating place in the city • Now, this issue can be resolved by including some randomness as proposed by [21]. They proposed the use of genetic algorithm in the perspective of information filtering • Sometimes, this is also a situation that a same incident is mentioned in different news articles. So, recommendation system should not provide these articles if the user has already seen something similar to that. Hence, some content-based recommendation techniques discard few items not only if those are exactly same to something the user has seen earlier, but also if they are too dissimilar from the preferences of the user • In an ideal situation, the user should not be provided with an identical set of recommendations. Instead user should get a range of alternatives [22]. Necessarily, it is not recommended to suggest all the movies directed by James Cameron to a person who enjoyed one of them
Limited content analysis	• One of the most important aspects of content-based recommender methods is the features which are linked with the items that are generally recommended by this technique • So, there are two ways to get an adequate number of features. Either the features must be assigned to items manually or the content should be parsed by a computer automatically • However, information recovery methods work fine in finding features from the text documents. But in automatic feature extraction method, few other fields have an intrinsic problem such as, it is very difficult to apply this technique to multimedia data, such as graphical pictures, audio, and video streams. • And normally, it is not feasible to allocate attributes manually because of limited resources [16] • There is another issue, if two different items are described by the same features, then it is impossible to differentiate between them • Consider the case of text-based documents, which are generally prepared with only few essential keywords. So, they generally use the same terms. Therefore, it is very difficult for content-based techniques to recommend the articles to users. Because the system is not able to find out which article is poorly written and which one is well written [16]

Linden et al. [17] proposed a unique and improved collaborative filtering technique in which the user specifies the input set of ratings. And those ratings are selected through various different techniques that remove redundancy, noise, etc. The results show high efficiency and high accuracy for model-based collaborative filtering methods. Now, among the latest researches, [18] also presented a new approach to collaborating filtering method in which the model-based and memory-based techniques are combined together. Basically, [18] suggested that recommendations can be calculated by using the mixture model with the help of user profiles stored in it.

Collaborative systems can handle any type of content and suggest any items, since they use other users' recommendations. It can suggest even those that are not

Table 2 Problems in collaborative-based recommender system

New item problem	• In this recommender system, an item can only be recommended to a user if it has been rated earlier by other users • But new items keep on adding to the system on a regular basis. This issue is known as new item problem • Therefore, it is necessary to get the ratings for the new items from a significant number of users. Then only, the new items would be recommended to the users
Sparsity	• In recommendation systems, the amount of ratings that are gathered previously is not enough to recommend accurately because those are very less as compared to the amount of ratings that requires to be predicted • Then it becomes a challenge to predict the ratings correctly from that small amount of ratings • Moreover, the success of collaborative recommendation method is directly related to the availability of a significant number of users • For instance, consider the case of movie recommender algorithm. There might be various movies that have been rated by very less number of people. That is why these movies would not be recommended very often, even if a movie got high ratings by those users • It will affect the recommendations for a particular user whose interests are not usual or similar to the other people [15] • This issue of sparsity can be resolved if we take help of user profiling while estimating similarity among the users. That means two different users would be taken as similar if they rate a same movie similarly and they fit into the same demographic section also • Chen and He [4] use the age, gender, education, employment information, and area code of people in their recommendation algorithm to provide accurate recommendations for restaurants. This technique is known as "demographic filtering" sometimes
New User Problem	• For suggesting perfect recommendations, a particular number of items need to be rated by a user before the recommender system can really understand the preferences of the user • To overcome this issue, various techniques have been proposed such as, hybrid recommendation technique. It integrates both the techniques—collaborative and content-based techniques into one recommendation algorithm • An alternative technique is also described in [23], where various methods are proposed for determining the best items to rate based upon item entropy, item popularity, and user personalization [23]

similar to the ones seen earlier in the past. But, still there are various problems in collaborative systems [19], as mentioned below in Table 2.

2.3 Hybrid Approaches

When collaborative and content-based techniques are integrated together, then that newly developed approach is known as a hybrid recommendation technique. It helps to reduce the problems which are found in collaborative-based and content-based methods [6, 7]. There can be three different ways to integrate the two approaches into a hybrid recommendation system. One is to implement both the methods individually and then combine their predictions. Another way is to incorporate few collaborative features into a content-based approach or vice versa. And the last one is to develop a general approach that integrates the characteristics of both the techniques— collaborative-based method as well as content-based method.

Most of the hybrid approaches are based on conventional collaborative filtering methods but they also maintain the content-based profiles for each and every user. Then the similarity between two users is calculated with the help of these content-based profiles which are not the usually rated items. This allows to prevail over some sparsity-related issues of an entirely collaborative system because not several pairs of users will have a considerable amount of usually rated items.

Furthermore, various papers, such as [15, 20], compare the recommendations suggested by the hybrid technique with the recommendations suggested by pure collaborative filtering methods and content-based methods and show that we get more precise recommendations results by using hybrid approaches.

3 Extensions to Recommender System's Capabilities

There are many ways in which recommender systems can be extended including, integration of the contextual information into the recommendation process, improving the understanding of users and items, providing less intrusive more flexible kinds of recommendations, and supporting multi-criteria ratings. Such improved models of recommendation techniques can give improved recommendation results. These potential extensions will be discussed in the rest of this section.

3.1 Complete Understanding of Items and Users

Current recommendation systems can never understand the items and users completely, since they only have item and user profiles to extract all the relevant information related to items and users. Most of the recommender systems produce ratings depending on that inadequate understanding of items and users. They do not extract the full information from the other accessible data which is stored in the user's transactional histories. Such as, typical collaborative recommendation techniques [16] do not utilize item and user profiles at all to make recommendations because they solely depend upon the ratings information. Though there have been some improvements made on integrating item and user profiles into few methods since the times of old recommender systems, these profiles still do not make use of the more developed profiling approaches. Apart from using conventional profile characteristics, such as keywords and regular demographics of a user, we can develop user profiles by using more developed profiling methods based on sequences, signatures, and data mining rules that illustrate the taste and interest of a user. Moreover, apart from using the conventional item profile characteristics, like keywords, similar developed profiling methods can also be used to make comprehensive item profiles.

3.2 Multidimensionality of Recommendations

Currently, all the recommendation systems work in the 2-D space, i.e. *User ×Item,* which means, they provide the recommendations that depend only on the item and user information. They do not consider other contextual information that could be important in various situations. Though, in many cases, the usefulness of a particular item for a person may vary according to time (such as, day of the week, a particular month or a season, or time of the year). It may also vary for a person to person, such as who will consume the product or with whom it will be shared and under which situation. Then, it might not be appropriate to just suggest a particular item to user. Other relevant information such as time, place, and the company of the user must be considered before making the recommendations. Consider the case of a user who wants to see a movie when she is with her family at home on a Tuesday evening and in another case, she is going out with her friends to a movie theatre on a Friday night. So she must have different choices for the movies she wants to see in both the situations. Consider another example, in tour package recommendations; it should also take into account few things such as, the time of the year, travelling environment and restrictions at that time, with whom the user wants to travel, and other contextual information.

Now, the concern is that the two-dimensional recommender systems cannot be directly converted to a multidimensional system. A reduction-based recommender system can be used which will use the ratings that are relevant to the context in which a recommendation is made. Suppose, a user wants to watch a movie in a movie theatre on a Friday night, then the reduction-based recommender system would use only the available ratings of the movies which are already seen in the movie theatres by the user over the weekends. By choosing only the ratings that are significant to a recommendation context, this approach reflects the multidimensional cube of ratings on the two primary dimensions, i.e. item and user.

3.3 Multi-criteria Ratings

Generally, all the current recommendation methods take single-criterion ratings into account, for example, movies ratings or books ratings. But, in few cases, such as recommender system for restaurant, it is very important to consider multi-criteria ratings into recommender systems. For example, various restaurant apps like Zomato provide three criteria for restaurant ratings: food, service, and decor. But then, the problem of multi-criteria optimization comes into the picture. So, we can have following easy solutions for the optimization problem:

1. We can create a linear combination of multiple criteria and then convert it to a single-criterion problem. And finally find the optimize solution for that single-criterion problem.
2. Or, we can optimize the most crucial criterion and convert the other less important criteria to constraints.
3. Lastly, we can consecutively optimize one criterion at a time and convert an optimal solution to constraints. And then repeat the procedure for other criteria also.

3.4 Flexibility

Since every recommendation system supports only a fixed and predefined set of recommendations, therefore none of the system is flexible. So, it does not allow the end-user to customize the recommendations as per his/her requirements in real time. Moreover, almost all the recommendation algorithms recommend only particular items to particular users and do not consider aggregation. But, in some cases, it is necessary to give aggregated recommendations. For instance, a recommendation application which deals in tours and travels may want to recommend few vacation packages in Singapore (product category) to the postgraduate students from India (user segment) during the summer break. In this case, the recommender system cannot recommend the same package to students which it provides to a newly married couple for their honeymoon package. Few modifications and customization is necessary in suggesting an attractive deal to them. OLAP-based systems can be used to maintain the aggregated recommendations since it supports aggregation hierarchies.

3.5 *Nonintrusiveness*

Almost all of the current recommender systems need a high level of user involvement and precise feedback from them. For instance, before recommending any newsgroup article to a user, system generally obtains the ratings of earlier read articles from that particular user. But, it is unfeasible to extract all the possible ratings from the users. Therefore, we suggest the use of nonintrusive rating determination techniques in recommendation systems to calculate the real ratings with the help of some proxies. Such as, the time required by a user to read an article can be a proxy of the article's rating given by that particular user. But, these nonintrusive ratings are generally not accurate and cannot exactly replace the ratings given by the user. Thus, the recommendation algorithm developers have to resolve the issue of minimizing the intrusiveness and maintaining a particular level of precision of recommendations as well. One way to resolve this issue is to obtain an optimal amount of ratings the system should take from a new user. Such as, MovieLens.org initially needs a user to rate a particular number of movies before recommending any movies. This will incur some costs on the user, but every single addition to the ratings provided by the user will enhance the correctness of recommendations and will result in some benefits for the user.

4 Conclusions and Future Scope

Recommendation systems made considerable growth and development over the time period when various collaborative, content-based, and hybrid approaches were proposed and developed. Even after all of these advancements, the current recommendation systems still need few enhancements to make better use of them. In this paper, we studied various shortcomings of the current recommender systems and suggested few potential enhancements that can improve the capabilities of the recommendation techniques as shown in Table 3. These potential extensions include integration of contextual information into the recommendation method, an improvement in understanding of items and users, developing less intrusive recommendation approaches, utilization of multi-criteria ratings, and providing more flexible types of recommendations.

In most of the recommendation systems, their performance is generally evaluated in terms of accuracy and coverage metrics. Though popular, these metrics have few drawbacks. One drawback is that these metrics are usually performed on user selected test data. But, items that users select to rate are expected to have a skewed sample. Developing the metrics that would resolve these drawbacks would be an important research area.

Table 3 List of potential extensions and their impact on the current recommender systems

Potential extension	Potential improvement in the recommendation systems
Comprehensive understanding of items and users	Recommendation systems can produce ratings based on the adequate understanding of items and users by extracting all the relevant information of items and users and recommend more accurate items to the users
Multidimensionality of recommendations	Recommendation systems can consider the other contextual information that could be relevant in various situations while making recommendations. It also adds up to accurate recommendations
Utilization of multi-criteria ratings	Nowadays, single-criterion ratings are not enough, since we have more complex applications and systems. Therefore, it is very important to consider multi-criteria ratings into recommender systems. It will improve the recommendation process by considering all the criteria which are important in the given scenario
Providing more flexibility	By allowing the end-user to customize the recommendations as per his/her requirements in real time, recommendation systems can really improve their efficiency. One of such customizations is providing aggregated recommendations
Nonintrusiveness	We suggest the use of nonintrusive rating determination techniques to calculate the real ratings with the help of some proxies. In this way, recommender systems will not require a high level of user involvement and can provide recommendations in less time

References

1. Ricci F, Rokach L, Shapira B (2011) Introduction to recommender systems handbook. Recommender Systems Handbook, Springer, pp 1–35
2. Kim HW, Han K, YI MY, Cho J, Hong J (2012) Moviemine: personalized movie content search by utilizing user comments. J IEEE Trans Consum Electron 58
3. Huete JF, Fernandez JM, Campos LM, Rueda-Morales MA (2012) Using past-prediction accuracy in recommender systems. J Inf Sci 199(12):78–92
4. Chen T, He L (2009) Collaborative filtering based on demographic attribute vector. In: Proceedings of international conference on future computer and communication. IEEE
5. Burke R (2000) Knowledge-based recommender systems. Encyclop Lib Inf Syst 69 (supplement 32):175–186
6. Liu DR, Lai CH, Lee WJ (2009) A hybrid of sequential rules and collaborative filtering for product recommendation. J Inf Sci 179:3505–3519
7. Nilashi M, Ibrahim OB, Ithnin N (2014) Hybrid recommendation approaches for multi-criteria collaborative filtering. J Expert Syst Appl 41:3879–3900
8. Liu Q, Chen E, Xiong H, Ge Y, Li Z, Wu X (2014) A cocktail approach for travel package recommendation. J IEEE Trans Knowl Data Eng 26
9. Gavalas D, Konstantopoulos C, Mastakas K, Pantziou G (2014) Mobile recommender systems in tourism. J Netw Comput Appl 39:319–333
10. Hyung Z, Lee K, Lee K (2014) Music recommendation using text analysis on song requests to radio stations. J Expert Syst Appl 41:2608–2618
11. Lin C, Xie R, Guan X, Li L, Li T (2014) Personalized news recommendation via implicit social experts. J Inf Sci 254:1–18

12. Agarwal V, Bharadwaj KK (2012) A collaborative filtering framework for fiends recommendation in social networks based on interaction intensity and adaptive user similarity. Springer
13. Esslimani I, Brun A, Boyer A (2010) Densifying a behavioural recommender system by social networks link prediction methods. Springer
14. Arias JP, Vilas AF, Redondo RP (2012) Recommender systems for social web. Springer (2012)
15. Balabanovic M, Shoham Y (2007) Fab: content-based, collaborative recommendation. ACM 40:66–72
16. Shardanand U, Maes P (2005) Social information filtering: algorithms for automating 'word of mouth'. In: Conference of human factors in computing systems
17. Linden G, Smith B, York J (2003) Amazon.com recommendations: item-to-item collaborative filtering. IEEE Int Comput
18. Miller BN, Albert I, Lam SK, Konstan JA, Riedl J (2003) Movielens unplugged: experiences with an occasionally connected recommender system. In: International conference of intelligent user interfaces
19. Billsus D, Brunk CA, Evans C, Gladish B, Pazzani M (2002) Adaptive interfaces for ubiquitous web access. ACM 45:34–38
20. Melville P, Mooney RJ, Nagarajan R (2002) Content-boosted collaborative filtering for improved recommendations. In: 18th national conference of artificial intelligence
21. Sheth B, Maes P (2003) Evolving agents for personalized information filtering. In: Ninth international conference of artificial intelligence for applications. IEEE (2003)
22. Zhang Y, Callan J, Minka T (2002) Novelty and redundancy detection in adaptive filtering. In: 25th annual international ACM SIGIR conference, pp 81–88
23. Rashid AM, Albert I, Cosley D, Lam SK, McNee SM, Konstan JA, Riedl J (2002) Getting to know you: learning new user preferences in recommender systems. In: International conference of intelligent user interfaces

Comparative Analysis of Application Layer Internet of Things (IoT) Protocols

Himadri Chaudhary, Naman Vaishnav and Birju Tank

Abstract IoT devices will be everywhere, will be context-aware and will enable ambient intelligence. This provides a possibility of a framework that would allow direct machine-to-machine communication. This vision has produced an exemplar which is often referred to as the Internet of things (IoT). We have studied the application level IoT protocols namely, The Constrained Application Protocol (CoAP), Message Queuing Telemetry Transport (MQTT) and the Hypertext Transfer Protocol (HTTP)/REpresentational State Transfer (REST). We have used the cooja simulator for carrying out the simulation. We have simulated the above protocols in the cooja simulator demonstrating a simple scenario of three motes. We have read the sensor information from the server with the help of browser. We can also make use of different topologies and simulate the scenario according the requirement.

Keywords Internet of things (IoT) · Constrained Application Protocol (CoAP) Machine-to-machine (M2M) · Message Queuing Telemetry Transport (MQTT) Hypertext Transfer Protocol (HTTP) · REpresentational State Transfer (REST)

1 Introduction

While the Internet is formed primarily by interconnecting homogenous devices (i.e. computers), there have been recently several paradigms in networking such as mobile, grid and cloud computing which enabled a purposeful interconnectivity between various semi-homogenous devices such as computers, cameras, smart-

H. Chaudhary (✉) · N. Vaishnav · B. Tank
GTU PG School, Ahmedabad, Gujarat, India
e-mail: himadrichaudhary1314@gmail.com

N. Vaishnav
e-mail: namanmvaishnav@gmail.com

B. Tank
e-mail: birjutank27@gmail.com

© Springer Nature Singapore Pte Ltd. 2018
D.K. Mishra et al. (eds.), *Information and Communication Technology for Sustainable Development*, Lecture Notes in Networks and Systems 9, https://doi.org/10.1007/978-981-10-3932-4_18

phones, sensors and other instrumentation (i.e. satellites). Nowadays, there has arisen a need for the machine-to-machine (M2M) communication that requires a need for a new technology such as the Internet of things (IoT) to come into existence keeping in mind the future of the Internet. This is connecting the various objects around us to the Internet. If the objects could be identified with unique identifier, we can interconnect various objects [6]. This would result into the Internet of things, which is an enhanced version of the traditional Internet. Various objects such as cell phones, household appliances and vehicles can be included in these devices. This can be done using the lightweight devices such as the wireless sensors and RFID tags. These small devices provide us with a fast-paced interactions with the inter devices as well as to the other devices and human beings at any time. If we are able to do so, there arises one important question: Are these devices secure enough to depend on? Smart devices not only pass on information but they also interact with human beings [5]. We in return provide these devices with the details of our daily routine, which can be easily tracked down.

We are, therefore, exposed to the world regarding to our personal, professional information. These devices can be designed to think and react to various situations and scenarios [7].

The concept where the simple devices that could communicate through the Internet is what IoT is all about. These devices are very simple that we use in our day-to-day life, such as car, television, washing machine, a/c, even lights and fans. All these devices with a little modification can be made smart enough to communicate through the Internet or other similar devices. We need actuators along with sensor nodes to make this possible. Actuators are the devices that can react when the event takes place. We can also make use of the RFID, sensors and actuators to turn normal devices into the smart devices [8, 9].

2 Related Work

2.1 Constrained Application Protocol (CoAP)

The Constrained Application Protocol (CoAP) is an application layer protocol in Internet of things (IoT) protocol stack. It is a specialized Web transfer protocol used for constrained nodes. CoAP gives request/response communication model for the endpoints of IoT application [1]. So many protocols are available for communication at the application layer. But those protocols are too heavy for the IoT networks. So Internet Engineering Task Force (IETF) designed one lightweight protocol called Constrained Application Protocol (CoAP) for the communication at application layer in IoT. It is mainly designed for the small, low-power and constrained devices. As the requests and responses are exchanging asynchronously, the use of CoAP is relatively easy [2]. The packet size of CoAP is smaller than HTTP TCP packet size. Generation and parsing of packets are simple that they will not be consuming RAM in constrained devices. CoAP uses UDP as transport layer protocol instead of TCP. So, the communication done between sender and receiver is connectionless. CoAP uses a client/server model. Client may have different access to the data of server. Messages are marked confirmable or non-confirmable. Acknowledgement of confirmable message will be send to the receiver by ACK packet. Non-confirmable messages are discarded. Content negotiation is also supported by CoAP. Security is not provided by SSL/TLS as CoAP is in the top of UDP and not TCP. Datagram Transport Layer Security (DTLS) provides same security as TLS but data transfer should be over UDP.

2.2 Hypertext Transfer Protocol (HTTP)/REpresentational State Transfer (REST)

Communication between independent systems is done in simple way through REST. Whereas sending and receiving of data on the Web is done through Web. We can communicate with minimal overhead in REST [3]. REST is almost same as HTTP. HTTP uses TCP as a transport layer protocol. HTTP works as a client/server or request/response model. The example of that is Web browser. REST supports scalability. So that large number of components can communicate with each other. REST has many features such as simple interface, versatility of component and transparency of communication between components. The architecture of REST applies constraints to the different components as well as data elements. REST is also a lightweight protocol which can be implemented in small devices. It can also be used in embedded system.

2.3 Message Queueing Telemetry Transport (MQTT)

In order to provide M2M communication, we have one messaging protocol named Message Queue Telemetry Transport (MQTT) that works on the publish/subscribe format. It follows a client/server model. Each node connects to the broker which is the server using TCP [4]. The messages in the MQTT are simply chunk of data that cannot be read by the broker. Messages are sent to an address which is called topic. Subscription to multiple topics by a single client is possible. As it is publish/subscribe model, client receives messages to all the topics it has subscribed to. The topics are structured hierarchically [5] (Fig. 1).

Three levels of QoS provided by MQTT are as follows:

1. **at most once**: Also referred as "fire and forget", as no care is taken for the reception of acknowledgement. The messages are not even stored. The messages are delivered at the most once to the subscriber or not at all. There is a

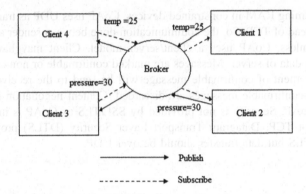

Fig. 1 Client and broker architecture

Table 1 Comparison of CoAP, MQTT and HTTP/REST

Features	CoAP	HTTP	MQTT
Transport	UDP	TCP	TCP
Messaging	Req/Resp	Req/Resp	Pub/Sub
Security	Medium-optional	Low-optional	Medium-optional
Architecture	Tree	Client–server	Client–server
Applications	Utility field area networks	Smart energy phase II	IoT messaging
Type	State transfer		Event-based

    possibility of the messages getting lost if the subscriber disconnects or if the server fails [6].

2. **at least once**: Here, the messages are to be delivered at least once to the subscriber. If the receiver fails to send the acknowledgement, the messages can be received more than once. The messages are locally stored by the receiver until the acknowledgement is received or also in case if it is to be sent again.

3. **exactly once**: As the name suggests the message is to be received exactly once. Local storage of the message is done until sender gets an acknowledgement. It is a safe mode of transfer while being very slow. It follows handshake method to keep a check no duplicate messages are being sent (Table 1).

3 Implementation

We have implemented these protocols in contiki 2.7 OS. Contiki is an open source operating system for the Internet of things (IoT). It provides low-power communication over Internet. It provides easy and fast development. Applications of contiki are written in C language. Contiki runs on low-power, small tiny devices. Contiki runs on virtual box or VMware.

Here, we have taken three motes. All of them are sky motes. Initially one node is added, this mote is configured as the REST border router. As shown in the following Fig. 2.

Then, we have added two more motes. These are configured using the IPv6 border router. We then set a mote as server. It is also shown in Fig. 3.

Then, we make one of the motes as the server mote and others will be considered as the client motes. One of the client motes is in the vicinity of the server mote, while the other is not. So, the packet will be transferred from the server via the second mote to the end mote. This is as illustrated in the following Fig. 4.

All the motes will have the IPv6 type of address like:

aaaa::212:7401:1:101 for the border router.

Finally, all the motes are bridged to each other using the "make connect-router-cooja" command.

We can read the sensors information using the Firefox browser. In the address bar, we can input the address of the mote as: coap: //[aaaa::212:7401:1:101]

As this is the border router, it will print the information of the neighbour motes and routes from the border router. Similarly, we can print the information of the other motes providing their addresses [10]. The Firefox output is shown in the following Fig. 5.

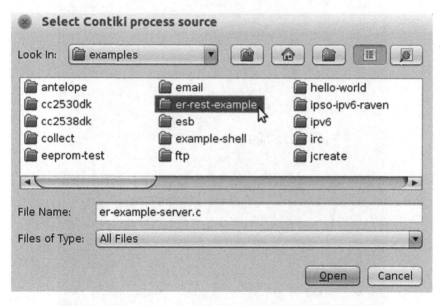

Fig. 2 REST border router

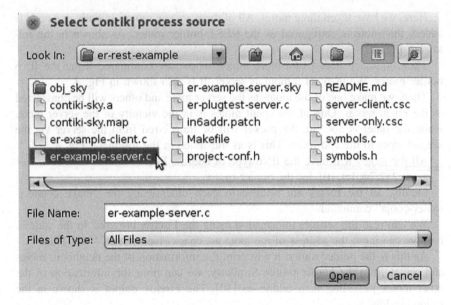

Fig. 3 IPv6 border router

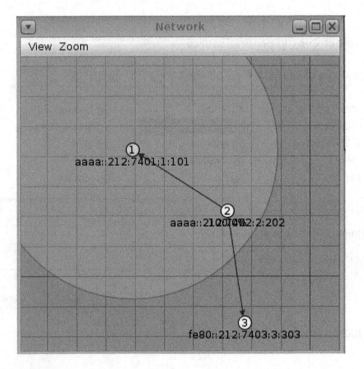

Fig. 4 Motes topology

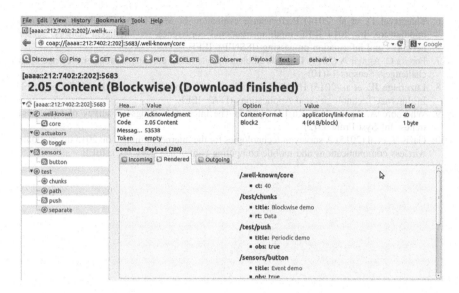

Fig. 5 Reading sensors using browser

4 Conclusion

We have studied the three application layer IoT protocols, i.e. CoAP, MQTT and HTTP/REST. We came to a conclusion that these protocols are really very useful in context of the M2M communication. Here, we have successfully simulated a scenario using the CoAP and REST protocols. As a context of the future work, we would like to configure the CoAP, REST and MQTT protocols using different motes and different topologies.

References

1. Ugrenovic D, Gardasevic G (2015) CoAP protocol for web-based monitoring in IoT healthcare applications. In: 2015 23rd Telecommunications Forum Telfor (TELFOR). IEEE
2. Potsch T, et al (2012) Performance evaluation of CoAP using RPL and LPL in TinyOS. In: 2012 5th international conference on new technologies, mobility and security (NTMS). IEEE
3. Teklemariam GK, et al (2014) Simple RESTful sensor application development model using CoAP. In: 2014 IEEE 39th conference on local computer networks workshops (LCN Workshops). IEEE
4. Govindan K, Azad AP (2015) End-to-end service assurance in IoT MQTT-SN. In: 2015 12th annual IEEE consumer communications and networking conference (CCNC). IEEE
5. Singh M, et al (2015) Secure MQTT for internet of things (IoT). In: 2015 5th international conference on communication systems and network technologies (CSNT). IEEE

6. Thangavel D, et al (2014) Performance evaluation of MQTT and CoAP via a common middleware. In: 2014 IEEE ninth international conference on intelligent sensors, sensor networks and information processing (ISSNIP). IEEE
7. Pereira C, Aguiar A (2014) Towards efficient mobile M2M communications: survey and open challenges. Sensors 14(10)
8. Luzuriaga JE, et al (2015) Handling mobility in IoT applications using the MQTT protocol. In: Internet Technologies and Applications (ITA). IEEE
9. Whitmore A, Agarwal A, Xu LDa (2015) The internet of things—a survey of topics and trends. Inf Syst Front 17(2)
10. Gazis V, et al (2015) A survey of technologies for the internet of things. In: 2015 international wireless communications and mobile computing conference (IWCMC). IEEE

Machine-Learning-Based Android Malware Detection Techniques—A Comparative Analysis

Nishant Painter and Bintu Kadhiwala

Abstract Today, Smartphones can handle myriad of programs and applications that perform a wide varieties of functions. In recent years, Android has been globally anticipated open source operating system for Smartphones. However, rapid advancement of Android is marred with augmenting threats of Android malwares that performs pernicious activities on Smartphones. Malwares exercising different techniques to dodge existing detection methods offers uncommon challenges for their accurate detection. Signature-based detection approach and machine-learning-based detection approach are the broad classifications for existing Android malware detection techniques. Researchers and antimalware companies have identified the inefficiency of signature-based detection approach and shifted to machine-learning-based detection approach to overcome the limitations of signature-based detection approach. This paper disserts existing machine-learning-based Android malware detection techniques and presents parametric comparison of discussed malware detection techniques. Hence, this paper targets to study various machine-learning-based detection techniques and to establish probable future directions.

Keywords Android · Malware · Signature-based · Machine-learning-based Static analysis · Dynamic analysis · Hybrid analysis

N. Painter (✉) · B. Kadhiwala
Department of Computer Engineering, Sarvajanik College
of Engineering and Technology, Surat, India
e-mail: nishant.painter@gmail.com

B. Kadhiwala
e-mail: bintu.kadhiwala@scet.ac.in

© Springer Nature Singapore Pte Ltd. 2018
D.K. Mishra et al. (eds.), *Information and Communication Technology*
for Sustainable Development, Lecture Notes in Networks and Systems 9,
https://doi.org/10.1007/978-981-10-3932-4_19

1 Introduction

Android open source Smartphone operating system currently secured largest market share of Smartphone worldwide by 80.7%, withdrawing its opponent iOS, Windows, Blackberry and other OS [1]. However, the mammoth compliance of Android is coupled with growing Android malwares inclusive of adware, botnet, ransomware, spyware, backdoor, SMS Trojan and worm [2–5] that may perform various malicious activities on Smartphone [2]. Furthermore, malicious activities includes pilfering confidential user information [6], creating botnet [7], gaining Smartphone control by taking advantage of platform weaknesses [8] and sending premium rate messages and making premium rate calls [9].

Numerous research studies, concerning Android malware detection techniques, apply efforts to confront the threats of Android malwares [10, 11]. These techniques can be broadly classified into two major approaches namely signature-based detection approach and machine-learning-based detection approach [2]. Signature-based detection approach creates a unique regular expression called signature for a malware and detects the malware by matching the extracted data with signature [12]. However, a small alteration in malware leads to a new variant of malware that can easily bypass the signature-based detection [13], forcing the approach to update its signature database frequently [14]. Moreover, the signature-based detection approach fails against the unknown malware since it is capable of detecting the malware whose signature is included in signature database and also the signature extraction process being manual requires more time for implementation [10]. Henceforth, to overcome the discussed limitations of signature-based detection approach, researchers deviated to machine-learning-based detection approach that records behavior of known malwares, trains machine-learning algorithms and detects novel or unknown malware [12].

This paper aims to study and scrutinize various existing machine-learning-based Android malware detection techniques and conjointly draws attention to parametric evaluation of these techniques. This paper is organized as follows.

Section 2 discusses various machine-learning-based Android malware detection techniques and different analysis practiced by these techniques. In Sect. 3, the parametric evaluation of machine-learning-based Android malware detection techniques is presented. Finally, Sect. 4 concludes the paper and enlightens future research directions.

2 State-of-the-Art

In this section, we present theoretical background of various existing machine-learning-based Android malware detection techniques along with different analysis. Machine-learning-based detection approach, employed in various android malware detection techniques, records behavior of known malwares, trains machine-learning algorithms and detects novel or unknown malware [12].

First and the most important step in machine-learning-based detection approach is feature extraction. In Android application, features consist of permission, java byte code, network traffic, system calls and various others. Based upon the extracted feature, machine-learning-based Android malware detection techniques can be classified into static, dynamic and hybrid analysis [2, 10].

Static analysis: Static analysis analyzes an Android application source code and files, without executing the application. This analysis utilizes static features of an application including permissions, java byte code, intent filter, strings and network address [10, 12].

Dynamic analysis: Dynamic analysis executes an Android application in a controlled environment, monitors the executing code and inspects its interaction with the system. Dynamic features of an application includes system call, network traffic, system component and user interaction [2, 10].

Hybrid analysis: Hybrid analysis converges good of both the static and dynamic analysis methods [2].

Various techniques employing machine-learning-based detection approach for malware detection are as follows:

2.1 Lu et al.

In [15], Lu et al. implement a novel machine-learning-based Android malware detection technique that classifies Android application as benign or malicious by observing Android mobile device and applying Naïve Bayesian machine-learning classifier. Permissions describing malicious behavior incorporating RECEI-VE_MMS, RECEIVE_SMS, READ_CONTACTS, CALL_PHONE, ACCESS_-FINE_LOCATION, CHANGE_NETWORK_STATE and INSTALL_PACKAGES are extracted from an application and Chi-Square feature selection method is used to obtain the optimal subset of features. Naïve Bayesian classifier combined with Chi-Square filtering is employed to a corpus of total 477 applications for implementing the system.

2.2 STREAM

Amos et al. [16] present the assessment of existing machine-learning classifiers, evaluating a dataset containing real malware applications, by designing distributed framework STREAM, *a System for Automatically Training and Evaluating Android Malware Classifiers* that detects Android malware by profiling Android applications and examining machine-learning classifiers. STREAM is a system, executing on a single server or cluster of scattered servers, for distributing roles to Android devices or emulators and collecting configurable feature vectors. For every Android device and emulator, STREAM manages Android application, monitors feature vector collection and supervises training and evaluation of machine-learning classifiers. The implemented system collects information regarding memory, network, battery and permission to train six distinct machine-learning classifiers enclosing Random Forest, Naïve Bayes, Multilayer Perceptron, Bayes Net, Logistic and J48 classifier. Results are tested upon a total of 1738 unique applications covering 1330 malicious applications and 408 benign applications.

2.3 MADAM

In [17], Dine et al. present a *Multi-level Anomaly Detector for Android Malware*, MADAM. This system detects real malware applications by supervising Android mobile devices at kernel-level and user-level simultaneously. System calls accurately describe functioning of a device in context to user activity, files and memory access, incoming/outgoing traffic, energy consumption and sensor status etc. Furthermore, assuming that the attacker needs to execute one or more system calls, they can also be used for intrusion detection; therefore MADAM initially monitors system calls. Features extracted at user-level decide the state of user, idle or not, and check the numbers of SMS sent from the Android mobile device. Machine-learning classifier, K-NN, is utilized by MADAM for differentiating malicious and benign behavior of applications and classification of new applications. Implemented system is also capable of detecting unnecessary outgoing SMS, maliciously sent by Android malware applications.

2.4 Huang et al.

In [18], Huang et al. propose a simple logic to identify malware applications based on requested permissions. Initially, source code of an Android application is analyzed, permissions required by the application are identified and features for malware detection are extracted. Feature vectors are then constructed containing the sequence of extracted features values, separated by comma. Along with the extracted features, end of each feature vector is appended with a BoM label that identifies the vector association to either benign or malicious application. Accurately labeling an application, with value benign or malicious for BoM label, is a critical task. *Site-based labeling, scanner-based labeling* and *mixed labeling* are the three strategies used to label the obtained feature vectors. The accuracy of this system is evaluated by extracting permission feature vectors from a total of 125,249 applications, consisting of 124,769 benign and 480 malicious applications. Finally, AdaBoost, Naïve Bayes, C4.5 (J48) and Support Vector Machine (SVM) classifiers are applied for classification to retrieved datasets.

2.5 Sahs et al.

Sahs et al. [19] develop a machine-learning-based Android malware detection technique based on static analysis. Permission and Control Flow Graph (CFG) features are extracted from packaged Android applications, by means of an open source project Androguard. A CFG is a graphical representation of program conceptual logic, containing non-jump instructions as vertices of the graph and the available paths of program flow as edges. Later, using benign applications, the system trains a One-Class SVM in an offline manner by Scikit-learn framework. As benign applications are easily available, thereupon constructing a classifier, classifying most of the training data as positive and testing data as negative by its appropriate divergence from training data is the fundamental scheme of the proposed system. Experiments are conducted on the collection of 2081 benign and 91 malicious Android applications, performing k-fold cross validation by selecting random subset of benign applications for each datapoints.

2.6 Shijo et al.

Shijo et al. [20] develop an integrated static and dynamic analysis technique for analysing and classifying Android applications. Identified malware applications and known benign applications are used to train machine-learning classifiers. Android application source code is analyzed and the dynamic characteristics of the applications are noted for constructing feature vectors. Computational strength and classification accuracy of the system is improved by leveraging the benefits of both static analysis and dynamic analysis. Printable strings information (PSI) is extracted as static feature by examining the binary code of the applications. Cuckoo sandbox is used for dynamic analysis that primarily focused on the extraction of system call sequences triggered by the applications. Finally, Support Vector Machine and Random Forest machine-learning classifiers are applied for classifying Android applications.

2.7 Crowdroid

Burguera et al. [21] develop Crowdroid, a behaviour-based detection system for detecting Android malware. Crowdroid is a lightweight Android application that can be downloaded and installed from Google Play Store. Crowdroid application obtains system calls, by monitoring Linux kernel, and transmits them to centralized server after applying pre processing. The received data is parsed by the remote server and for each interaction of the user within their application a new system call vector is generated by the system. Lastly, K-means partitional clustering algorithm is applied to differentiate benign and malware applications. Benign applications show identical patterns of system calls, while the malware applications behave differently in terms of distance between the application system vectors. K-means clustering algorithm considers k = 2 clusters, as the classification result can either be benign or malicious. The developed system is tested by artificially created malware and real malware applications.

2.8 DroidMat

In [12], Wu et al. develop static analysis based Android malware detection technique *DroidMat*. Packaged Android application is disassembled into its constituent files and then AndroidManifest of the application is processed to extract features including permission, API call, deployment of component and Intent message passing. Malware modeling capability of the system is improved by applying K-means and EM clustering algorithms. *Singular Value Decomposition* (SVD) method using low rank approximation is used to decide the number of clusters during clustering. Finally, DroidMat applies K-NN classification algorithm with k = 1 to evaluate the applications as benign or malicious. Total 1,738 applications, including 238 malware applications and 1,500 benign applications, were collected, extracted and analyzed.

2.9 Kim et al.

In [22], Kim et al. propose a robust feature extraction tool for extracting Android application features and a detection framework on android market that efficiently detects Android malware applications. Formerly the applications are rapidly classified by static analysis and subsequently, dynamic analysis is applied to only those applications that are termed as suspicious. Static analysis acts as a filtering method that limits the number of applications for dynamic analysis and reduces the time required for dynamic analysis of the applications. Primarily classification of applications by static analysis also allows inclusion of time-consuming dynamic mechanism, for thorough analysis of application in the existing framework maintaining its efficiency. To evaluate the system, feature extraction tool is used to extract permissions and methods of API from 893 benign and 110 malicious applications. Finally, J48 Decision Tree classifier is utilized with ten fold cross validation scheme to train and test data for classifying Android applications.

2.10 Yerima et al.

Yerima et al. [23] present a system for quick detection of Android malware by implementing parallel machine-learning classification model. A complex model for classification of malware and benign applications is generated from parallel combination of distinct machine-learning classifiers. API call, permission and command are extracted from an Android application as features during learning phase of the system. Computational efficiency of heterogeneous machine-learning classifiers is aggregated in various parallel combinations for constructing a single classification model that classifies new or unknown applications. Parallel combinations are implemented by averaging probabilities, product of probabilities, maximum of probabilities and majority vote considering Decision Tree, Simple Logistic, Naïve Bayes, PART and RIDOR machine-learning classifiers. Overall 6,863 applications, containing 2925 malicious applications and 3,938 benign applications, were collected to evaluate the system.

3 Parametric Evaluation

As mentioned in Sect. 2, machine-learning-based Android malware detection techniques can be further classified into static analysis, dynamic analysis and hybrid analysis based upon the features extracted. Additionally, required optimal subset of features for accurate classification should be generated by employing feature selection methods. Lastly, classification and clustering algorithms evaluate the system and categorize Android applications as benign or malicious. Machine-learning-based detection techniques for malware detection can be entirely implemented on a workstation (off-device), also it can completely be executed on Smartphone (on-device) or it can be performed partly on workstation and Smartphone simultaneously (distributed). Parametric evaluation of these detection techniques mentioned in Sect. 2 is described in Table 1.

Table 1 Parametric evaluation of machine-learning-based Android malware detection techniques

Technique	Analysis	Feature	Feature selection method	Classification algorithm(s)	Clustering algorithm(s)	Deployment	Evaluation metrics	Number of applications in dataset
Lu et al. [15]	Static	Permission	Chi-square	Naïve Bayes	–	Off-device	FPR, FNR, ACC	477
STREAM [16]	Dynamic	Battery, binder, memory, network information	–	RF, Naïve Bayes, multilayer Perceptron, Bayes net, logistic, J48	–	Distributed	TPR, FPR	1,738
MADAM [17]	Dynamic	System call, running process, free RAM, CPU usage	–	K-NN	–	Distributed	FPR, ACC	–
Huang et al. [18]	Static	Permission	–	AdaBoost, Naïve Bayes, J48 and SVM	–	Off-device	TPR, FPR, precision, Recall, F-measure	125,249
Sahs et al. [19]	Static	Permission, CFG	–	One-class SVM	–	Off-device	Precision, Recall, F-measure	2,172
Shijo et al. [20]	Hybrid	Print string information, API call sequence	N-gram creation	SVM, RF	–	Off-device	ACC, FPR, TPR	1,487
Crowdroid [21]	Dynamic	System call	–	–	K-means	Distributed	ACC	–
DroidMat [12]	Static	Permission, deployment of component, intent message, API call	–	K-NN	K-means and EM	Off-device	ACC, recall, precision, F-measure	1,738
Kim et al. [22]	Hybrid	Permission, methods of API	–	J48	–	Off-device	ACC	1,003
Yerima et al. [23]	Static	API call, commands, permission	–	DT, simple logistic, Naïve Bayes, PART, RIDOR	–	Off-device	TPR, TNR, FPR, FNR, ACC, ERR, AUC	6,863

RF Random Forest, *K-NN* K-nearest neighbor, *SVM* Support Vector Machine, *DT* Decision Tree, *RIDOR* Ripple Down Rule Learner, *FPR* false positive rate, *FNR* false negative rate, *ACC* accuracy, *TPR* true positive rate, *TNR* true negative rate, *ERR* error ratio, *AUC* area under curve

4 Conclusion and Future Work

In this paper, we have discussed the existing machine-learning-based Android malware detection techniques and also the feature selection of these techniques has been classified into static analysis, dynamic analysis and hybrid analysis. Static analysis extracts the features required for classification by only analyzing the source code and files that is easy as compared to dynamic analysis that executes the applications and extracts the features. Moreover, applications for dynamic analysis are executed in secure environment and monitored with respect to their behavior that makes the detection process time consuming. Also, for dynamic analysis, the applications are interacted with the help of programs imitating human behavior that may not trigger all events of the application; as a result code coverage of malware detection technique is not satisfying. Furthermore, certain malware behaves differently in secure environment and real runtime environment that fools the detection technique to record inaccurate information and decreases its efficiency. However, dynamic analysis and hybrid analysis provides protection against code obfuscation that static analysis fails. Hybrid analysis, leveraging the best of both static and dynamic, also inherits all limitations of dynamic analysis.

Static analysis is simple to implement and advantageous over dynamic analysis as discussed. Future research directions may include creating malware detection technique based upon static analysis, capable of protecting the system against code obfuscation thereby overcoming the limitation of analysis and utilizing its benefits. Moreover, for extracting effective features that accurately classify Android application, different feature selection methods can be employed. Furthermore, various other static features should be extracted to improve the code coverage of the malware detection techniques. Finally, with the help of different clustering and classification algorithms, performance of the system can be measured and compared with the performance of the existing Android malware detection techniques.

References

1. G Inc, Gartner says worldwide smartphone sales grew 9.7 percent in fourth quarter of 2015. http://www.gartner.com/newsroom/id/3215217
2. Feizollah, A, Anuar, N, Salleh, R, Wahab, A (2015) A review on feature selection in mobile malware detection. In: Digital investigation, vol 13, pp 22–37
3. Castillo, C (2012) Android malware past, present, future. In: Mobile working security group McAfee, Santa Clara, CA, USA, Technology Report
4. Zhou, W, Zhou, Y, Jiang, X, Ning, P (2012) Detecting repackaged smartphone applications in third-party android marketplaces. In: 2nd ACM conference on data and application security and privacy, New York, NY, USA, pp 317–326
5. Android Malware Genome Project. http://www.malgenomeproject.org/
6. Android.Bgserv. http://www.symantec.com/securityresponse/writeup.jsp?docid=2011-031005-2918-99
7. Android/NotCompatible looks like piece of PC botnet. http://blogs.mcafee.com/

8. RageAgainstTheCage. https://github.com/bibanon/android-development-codex/blob/master/General/Rooting/rageagainstthecage.md
9. Android Hipposms. http://www.csc.ncsu.edu/faculty/jiang/HippoSMS/
10. Faruki P, Bharmal A, Laxmi V, Ganmoor V, Gaur M, Conti M, Rajarajan M (2015) Android security: a survey of issues, malware penetration, and defenses. IEEE Commun Surv Tutor 7 (2):998–1022
11. Sufatrio, Tan, D, Chua, T, Thing, V (2015) Securing android: a survey, taxonomy, and challenges. In: ACM Comput Surv 47(4):58
12. Wu, D, Mao, C, Wei, T, Ming, H, Wu, K (2012) DroidMat: android malware detection through manifest and api calls tracing. In: 7th Asia joint conference on information security. IEEE, pp 62–69
13. Teodoro P, Verdejo J, Fernndez G, Vzquez E (2009) Anomaly-based network intrusion detection: techniques, systems and challenges. Comput Secur (Elsevier) 28(1–2):18–28
14. Fedler, R, Schütte, J, Kulicke, M (2013) On the effectiveness of malware protection on android. In: Fraunhofer AISEC, Berlin, Germany, Technology Report
15. lu, Y, Zulie, P, Jingju, L, Yi, S (2013) Android malware detection technology based on improved Bayesian classification. In: 3rd international conference on instrumentation, measurement, computer, communication and control. IEEE, pp 1338–1341
16. Amos, B, Turner, H, White, J (2013) Applying machine learning classifiers to dynamic android malware detection at scale. In: 9th international wireless communications and mobile computing conference (IWCMC). IEEE, pp 1666–1671
17. Dini, G, Martinelli, F, Saracino, A, Sgandurra, D (2012) MADAM: a multi-level anomaly detector for android malware. In: 6th international conference on mathematical methods, models and architectures for computer network security. Springer
18. Huang, C, Tsai, Y, Hsu, C (2013) Performance evaluation on permission-based detection for android malware. In: Advances in intelligent systems and applications. Springer, pp 111–120
19. Sahs, J, Khan, L (2012) A machine learning approach to android malware detection. In: European intelligence and security informatics conference. IEEE, pp 141–147
20. Shijo, P, Salim, A (2015) Integrated static and dynamic analysis for malware detection. In: International conference on information and communication technologies (ICICT 2014). Proc Comput Sci (Elsevier) 46:804–811
21. Burguera, I, Zurutuza, U, Tehrani, S (2011) Crowdroid: behavior-based malware detection system for android. In: Security and privacy in smartphones and mobile devices. ACM, pp 15–26
22. Kim, D, Kim, J, Kim, S (2013) A malicious application detection framework using automatic feature extraction tool on android market. In: 3rd international conference on computer science and information technology (ICCSIT)
23. Yerima, S, Sezer, S, Muttik, I (2014) Android malware detection using parallel machine learning classifiers. In: 8th international conference on next generation mobile applications, services and technology. IEEE, pp 37–42

Sensing Technology for Detecting Insects in a Paddy Crop Field Using Optical Sensor

Chandan Kumar Sahu, Prabira Kumar Sethy
and Santi Kumari Behera

Abstract This paper proposed a system which is to detect insects in a paddy crop field. Today we are living in the twenty-first century where computer vision is playing important role in human life. Computer vision provides image acquisition, processing, analyzing, and understanding images and, in general, high quality image from the real world in order to produce numerical or symbolic information, in the forms of decisions. It provides not only comfort but also efficiency and time saving. Today satellites are used as computer vision technology; by analyzation of the satellite images, it gives the information to the user. But this is only applicable for scientific level research laboratory because the cost of this type of devices is very high and not suitable for using in a farm field. So here we design a system, which detects insects in a farm filed and population estimation of insects in a farm field. The objectives of this paper are to control pests in a farm field and a healthy crop yielding for increased food production.

Keywords MATLAB image-processing tool · Object detection
Object extraction · Paddy field insects

1 Introduction

In our country, agriculture is a major source of food production for the growing demand of human population. But crops production losses due to the effect of insects in agricultural fields [1]. Farmers generally visit their lands periodically to

C.K. Sahu (✉) · P.K. Sethy
Sambalpur University, Sambalpur, India
e-mail: chandan.sahu@suiit.ac.in

P.K. Sethy
e-mail: psethy@suiit.ac.in

S.K. Behera
Veer Surendra Sai University of Technology, Burla, India
e-mail: b.santibehera@gmail.com

© Springer Nature Singapore Pte Ltd. 2018
D.K. Mishra et al. (eds.), *Information and Communication Technology
for Sustainable Development*, Lecture Notes in Networks and Systems 9,
https://doi.org/10.1007/978-981-10-3932-4_20

check the condition of crops, whether it is infected by insects or not. This takes a lot of time, if the farmers go around the fields and visualize the density of insects nevertheless farmers need to manage their agricultural activities along with other occupations. So computer vision takes an important role here, by visualizing the farm fields using cameras and it automatically analyze the density of insects in farm field. Computer vision is based on automatic analyzation of farm field system, as it provides optimizing solution to farmers where the presence of farmer in field is not compulsory [2]. A digital camera is used for capturing images from farm filed, and a small processor programmed for processing the images and measuring the density of insects. Generally, Indian farmers need cheap and simple user interface for analyzing the condition of farm field whether the farmer is far away from the field. Using Internet, farmer knows about the processed data of farm field image. This helps farmer to know the condition of insect-affected area.

In this paper, we represent a prototype, for fully automatic device for detecting insects using camera and image-processing tool and counting the density of insects in a farm field. The camera would be integrated with a quad-copter and also with a tiny processor Raspberry-Pi [3]. The Raspberry-Pi is used to process the image and analyze the required data for a farmer. The processor is also connected with Internet to store the data in cloud and give information to the farmer. For experimentation, we have used the quad-copter, as it moves randomly above the farm field finding insects-affected area of that farm field. The quad-copter automatically starts moving everyday for checking insects; Raspberry-Pi processes all captured images and calculates the density of bugs in field; and notifications as SMS are sent to the registered mobile phone which is registered in Raspberry-Pi through Internet.

2 Methods of Insects Analyzing

There are different types of methods for analyzing insects from various crop fields. Basically, Indian farmer uses visualization method for finding insects in farm fields. But by only using visualization method, it is difficult to analyze the density of insects. Whether a processor is used to easily calculate the density of insects in farm field and analyze which type of insect affects to that farm field.

2.1 Visualization Method for Insect Analyzing

This system is a common method for analyzing insects of a farm field, as this method is very simple and farmer by himself visits the farm field and analyzes the insects. In a large area farm field, the farmer has no much more time to visit the whole field in a single day. So the farmer starts moving whether one side of the field or may be other side of the field to check the damaged due to insects (Fig. 1).

Fig. 1 Visualization method

Fig. 2 Sensor-based insects detection

2.2 Optical Sensors (Camera) Used to Detect Insects in Farm Field

In this technique, camera captures the leaf of the crop and analyzes the color of leaf and detects the infected part of the leaf. The camera is used for capturing the image of crops and sends that image to a processor which processes the image and detects the disease of crops. The camera is fitted with a drone which randomly moves surrounding the agricultural field and takes images. The Raspberry-Pi controls the drone and sends those images to cloud. Local server interprets the data from cloud and processes the image to analyze the disease and count the density of insects in the farm field (Figs. 2 and 3).

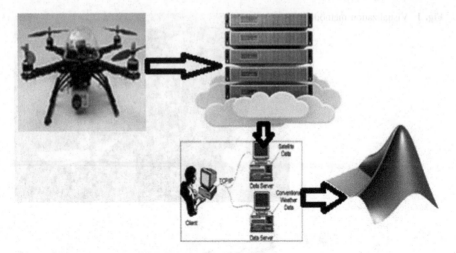

Fig. 3 System components and operation

3   System Overview

A. Drone:

The drone, driven by a Raspberry-PI, is equipped with a camera. The camera will take pictures from different angles and by using the Raspberry-Pi the pictures are sent to the cloud server. Then from the cloud server, the images are sent to the local server and then to MATLAB, where it is processed using the image-processing tool and the user gets the output.

B. Optical Sensor (Camera):

The camera is used to capture images of the site from different angles and then send those images to the Raspberry-Pi.

C. Raspberry-Pi:

It is used to control the drone, and the images captured by the drone which are stored and then sent through the Raspberry-Pi. The Raspberry-Pi also processes the image using Open-CV tool. But in our technology, farmer also sees the agricultural field image inside a room.

D. Cloud Server:

The images captured by the drone are sent to the cloud and stored for retrieval as per requirement.

E. Local Server:

Required images from the cloud are taken, and the local server feeds the images as an input to MATLAB and the result is displayed.

4 Result and Analysis

The local server processes the image using MATLAB image-processing tool. The MATLAB tool is linked with cloud server, and when image data is sent to cloud, the local server downloads that image from cloud and again sends it to MATLAB for processing.

Processing Steps:

- Original Image (RGB), i.e., Fig. 4 converted to gray scale image, i.e., Fig. 5. Gray image = 0.2989 * R + 0.5870 * G + 0.1140 * B

Fig. 4 Original image

Fig. 5 Gray scale image

- Use morphological opening to estimate the background.
  Background = imopen(I, strel('disk', 20));
  strel: flat structure of neighbor elements having radius 20 elements. In which, the true pixels are included in the morphological computation; the false pixels are not.
- Subtract the background image from the original image new image = Original Image—Background (i.e., Fig. 6).
- Increase the contrast of new image (i.e., Fig. 7).
- Threshold the Image
  The thresholding method used to replace each pixel in an image with a black pixel if the image intensity $I_{i,j}$ is less than some fixed value T(i.e., $I_{i,j<T}$) or a white pixel if the image intensity is greater than that constant.
  This is the best way to remove unwanted region from image technique for clearly counting the insects in crop field. This technique removes all unwanted part of an image, i.e., Fig. 8.
- Identify objects in the image
  After removing the unwanted part from an image, measure the number of sets specified for each connected components in the binary image. The number of sets is the required objects which are insects of an image.

Fig. 6 Subtract the background image from the original image

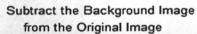

Fig. 7 Contrast-increased image

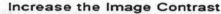

Fig. 8 New image after thresholding

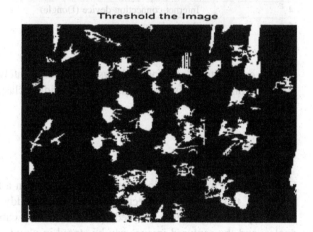

Connectivity: 8
Image Size: [550 403]
Num Objects: 37
PixelIdx List: {1x37 cell}

In this picture, a number of insects are approximately 37. Using this technique, count insects from the all captured images. And count the mean of the insects from the number of images to get approximate value of insects in a farm field.

$$\text{Approximate value of insects from a field} = \frac{\text{values of}\,(I1 + I2 + I3 + \ldots + In)}{In}$$

I1, I2, and In are number of images.

5 Cost Analysis

The proposed system is average cost which is easy to use and get the appropriate value of insects. In the table below, we show the cost of proposed system.

Sl no	Name of device	Price
1	Drone	5000.00
2	Raspberry-Pi	3000.00
3	Camera	1000.00
4	Internet connection device (Dongle)	1000.00
5	Personal computer	30000.00
TOTAL		40000.00

The expenditure of proposed system approximately budgets nearly around 40000.00 only which is easy to install for a middle-class farmer.

6 Conclusion

For increased crop productions by controlling pests in a farm field, we proposed a system in this paper, for insects detection from paddy crop field which is economically beneficial. Here we used optical sensor for capturing images in different angles, and the captured images can be stored in cloud server for further use. The farmers can get the images from cloud by local server, and they get the result of insect density. So that according to the density, they can apply that amount of pesticides for useful effect so that the environment is also not much more depleted and the farmers can assign him to multiple work.

7 Future Work

In the future times, we are going to work on using images of agriculture analysis to the structure of insect and classify different types of insects and get knowledge about the type of insect. Using regression method, predict disease severity and calculate the amount of pesticide apply to the field. For which, applying of pesticide amount will decrease and environment becomes eco-friendly due to less amount of pesticides applied to the fields.

References

1. Azfar S, Nadeem A, Basit A (2015) Pest detection and control techniques using wireless sensor network: a review 3(2):92–99. E-ISSN: 2320-7078 P-ISSN: 2349-6800JEZS © 2015 JEZS
2. Lee WS, Alchanatis V, Yang C, Hirafuji M, Moshoue D, Li C (2010) Sensing technologies for precision specialty crop production. Comput Electron Agric 74(2–33): 0168–1699/$—see front matter© 2010 Elsevier B.V. All rights reserved. doi:10.1016/j.compag.2010.08.005
3. Raspberry-pi. https://www.raspberrypi.org/products/camera-module/
4. Miranda JL, Gerardo BD, Tanguilig III BT (2014) Pest detection and extraction using image processing techniques. Int J Comput Commun Eng 3(3). doi:10.7763/IJCCE.2014.V3.317
5. Johnny L. Miranda, Bobby D. Gerardo, Bartolome T. Tanguilig (2014) Pest identification using image processing technique in detecting image pattern through neural network. In: Conference on advances in computer and electronics technology-ACET 2014. ISBN: 978-1-63248-024-8. doi:10.15224/978-1-63248-024-8-10
6. Akriti P, Sonal K, Nandhini V. Real time pest detection and identification using image processing. Comput Sci Eng BMSCE. Karnataka, India. ISSN: 2277 128X
7. Chitade AZ, et al (2010) Colour based image segmentation using k-means clustering. Int J Eng Sci Technol 2(10):5319–5325
8. Deshmukh KS (2012) Disease detection of crops using hybrid algorithm. Int J Eng Res Technol (IJERT) 1(10). Tulsiramji Gaikwad Patil College of Engineering & Technology Nagpur. ISSN: 2278-0181
9. Wang Y-H (2010) Tutorial: image segmentation. Graduate Institute of Communication Engineering National Taiwan University, Taipei, Taiwan, ROC
10. Liu ZY, Huang JF, Shi JJ, Tao RX, Zhou W, Zhang LL (2007) Characterizing and estimating rice brown spot disease severity using stepwise regression principal component regression and partial least-square regression. J Zhejiang Univ Sci B. 8(10):738–44

Novel Approach to Image Encryption: Using a Combination of JEX Encoding–Decoding with the Modified AES Algorithm

Sneha Birendra Tiwari Sharma, Manjeet Kantak
and Nagaraj Vernekar

Abstract Many research works have been conducted so as to design an efficient image encryption algorithm. In this paper, we propose an algorithm that combines the image space requirement with image encryption. The proposed encryption technique uses the JEX encoding–decoding in combination with the Modified AES algorithm for encryption. In JEX encoding, the image is divided in pixel blocks, and on each pixel block column–row permutations are performed resulting into the generation of a header file which is then added to the image file. The resultant image then passes through the remaining steps of basic JPEG compression. The JEX-coded image is then encrypted using the Modified AES encryption algorithm. The modification in AES algorithm is performed on the round key generation procedure wherein a circular shift and mirror operations are performed on the 128-bit cipher key matrix. On the decryption unit, the encrypted image is decrypted using the same key obtained from the Modified AES decryption algorithm which is then followed by the JPEG decompression. On the decoder unit, the JPEG decoder decodes the image file by extracting the information from the header and performing the inverse transformation operation to decode the coded image so as to recover the uncompressed image.

Keywords Conventional JPEG · JEX encoding–decoding · Quantization
Permutation · Header · M.A.E.S algorithm · Circular shift
Mirror operation

S.B.T. Sharma (✉) · M. Kantak · N. Vernekar
Computer Science and Engineering Department, Goa College of Engineering,
Farmagudi, Ponda, Goa, India
e-mail: snehabirendra@gmail.com

M. Kantak
e-mail: Manjeet959@gmail.com

N. Vernekar
e-mail: nkv_2447@yahoo.com

© Springer Nature Singapore Pte Ltd. 2018
D.K. Mishra et al. (eds.), *Information and Communication Technology
for Sustainable Development*, Lecture Notes in Networks and Systems 9,
https://doi.org/10.1007/978-981-10-3932-4_21

1 Introduction

The recent advances in the last decade have shown growth in the numbers of image pixels that are used in digital cameras or the cellular phones. In order to handle such an increase in the amount of information, JPEG and JPEG 2000 compression techniques are most widely used. The basic JPEG [1, 2] is a lossy DCT-based image compression technique. Being a lossy image compression technique, JPEG is not suited for images with sharp edges as it can result to artifacts. The other drawback that arises is the need to store the preprocessing steps on the JPEG coder unit that leads to the increase in the computational complexity. JPEG 2000 [3] is a wavelet-based image compression technique. In spite of better compression performance; JPEG 2000 is not preferred as standard technique for image compression due to the complexity of the encoder and decoder unit when compared to JPEG. Image encryption techniques are used to provide confidentiality to the images [4, 5]. A robust image encryption technique is proposed in this paper.

In this paper, we propose an image encryption technique that uses a different approach to image encryption. The image encryption procedure is first preceded by JEX coding which is an enhancement to JPEG compression algorithm. The proposed compression format decodes the image using only the compressed image file header with a JPEG decoder. The resultant JEX-coded compressed image is then encrypted using the proposed Modified AES encryption algorithm. The existing AES algorithm is modified, and the modification is performed on the round key generation operation. A round key is generated by performing various operations on the cipher key. On the decryption unit, the encrypted image is decrypted using the same key obtained from the Modified AES decryption algorithm which is then followed by the JPEG decompression followed by JEX decoding. To perform the JEX decoding, the JPEG decoder examines the header section, extracts the information related to the image data and thus decodes the coded data to obtain transform coefficient in spatial frequency domain so as to recover the encoded image data.

The paper organization consists of the following sections; the second section provides a description of some of the "research work related to Image compression and Image encryption," the third section provides a detailed view of our "proffered research algorithm," the fourth section shows the "test results generated from the experiments performed on our proffered algorithm," followed by "conclusion" in Sect. 5 and "references."

2 Related Work

This section gives a brief overview on the related work done on image compression related to the various modifications done in the conventional JPEG compression and image encryption algorithm so as to improvise the algorithms.

In the research paper [6], the author has analyzed the Huffman coding of JPEG compression and proffered a small modification to the Huffman coding of the JPEG baseline compression algorithm wherein the author exploits the redundancy in Huffman coding. To perform this modification, the author segments the DCT blocks into number of separate bands and then codes them using a separate code table. The proposed modification is performed in three different methods. The experimental results have shown that the proposed algorithm obtains 4% of an average code reduction rate to the total image code size when compared to the JPEG with sequential Huffman coding and arithmetic coding method.

In the research paper [7], authors have analyzed the Advanced Encryption Standard (AES), and in their image encryption technique they add a key stream generator (A5/1, W7) to AES to ensure improving the encryption performance.

3 Proposed Algorithm

This section provides the complete specification of the proffered algorithm. Block diagram in Figs. 1 and 2 provides the flow of the proffered algorithm which is followed by the step-by-step description of the algorithm.

Algorithm 1: Proposed JEX Encoding
The conventional JPEG image data compression method is enhanced by adding the JEX coding method before performing the quantization step. The resultant encoded image is then encrypted using Modified AES encryption. The algorithm procedure starts as follows:

Step1: Divide the image into 8 * 8 pixel block;
Step2: Perform JEX encoding using following steps:

Step2.1: For every pixel block do:
$sc(i) = \sum_{l=0}^{7} f(i,l)$ and $sr(j) = \sum_{k=0}^{7} f(k,j)$ //Sum of pixels in column and row-wise;
Next; Calculate nc(i) and nr(j) //the pixel position;
Step2.2: If(sc(nc(0))-sc(nc(7)))>Thc*8 AND Chc>=Nc; Then, Set: Pc=1 //Perform column-wise permutation; Else, Set: Pc=0; // No column-wise permutation

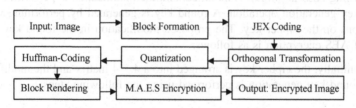

Fig. 1 Steps in image encryption

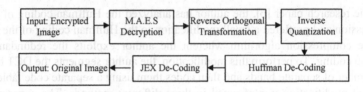

Fig. 2 Steps in image decryption

Step2.3: If(sr(nr(0))-sr(nr(7)))>Thr*8 AND Chr>=Nr; Then, Set: Pr=1 //Perform row-wise permutation; Else, Set: Pr=0; //No Row-wise permutation.

Step3: Generate the header information Ph(n); Output the resultant JEX-encoded image data with the header portion combined to the permuted image;

Step 4: The resultant image is then quantized using DCT followed by run-length encoding and Huffman coding. The resultant image is encrypted using Modified AES encryption algorithm.

Algorithm 2: Proposed JEX Decoding

The encrypted image is then decrypted which is followed by the conventional JPEG decompression and the JEX decoding procedure. The algorithm procedure starts as follows:

Step1: Perform Modified AES decryption;

Step2: Perform Huffman decoding followed by inverse DCT

Step3: Perform JEX decoding using following steps:

Step3.1: For every pixel block extract the header and check the following conditions;

Do: If (Pc = 1): Acquire {nc(0)...nc(7)} values from Ph(n); Perform inverse column-wise operation; Else Do: No transformation; goto L1:

L1: Do: If (Pr = 1): Acquire nr(0)...nr(7) values from Ph(n); Perform inverse row-wise operation; Else Do: No transformation; goto L2:

L2: Output the JEX-decoded image;

Step 4: Generate the synthesized image and output the resultant decrypted-decoded image.

Algorithm 3: Proposed Modified A.E.S Encryption

The existing AES algorithm is modified, and the modification is performed on the round key generation operation. A round key is generated by performing various operations on the cipher key. The proposed modification in the round key generation of AES encryption is as follows:

Step1: Initially, the cipher key is arranged into a 4 * 4 matrix and the modification to the round key generation is performed using the following steps:

Step1.1: Select the third row of the key matrix and convert every value to its binary equivalent form which is then grouped into eight groups having four bits. The binary bits are XOR-ed using the following conditions to obtain the binary result:

Step1.1.1: The bits of group1 starting with position a_0 are XOR-ed with those starting at position a_4 to obtain a 4-bit binary result say "L." Using the same XOR operation, the remaining groups say "M, N, O" are obtained, i.e., [a1, a5], [a2, a6] and [a3, a7], respectively.

Step1.2: To perform circular shift the following condition is considered:
Step1.2.1: First two bits of L and N represent the row numbers which are to be circular shifted and the number of one's in M and O defines the number of circular shift
Step1.2.2: To perform circular shift for row L, the row one less than the number of one's in M is calculated. Similarly to perform circular shift of row N, the row one less than the number of one's in O is calculated.
Step1.2.3: While performing the circular shift of N if the same row is represented by the first two bits of N as that of L that where shifted then the next immediate bits are considered in N for circular shift. After performing a check of all the bits if the row number comes out to be the same in N, then the same row is shifted by one less than the number of one's in O.

Step2: Next, perform the mirror operation on the remaining row. In mirror operation, the individual data are converted from each row into its binary form and read it from right to left, finally converting it back to hexadecimal value. The resultant matrix is used as a round key which is fed to the AddRoundKey operation. This round key modification loops till the N_r-1 rounds.

4 Test Results

The proffered image encryption algorithm is materialized by using java programing and the test outcomes are obtained using MATLAB. The experiment is conducted on a total of 50 images obtained from different image capturing devices and SIPI miscellaneous database of images. The performance comparison is based on the average values of PSNR, MSE, entropy, mean and standard deviation, skewness and kurtosis, joint entropy and mutual information of the JEX-coded image with the JPEG compression image. The key space and statistical analysis is performed to check the security of the proposed modification to the AES algorithm (Figs. 3, 4, 5, 6, 7, 8, 9 and 10, Tables 1 and 2).

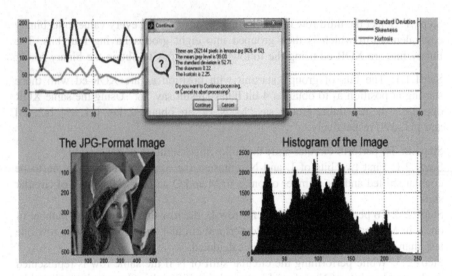

Fig. 3 Histogram plot of JPEG image with the various test parameter values

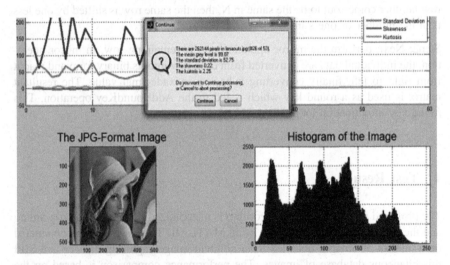

Fig. 4 Histogram plot of JEX-encoded image and the various test parameter values

Fig. 5 Mutual information and joint entropy value of JPEG decompressed image and JEX-decoded image

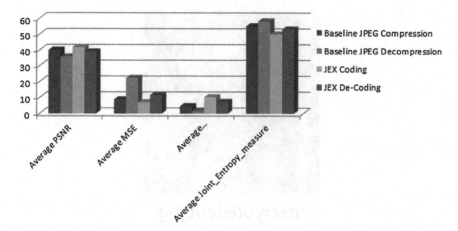

Fig. 6 Overall performance comparison chart of existing JPEG compression–decompression algorithm with the proposed JEX encoding–decoding algorithm

Fig. 7 JEX-encoded image

lenaouts.jpg

Fig. 8 Encrypted image

encryptedout.jpg

decryptedout.jpg

Fig. 9 Decrypted image

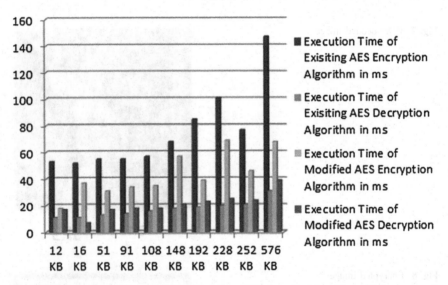

Fig. 10 Performance comparison chart of existing AES encryption algorithm with the proposed Modified AES encryption algorithm

Table 1 Performance comparison

Test module	Average PSNR	Average MSE	Average mutual information	Average joint entropy
JPEG compression	39.97	8.74	4.5124	54.9548
JPEG decompression	35.76	22.36	1.47	58.13
JEX encoding	41.79	6.94	10.24	49.92
JEX decoding	38.89	11.50	7.20	52.99

Table 2 Execution time calculation of AES and Modified AES encryption and decryption algorithm

Sr. no	Image size (KB)	Execution time of existing AES encryption (ms)	Execution time of existing AES decryption (ms)	Execution time of Modified AES encryption algorithm (ms)	Execution time of Modified AES decryption algorithm (ms)
1	12	11	11	11	17
2	16	11	11	12	17
3	51	13	13	13	17
4	91	14	14	13	18
5	108	16	16	16	18
6	160	18	18	16	21
7	208	19	19	19	23
8	240	20	20	21	25
9	252	21	21	21	24
10	576	31	31	33	39

4.1 Key Space Analysis

The proposed modification has a key space of size 10^{128} which is very large enough to avoid any kind of brute-force attack. Also because of the use of cyclic shift operations, the round keys obtain high degree of randomness which makes it difficult for any intruder to obtain the correct key.

5 Conclusion

In this paper, the proposed algorithm is based on the idea which considers image storage and image security concept as one. To overcome the aforementioned drawbacks of the concerned areas, the proposed JEX method is designed with the appropriate mathematical equations so that the image can be decoded using the generated header data attached that is attached to the image data file alone, thus avoiding the need to store any information related to the compression of the image data on the JPEG decoder unit reducing the computational complexity to the minimum. To provide security to the image, the image is encrypted with Modified AES encryption algorithm. The experimental results have shown that the JEX encoding–decoding gives a better performance compared to the JPEG compression algorithm. Also the modification to the AES algorithm gives a better security as shown by the execution time and the key space analysis.

References

1. Ashourian M, Afzal A, Moallem P (2012) Using reversible variable-length codes for JPEG image transmission in a noisy channel. Analog Integr Circ Signal Process 71:337–341
2. Neelamani R, Fan Z (2006) JPEG compression history estimation for color images. IEEE Trans Image Process 15(06):1365–1378
3. Khademi A, Krishnan S (2005) Comparision of JPEG 2000 and other lossless compression scheme for digital mammograms. In: 5th IEEE conference on engineering in medicine and biology, Shanghai, China, pp 201–210, 1–4 Sept 2005
4. Gilani SAN, Bangash MA (2008) Enhanced block based color image encryption technique with confusion. In: The 12th IEEE conference on international multitopic conference, pp 200–206, 23–24 Dec 2008
5. Song O, Kim J (2011) Compact design of advanced encryption standard algorithm for IEEE 802.15.4 devices. J Electr Eng Technol 6:418–422
6. Lakhani G (2003) Modified JPEG Huffman coding. IEEE Trans Image Process 12(2):159–169
7. Zeghid M, Machhout M, Khriji L, Baganne A, Tourki R (2007) A modified AES based algorithm for image encryption. J World Acad Sci Eng Technol 27:206–211

A Survey on Video Smoke Detection

Princy Matlani and Manish Shrivastava

Abstract Fire destroys human lives and property. Therefore, there is a huge need for a reliable and probable fire detection technique. This paper provides a review on various methods developed to detect smoke through videos. The study basically categorizes techniques of smoke detection on the basis of feature extraction method (static/dynamic characteristics), locating region of interest (ROI), etc. It also discusses the nature of camera, color model used for detection and so on. A basic method of smoke detection is described stepwise with different types of algorithms used in each step. The pros and cons of each method are also discussed briefly in this paper.

Keywords Smoke · Detection · Image · Processing · Video

1 Introduction

Smoke is considered as a signal of fire. Fire can result in a damage of crucial property. Therefore, we need to have an early solution of fire detection, so that it can cause hazardous to a minimum. In recent years, computer vision technology has come into existence to overcome the drawbacks that we faced in sensor-based smoke detectors. It makes it possible in field of surveillance to keep a constant eye on every camera. Installing cameras to capture video and then using computer vision technology to detect smoke has made it all easy to do the surveillance job. Entire process has the advantage of being automated, and it also has negligible transportation delay, which was a great disadvantage of sensor-based fire alarm.

P. Matlani (✉) · M. Shrivastava
Computer Science and Engineering Department, Guru Ghasidas University,
Bilaspur, Chhattisgarh, India
e-mail: princy.matlani@gmail.com

M. Shrivastava
e-mail: manbsp@gmail.com

© Springer Nature Singapore Pte Ltd. 2018   211
D.K. Mishra et al. (eds.), *Information and Communication Technology
for Sustainable Development*, Lecture Notes in Networks and Systems 9,
https://doi.org/10.1007/978-981-10-3932-4_22

To avoid huge fire and its consequent damage, video processing technique for smoke detection and analysis of fire are being performed. As soon as smoke occurs in any one of the camera installed, it detects immediately and notifies the user. In this article, a revised review of smoke detection is presented. The basic difference in methods of detection is in feature extraction method, whether to use wavelet transform or ROI or clustering or color-based feature extraction, etc.

Generally, we see the first step in whole process of smoke detection involves detection of moving regions in the video. This is performed traditionally by background subtraction algorithm, in which the current video frame is subtracted from the background frame to get the moving areas. Whether it may be Gaussian mixture model, frame difference method, or optical flow method, all these algorithms are used as a first step to find the moving regions in the frame. This step often involves high computational cost and is also sensitive to noise. Next, after background subtraction, it is needed to distinguish smoke from other objects detected in first step. For that, features of smoke are studied and are extracted. Therefore, feature extraction is the soul of the whole smoke detection process. Some algorithms involve use of static characteristics of smoke such as color, texture, and contrast, while others use dynamic characteristics of smoke such as area of smoke, its direction, and growth of region and shape. Some use LBP (local binary pattern) to learn features of smoke. But this method has a drawback of LBP being sensitive to changes in the background or foreground.

After the feature is extracted in each block of image, positive samples and negative samples are used for training the classifier that classifies the given block contains smoke or not.

Camera for recording the smoke video can be still where the background subtraction method is involved and can be moving in other cases. Each algorithm takes image in different color model. The three basic models in which smoke images are taken are YUV, RGB, HSI.

Recent methods of smoke detection basically vary in the technique they use for feature extraction and classification of smoke. A number of smoke detection methods have been come into existence. Not only smoke, but flame is also used for fire detection. Firstly, taking smoke into account, we have huge algorithms. Using motion as a key for identifying smoke areas, background subtraction has been continuously in trend. Gaussian mixture model was used for preprocessing purpose very commonly in [1, 2]. Then in contrast, optical flow was used for detection of movement of smoke [3]. In [4], Kalman filtering for motion detection provided an efficient way of background estimation considering its nonlinear property, while [5] used combination of Kalman filter with MHI (motion history image) for extracting motion regions from image.

Some algorithms used static characteristics of smoke in feature extraction phase. Ma et al. and Xiong et al. [5, 6] use color information for identifying smoke in given video sequence. Another smoke detection is performed by making color

Fig. 1 Basic steps for video-based smoke detection

histogram for measuring color similarity features with reference to histograms of sampled smoke templates [7]. For dynamic features of smoke, [7–9] consider shape irregularity of smoke, [10] use texture information, [11, 12] use temporal wavelet transformation and discrete wavelet transformation, respectively. An approach that performs detection of region of interest (ROI) using stationary wavelet transform (SWT) is made in [13]. A four-stage algorithm for smoke detection that involves fuzzy c-means clustering to cluster candidate smoke regions is given in [14]. Another research was identifying ROI by connected component analysis and calculating area of ROI by convex hull algorithm after detecting area of change was proposed in [15]. These are some recent contributions made in the area of smoke detection (Fig. 1).

2 Overview of Visual Smoke Detection

There are many techniques for detecting smoke in the field of computer vision. And most of the techniques even use combination of several approaches to improve performance and reliability. Some of the steps are common in most smoke detection systems; they are motion detection, region analysis, dynamic analysis, and lastly smoke classification stages. The difference lies in algorithms used in these separate stages. Next, we will discuss each algorithm, its benefits, and drawbacks, so that one can choose the optimal algorithm for fire detection in future to improve the system performance.

2.1 Smoke Detection Based on Color

Mainly, RGB, HIS, or YUV model is used for color-based smoke detection. Nearly, all visible range cameras have sensors which detect video in RGB format. Although using RGB indicates very low computational complexity but in smoke pixels, RGB values are very close to each other. HIS is often adapted, because of its suitability of providing more people-oriented way of describing the colors [25].

YUV on the other hand describes luminance and chrominance values of a particular pixel.

(a) video sequences   (b) GMM background image

(c) current image   (d) Foreground image

Fig. 2 Foreground segmentation using background subtraction (GMM) [26]

2.2 Moving Region Extraction Method

Well-known moving region algorithms are background subtraction, temporal differencing, optical flow analysis, and Gaussian mixture model. Background subtraction is easy to understand but is very sensitive to noise, lightning, etc. Optical flow technique uses motion field but is computationally complex [26].

Temporal differencing has advantage of quick adaption of change in environment but has disadvantage of being incapable of extracting complete contours [27] (Fig. 2).

2.3 Feature Extraction Method

Every other smoke detection system mainly differs by the algorithm used for feature extraction method of smoke. Some of the methods are listed below

1. **Using Static Characteristics**: Static characteristics of smoke refers to some component which has some fixed value, for example, color, intensity, etc.

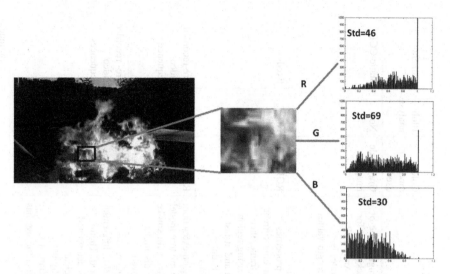

Fig. 3 Spatial difference analysis: in case of flames, the standard deviation σ G of the *green* color band of the flame region exceeds $\sigma = 50$ (Borges [30])

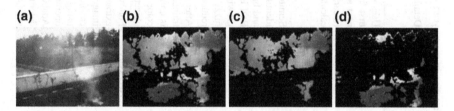

Fig. 4 Segmentation of smoke color using the fuzzy c-means algorithm: **a** original image, **b** moving regions, **c** smoke regions and **d** non-smoke regions [14]

2. **Using Dynamic Characteristics**: Refers to some uncertain characteristics or in which value is uncertain such as smoke area, moving direction, shape, and growth of region [15].
3. **Spatial and Temporal Analysis (Flicker Analysis)**: Since smoke is semi-transparent, therefore, the edges can lose their sharpness; this can lead to a decrease in high-frequency content of an image. This decrease in high frequency energy was used in spatial wavelet transform.

Also, it is very important to distinguish between fire smoke and other fire smoke-colored object. The key to do this is to observe their motion. To study such characteristic of smoke which changes with time refers to temporal analysis. One of the most common is flicker analysis that says that at any time in any pixel, fire flames may be present for a fraction of time. The candidate regions are checked for the presence of flickers (Fig. 3).

Table 1 A tabular comparison of different methods of smoke detection in recent years

Paper	Technology	Moving region extraction	Background noise removal	Details	Color model used	Blob analysis	Camera
Liu YunChang, Yu ChunYu, Zhang YongMing [16], 2010	CCD infrared video image	Background Subtraction	Necessary	Uses segment length value of first frame as background for next and so on. And MCR (mean crossing rate) is calculated from length values	Gray scale image	No	Static CCD
Yuan De-fei, Hu ying, Bi feng-long [17], 2015	Semitransparent properties-based algorithm for detection of smoke in video	Background and current frame are recognized using optical model used in haze image restoration	Not necessary	Semitransparent properties identified by matching current and background frame, then ROI is obtained using region growing method and at last fuzzy clustering is done to reduce false alarms	RGB	No	fixed
Li Ma, Kaihua Wu, L. Zhu [5], 2010	Kalman filter (moving history image), MHI analysis, and Gaussian mixture color model	Kalman filter and MHI to extract moving regions	Necessary	After moving region extraction, trained Gaussian mixture model is compared online with smoke color model and dissimilar objects are removed	RGB	No	Pre-recorded video sequence
H Kim, D. Ryu, J. Park [18], 2014	Gaussian mixture model, and Adaboost algorithm	GMM is used as background estimation	Necessary	Following moving region extracted by GMM, differential image is calculated. Then candidate area is determined using morphological operations. Finally using Adaboost algorithm, it is determined whether the area has smoke or not		Yes	Pre-recorded video sequences

(continued)

Table 1 (continued)

Paper	Technology	Moving region extraction	Background noise removal	Details	Color model used	Blob analysis	Camera
D. Kim, Y Wang [19], 2014	Block–based approach is used for detection of ROI, and K-temporal information is used for smoke classification of each blob	Block-based background subtraction	Not necessary	First is to ensure whether the camera is moving or not. Then, ROI is extracted and final step is to classify each blob as smoky/non smoky using temporal information of color and shape	YUV	Yes	moving
G. Lee, I Ince, G Kim J Park [20]	Histogram technique for varying shape of smoke	Frame difference method	Necessary	Chrominance/Luminance intercorrelation of video is examined periodically. High luminous and less chromatic pixels are used as features of smoke	L * a * b	Yes	stationary
A Benazza, N hamouda, F Tilli, S Ouerghi [21]	DCT(discrete cosine transformation) coefficients	Image difference using Hurst exponent	Not necessary	This paper calculates DCT coefficients at each coordinate pixel. Next, the smoke blocks are supposed to have roughness characterized by a specific range of values of Hurst exponent, which is different of that of non-smoke blocks	RGB	No	Compressed videos
J Li, W Yuan, Y Zeng, Y Zhang [22], 2013	DWT(Discrete wavelet transform)	Improved Gaussian mixture model	Needed	After background subtraction, extraction of smoke fuzzy characteristics based on wavelet analysis is done and finally mobility characteristics are selected for detection criterion of smoke	RGB	No	fixed

(continued)

Table 1 (continued)

Paper	Technology	Moving region extraction	Background noise removal	Details	Color model used	Blob analysis	Camera
T Tung, J Kim [14], 2011	Fuzzy c-means for clustering, SVM(support vector machine) to classify smoke and non-smoke region	Approximate median method	Not necessary	Moving region is segmented using median method, FCM clustering is performed to identify candidate region, then it utilizes combination of color and dynamic features of smoke and an SVM	L * a * b color space	No	static
B Toreyin, Y Dedeoglu, A Cetin [23], 2006	Hidden Markov model (HMM), temporal and spatial wavelet analysis of object contours	Recursive threshold estimation	Not necessary	Decrease in high frequencies is considered as edges using spatial wavelet transform. Decrease in U, V channels is identified, and flicker analysis is carried out by HMM. Finally, wavelet analysis of object contours is carried out	YUV	No	stationary
H Tian, W Li, P Ogunbona, D Nguyen, C Zhan [24], 2011	Local binary pattern (LBP), non-redundant local binary pattern (NRLBP), non-redundant local motion binary pattern (NRLMBP)	Adaptive Gaussian Mixture Model including block-based processing	Not necessary	LBP captures local appearance information based on texture, but is also sensitive to changes in background and foreground. Thus, NRLBP was introduced to reflect contrast between background and foreground and do the job of LBP. Further for temporal texture descriptor NRLMBP was introduced	RGB	No	stationary

(continued)

Table 1 (continued)

Paper	Technology	Moving region extraction	Background noise removal	Details	Color model used	Blob analysis	Camera
C Yu, Z Mei, X Zhang, [25]	Frame differential method for detecting smoke motion edges, while image block processing for flame motion features	Frame differential method	Necessary	Block image processing and optical flow technique are combined to extract smoke features, and backpropagation neural network for smoke classification	HIS for flame detection and RGB for smoke detection	No	Real time video sequence through static cameras

4. **Fuzzy Clustering Method**: It is a method of clustering of data sets. It is used additionally with some color or dynamic features of smoke with an SVM to further classify the clusters. The FCM algorithm is an iterative method of clustering that classifies each and every piece of data to belong to two or more clusters [14].

Support vector machines (SVMs) are a set of supervised learning techniques given by Vapnik, which analyzes data and recognize patterns [28, 29] (Fig. 4 and Table 1).

3   Tabular Comparison of Smoke Detection Methods

Analysis

Overall, the various techniques for video-based smoke detection fall under one or combination of one or more of the categories mentioned: (Fig. 5).

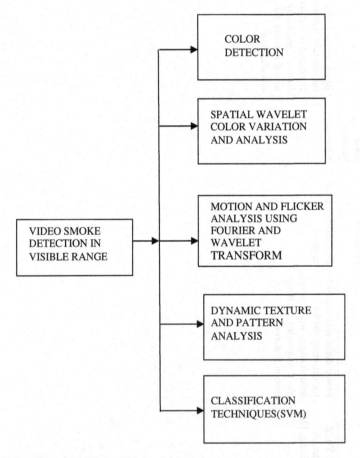

Fig. 5 Techniques of detecting smoke in visible range

4 Conclusion

In this paper, we tried to layout basic traditional methods of detecting smoke in video sequences in the field of computer vision. In spite of variation in smoke feature extraction, every algorithm has common steps as follows:

1. Foreground segmentation/moving region segmentation.
2. To analyze features of smoke and identify ROI.
3. Smoke/non-smoke classification of candidate regions.

Different features of smoke can also be integrated and used to detect fire smoke in a high accuracy. This paper reviewed smoke detection methods based on video images that are used in recent years. Algorithms can be opted for each part of detection in optimality, to improve the system performance.

References

1. Piccinini P, Calderara S, Cucchiara R (2008) Reliable smoke detection system in the domains of image energy and color, pp 1376–1379
2. Toreyin BU (2005) Wavelet based real-time smoke detection in video (2 0), pp 255–256
3. Comez-Rodriuez F (2003) Smoke monitoring and measurement using image processing: application to forest fires, pp 404–411
4. Rider C, Munkelt O, Kirehner H (1998) Adaptive background estimation and foreground detection using Kalman-filtering, vol 12, pp 193–199
5. Ma L, Wu K, Zhu L (2010) Fire smoke detection in video images using kalman filter and gaussian mixture color model, vol 1, pp 484–487
6. Xiong Z, Caballero R, Wang H, Finn A, Lelic MA, Peng P (2007) Video-based smoke detection: possibilities, techniques, and challenges. In: Suppression and detection research and applications
7. Chao-Ching H, Tzu-Hsin K (2009) Real time video-based fire smoke detection system, 1845–1850
8. Migliore DA, Matteucci M, Naccari M (2006) A revaluation of frame difference in fast and robust motion detection, pp 215–218
9. Kim D, Wang Y-F (2009) Smoke detection in video, pp 759–763
10. Maruta H, Kato Y (2009) Smoke detection in open areas using its texture feature and time series properties, pp 1904–1908
11. Toreyin BU, Dedeoglu Y, Cetin AE (2005) Wavelet based real-time smoke detection in video
12. Toreyin BU, Dedeoglu Y, Cetin AE (2006) Contour based smoke detection in video using wavelets
13. Gonzalez-Gonzalez R, Ramirez-Cortes J (2010) Wavelet-based smoke detection in outdoor video sequences
14. Tung T, Kim J (2011) An effective four stage smoke-detection algorithm using video images for early fire-alarm system
15. Surit S, Chatwiriya W (2011) Forest fire smoke detection in video based on digital image processing approach with static and dynamic characteristic analysis, pp 35–39
16. YunChang L, ChunYu Y, YongMing Z (2010) Nighttime video smoke detection based on active infrared video image

17. De-fei Y, Ying H, Feng-long B (2015) Video smoke detection based on semitransparent properties
18. Kim H, Ryu D, Park J (2014) Smoke detection uding GMM and Adaboost 3(2)
19. Kim DJ, Wang Y-F (2009) Smoke detection in video, pp 759–763
20. Lee G, Ince I, Kim G, Park J (2014) Patch-wise periodical correlation analysis of histograms for real—time video smoke detection
21. Benazza A, Hamouda N, Tilli F, Ouerghi S (2012) Early smoke detection in forest area from DCT based compressed video
22. Li J, Yuan W, Zeng Y, Zhang Y (2013) A modified method of video-based smoke detection for transportation hub complex
23. Toreyin BU, Dedeoglu Y, Cetin AE (2006) Contour based smoke detection in video using wavelets, pp 123–128
24. Tian H, Li W, Ogunbona P, Nguyen DT, Zhan C (2011) Smoke detection in videos using non-redundant local binary pattern-based features, 1–4
25. Yu C, Mei Z, Zhang X (2013) A real time video fire flame and smoke detection algorithm
26. Lee C, Lin C, Hong C, Su M (2012) Smoke detection using spatial and temporal analyses 8(6)
27. Valera M, Velastin SA (2005) Intelligent distributed surveillance systems 152(2):192–204
28. Vapnik V (1982) Estimation of dependences based on empirical data
29. Vapnik V (1982) Statistical learning theory. Springer, NewYork
30. Borges PVK, Izquierdo E (2010) A probabilistic approach for vision-based fire detection in videos 20(5):721–731

Understanding Intrafactor Relationships in Cyberloafing Using Predictive Apriori Algorithm

Soham Banerjee and Sanjeev Thakur

Abstract Modern business infrastructures are constantly evolving based on a fluctuating market scenario. It is very important for organizations to have a competitive edge for sustainance and market leadership. This is possible not only by planning risks ahead, but also by improving the productivity of the organization. Cyberloafing is a new trend across organizations that can hamper productivity to well. In this paper, we shall study a specific association rule mining algorithm known as Predictive Apriori algorithm and apply the same on a data set build from user responses. We will identify the best rules that will help us in future to develop a possible forecasting system that can identify when an employee can cyberloaf under certain conditions.

Keywords Cyberloafing · Association rule mining · Predictive Apriori Forecasting system

1 Introduction

Consider a simple scenario of an office, where most of the employees have access to unrestricted Internet, telephone minutes, and unsupervised flexible work hours. The amount of productivity that can be expected from such scenario provided there is no proper monitoring and an ineffective employee surveillance system is not promising. Cyberloafing has become one of the key issues each industry is facing. The problem with cyberloafing is that it relates to employee behavior and psychology, and employers cannot force employees to work all day round the clock [1, 2]. An employee will stop working and waste resources assigned to him if he is not

S. Banerjee (✉) · S. Thakur
Department of Computer Science and Engineering, Amity School of Engineering
and Technology, Amity University, Noida, Uttar Pradesh, India
e-mail: official.soham@gmail.com

S. Thakur
e-mail: Sthakur3@amity.edu

© Springer Nature Singapore Pte Ltd. 2018

D.K. Mishra et al. (eds.), *Information and Communication Technology
for Sustainable Development*, Lecture Notes in Networks and Systems 9,
https://doi.org/10.1007/978-981-10-3932-4_23

satisfied with certain office-oriented environment factors such as salary appraisal, perks, work hours [3]. Good organizations promote healthy and monitored work environment where employee surveillance is effective to penalize any employee wasting resources [4]. However, not every organization enjoys such luxury [2, 5]. So the question arises that can we relate those factors that promote cyberloafing by deriving a relationship among them? If certain parameters can help to classify an employee's satisfaction level toward his job, then it will be an effective technique to monitor and improve efficiency and productivity. In order to generate relationship among various work factors, association rule mining can really help us to generate association rules using the concept of CAR, which will help to assign a class to each association rule that is generated. Thus, we can generate rules that can be inherited by a learning system which will help to predict the class of employees who work or skip their work. However, the scope of this paper is to generate the association rules using Predictive Apriori algorithm that will generate best rules based on increasing support–confidence threshold over a data set generated by user responses. This paper is divided into five sections. The second section following this introduction is a detailed literature discussing the Apriori and the Predictive Apriori algorithm. The third section is a simple experimental setup along with the description of factors that have been taken into consideration while collecting user data followed by results in the fourth section. The final section will be a detailed discussion on the results along with plausible conclusion.

2 Literature Survey

In this section, we will discuss the fundamentals of association rule mining and basic terminologies related to it. Further, we will try to investigate the Apriori and Predictive Apriori association rule mining. We will proceed with the literature survey and have followed the work done by Banerjee et al. [3] and the understanding of data mining following the work done by Singh et al. [6].

2.1 Association Rule Mining

Let us understand what do we exactly mean by association rule mining. The simplest answer would be that it is an approach to determine the relationship among the items in a set of transaction of a database based on interestingness [3]. Association rule mining algorithms mine the various association rules based on support and confidence parameters by generating item sets [7]. At each iteration, the support of the selected item is compared with threshold support. If it is less than the predefined threshold support, then that item is removed from the item set. This process is iterated, and each association rule that can achieve the minimum confidence threshold is listed. The higher the confidence, the better chance of generating a good association rule [3]. Let us understand the terms *support* and *confidence* related to association rule mining.

Considering a database containing items {A, D, E, L, M, X, Y, W, W, J}, then:

1. Each tuple is a transaction.
2. Each transaction contains an attribute—value pair known as an item.
3. An association rule will be represented in the form of BODY => HEAD[support, confidence]. For example, eat(y, "chocolate") => sugar(y, "high") [0.3, 0.9].
4. *Support:* Consider two items A and D such that A => D: P(Q.W). Then, support is equal to number of transactions containing (AUD) divided by total number of transactions.
5. *Confidence:* Consider two items A and D such that A => D: P(W|Q). Then, confidence is equal to number of transactions containing (AUD) divided by number of transactions containing A.
6. So the objective is to determine the association rules that satisfy a predefined support and have high confidence.

Hence, in association rule mining, the item sets with minimum support are associated and each rule has been ranked on the basis of their confidence scores [7–9]. This simply means that higher the support, more will be the coverage of items, and confidence acts as a parameter for accuracy. For n item sets, the number of possible rules that can be generated is $2^n - 1$ [10].

It is clear that why association rule mining is important for real-world application since it can mine relationships among attributes for a specific data. The advantage is the ability to exercise each attribute-value pair as a separate class and learn the rules by incorporating other rules as an input and evaluate them on the basis of minimum threshold support and confidence parameters. However, the disadvantage with rule mining is that it generates too many classes for attribute–value pairs for which consequently too many rules are generated. It means that these algorithms are computationally intractable [11]. In order to understand the purpose and working of Predictive Apriori algorithm, we will discuss the most fundamental algorithm used for association rule mining which is basically the standard Apriori algorithm.

2.2 Apriori Association Rule Mining

Apriori rule is based on the fact that "*any subset of an frequent item set is frequent.*" To understand it's working, some of the related terms are discussed below:

1. *Frequent Item set* is the set of items that have fulfilled the minimum support and is represented by L_k for the kth item set.
2. *Join Operation* focuses on performing a join item L_{k-1} with itself to generate candidate item C_k.
3. *Prune Operation* basically confirms that if any k-1 item set is frequent, then it should belong to a subset of a frequent k item set.

The algorithm for implementing the Apriori association rule with reference to Fig. 1 is as follows:

C_j = Item sets of unary size in I;
Identify every large item sets with unary size, L_j;
j = 1;
Repeat
j = j + 1;
Cj = Apriori(Lj − 1);
Apriori(Lj − 1)

Generate candidates of size j + 1 from large item sets of size j.
Perform join operation on large item sets of size j if they agree on j − 1.
Perform pruning of candidates who have subsets that are not large.

Count Cj to determine Lj unless no large item sets are determined.

Assume a scenario where a specific database, contains some association rules with *sup = 0.5* and *conf = 0.8* [12]. To generate the association rules from frequent item set, first for all frequent item set L the nonempty subset of L is determined as shown in Table 1 such that a rule in the form of M => (L − M) is generated if and only if it satisfies Eq. 1.

$$Supp(L)/Supp(M) > = Min\ Conf \qquad (1)$$

2.3 Predictive Apriori Association Rule Mining

Predictive Apriori association rule mining is another confidence-based association rule mining algorithm that generates frequent item sets in the same fashion as

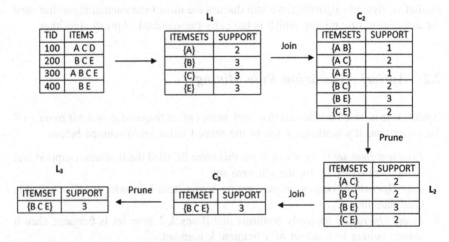

Fig. 1 Implementation of Apriori algorithm [12]

Table 1 Generation of association rules

Rules	Support (X Y)	Support (X)	Confidence
{A} => {C}	2	2	100
{B} => {C}	2	3	66.67
{B} => {E}	3	3	100
{C} => {E}	2	3	66.67
{B} => {C E}	2	3	66.67
{C} => {B E}	2	3	66.67
{E} => {B C}	2	3	66.67
{C} => {A}	2	3	66.67
{C} => {B}	2	3	66.67
{E} => {B}	3	3	100
{E} => {C}	2	3	66.67
{C E} => {B}	2	2	100
{B E} => {C}	2	3	66.67
{B C} => {E}	2	2	100

Apriori algorithm does with the only difference that Predictive Apriori estimates the confidence of an association rule differently [10, 13]. The objective of Predictive Apriori is to increase the correctness of the association rule rather than the correctness of data that is used as an input [10]. This accuracy is also known as expected accuracy or better known as predictive accuracy.

Let I be a database which contains some records t generated by process C, and let A => B be an associated rule. The predictive accuracy pra(A => B) is equivalent to the probability that t satisfies A such that it also satisfies B. This conditional probability of B being a subset of t given that A is also a subset of t is governed when process C distributes t. [13]. Where Apriori algorithm prefers generating more general rules, Predictive Apriori algorithm generates the best n rules based on the following criteria as described by Mutter et al. [10]

1. Among the n best rules, the predictive accuracy is calculated,
2. In case of same predictive accuracy, rules generated by Predictive Apriori are not subsumed.

When compared with Apriori algorithm, we observe that Predictive Apriori algorithm generates the association rules and puts them in the set of n best rules on increasing minimum support threshold. If a particular rule is to be discarded from the list of n best rules, then the algorithm has to rerun recursively again for generation of rules since it might be possible that the rule which could not make it to the set of n best rules at the first place might achieve its position in the set. Also, the classification of association rules becomes easier [10] when using Predictive Apriori algorithm. Also, the quality of rules generated with Predictive Apriori algorithm is better when using small data [14] sets which fits perfectly for our work. However, the drawback is that Predictive Apriori algorithm has higher time complexity as compared to Apriori algorithm [10]. Referring to Table 2, we have

Table 2 Comparative analysis of association rule mining algorithms

Factors	Apriori	FP-growth	Predictive Apriori
Time taken	Better time complexity than Predictive Apriori but worse than FP-growth	Takes only O(n) time to scan and generate rules using FP-tree	Has the worst time complexity among all association rules
Independence	Attributes are independent of each other	Attributes are dependent of each other	Attributes are independent of each other
Number of rules	Generate equal number of rules w.r.t to FP-growth but less than Predictive Apriori	Generate equal number of rules w.r.t to Apriori but less than Predictive Apriori	Generates maximum number of rules > 100
Quality measure	Support and confidence	Support and confidence	Predictive accuracy
Search strategy	Breadth-first search and Hash Tree with bottom-up approach	Depth first search with top-down approach	Breadth-first search and Hash Tree with bottom-up approach

outlined the comparison among various association rule mining algorithms. However, as explained earlier the focus is to understand Apriori algorithm and Predictive Apriori algorithm as covered in this section.

Now, we will move forward with Predictive Apriori algorithm and apply over the cyberloafing data set we have generated by collecting user responses in the next section.

3 Experimental Setup

In this part, we will discuss our cyberloafing data set and the factors that have been considered. We prepared a questionnaire for 627 participants out of which 500 responses were selected due to their completeness. 49.77% males and 40.23% of the females had participated between the ages of 18–60 years. Each participant is working in some organization with different work environment variables. The data set covers the following factors in the form of questions which can be answered by a participant in yes or no:

1. *Age between 18–30*: Our focus was to determine the extent to which junior level employees of any organization will work. Those who have responded yes belong to junior management, while the rest belong to senior management.
2. *Salary Appraisal*: Salary appraisal is considered as a key motivator for some employees.
3. *Promotion*: Promotion is considered as a key motivator for some employees.

4. *Prospective Work Env*: Most employees prefer a healthy work environment to learn and grow their skills.
5. *Perks*: Most employees favor noncashable income to uniquely identify their stand and power.
6. *Flexible Work Hours*: Employees often slack their work if they have the privilege to fulfill weekly hours of work.
7. *Designation*: The higher the designation, the more the authority an employee can exercise.
8. *Using office resources for personal use*: Employees do perform personal tasks using office equipment.
9. *Work From Home/Part-Time*: Employees who work from home or do part-time work remotely can slack.
10. *Attrition*: Has the employee experienced a high rate of releasing employees in the organization?
11. *Class*: Users were asked whether they were satisfied with their job or not.

4 Results

In this section, the Predictive Apriori algorithm has generated hundred rules based on *CAR* property as shown in Fig. 2. The top ten association rules are as follows:

1. Age Between 18 – 30 = Yes Salary Appraisal = No Prospective Work Env = No Designation = Junior Using Office Resources for personal use = No Attrition = No 13 ==> Class = Satisfied 13 acc:(0.98949)

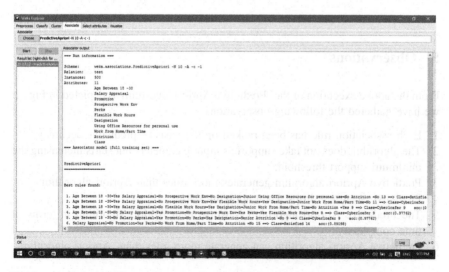

Fig. 2 Association rules generated by Predictive Apriori

2. Age Between 18 – 30 = Yes Salary Appraisal = No Prospective Work Env = Yes Flexible Work Hours = Yes Designation = Junior Work From Home/Part Time = No 11 ==> Class = Cyberloafer 11 acc:(0.98547)
3. Age Between 18 – 30 = Yes Salary Appraisal = No Flexible Work Hours = Yes Designation = Junior Work From Home/Part Time = No Attrition = Yes 9 ==> Class = Cyberloafer 9 acc:(0.97762)
4. Age Between 18 – 30 = No Salary Appraisal = Yes Promotion = No Prospective Work Env = Yes Perks = Yes Flexible Work Hours = Yes 9 ==> Class = Cyberloafer 9 acc:(0.97762)
5. Age Between 18 – 30 = No Salary Appraisal = Yes Promotion = No Perks = Yes Designation = Senior Attrition = No 9 ==> Class = Cyberloafer 9 acc:(0.97762)
6. Salary Appraisal = Yes Promotion = Yes Perks = Yes Using Office Resources for personal use = Yes Work From Home/Part Time = No Attrition = Yes 9 ==> Class = Cyberloafer 9 acc:(0.97762)
7. Salary Appraisal = Yes Promotion = No Prospective Work Env = Yes Perks = Yes Flexible Work Hours = Yes Attrition = No 9 ==> Class = Cyberloafer 9 acc:(0.97762)
8. Salary Appraisal = Yes Promotion = No Prospective Work Env = Yes Perks = Yes Designation = Senior Attrition = No 9 ==> Class = Cyberloafer 9 acc:(0.97762)
9. Salary Appraisal = Yes Prospective Work Env = Yes Perks = Yes Flexible Work Hours = Yes Designation = Senior Using Office Resources for personal use = Yes 9 ==> Class = Cyberloafer 9 acc:(0.97762)
10. Age Between 18 – 30 = Yes Salary Appraisal = No Prospective Work Env = No Designation = Junior Work From Home/Part Time = No Attrition = No 8 ==> Class = Satisfied 8 acc:(0.971)

5 Observations

From the above execution of the Predictive Apriori algorithm as depicted in Fig. 2, we have gathered the following observations.

1. Each association rule has been ranked on the basis of predictive accuracy.
2. The algorithm does not take support as input parameters but starts increasing the minimum support threshold.
3. Predictive Apriori algorithm generates more rules than Apriori algorithm.
4. Predictive accuracy drops when applying *CAR*.
5. The Predictive Apriori algorithm is worse than Apriori algorithm in terms of time complexity.
6. Predictive Apriori algorithm can generate more rules than Apriori algorithm.

7. Predictive accuracy can be used to classify the association rules.
8. Predictive accuracy is dependent on the quality of data set. The predictive accuracy is inversely proportional to the size of data set.

6 Conclusion

In this paper, we have discussed the use of association rule mining to identify the relationships among the factors that promote cyberloafing. A detailed discussion regarding Predictive Apriori algorithm and standard Apriori algorithm is discussed. The outcome from the experimental setup provided some observations that will help researchers to develop a forecasting system that can help organizations to identify whether an employee works according to the policies or slack from their work and waste organizational resources.

References

1. Banerjee S, Thakur S (2016) A critical study of factors promoting cyberloafing in organizations. In: Proceedings of the international conference on cyber security and digital forensic. ACM (2016). (In press)
2. Apriori rule. http://www3.cs.stonybrook.edu/~cse634/lecture_notes/07apriori.pdf
3. Jia H, Jia R, Karau S (2013) Cyberloafing and personality: the impact of the big five traits and workplace situational factors. J Leadersh. Organ. Stud. 20(3):358–365
4. Olguin D, Waber B, Taemie Kim MA, Ara K, Pentland A (2009) Sensible organizations: technology and methodology for automatically measuring organizational behavior. IEEE Trans Syst Man Cybern B. 39:43–55
5. Lim V, Chen D (2012) Cyberloafing at the workplace: gain or drain on work? Behaviour & Information Technology. 31(4):343–353
6. Singh A, Banerjee S, Shukla S, Singhal S (2016) Attenuation of broadband problems using data mining techniques. Proc First Int Conf Inf Commun Technol Intell Syst 2(2):313–322
7. Ceglar A, Roddick J (2006) Association mining. CSUR 38, 5-es (2006)
8. Helm B (2007) Fuzzy association rules an implementation in R, Dissertation, Vienna University of Economics and Business Administration
9. L.M.R.J, Lobo (2012) A comparative study of association rule algorithms for course recommender system in E-learning. Int J Comput Appl 39
10. Mutter S, Hall M, Frank E (2004) Using classification to evaluate the output of confidence-based association rule mining. Lect Notes Comput Sci 3339:538–549
11. Andreassen C, Torsheim T, Pallesen S (2014) Predictors of use of social network sites at work —a specific type of cyberloafing. J Comput-Mediat Commun 19(4):906–921
12. Apriori Algorithm-CodeProject. http://www.codeproject.com/Articles/70371/Apriori-Algorithm
13. Scheffer T (2001) Finding association rules that trade support optimally against confidence. Princ Data Min Knowl Discov 2168:424–435
14. Gyenesei A, Teuhola J (2001) Interestingness measures for fuzzy association rules. Princ Data Min Knowl Discov 2168:152–164

Fault Detection and Recovery for Automotive Embedded System Using Rough Set Techniques

Pattanaik Balachandra

Abstract The aim of this chapter is to propose the two main features, fault detection and recovery in a fault-tolerant system. These are to be modeled and designed through the use of correct mathematical approaches. The fault finding approach through rough sets using fuzzy logic control to detect exact location and nature of fault in terms of states in the automotive system. The clear distinction between thermal, mechanical, electrical, electronics, communication, and computing subsystems is another challenge in the design of fault-tolerant automotive embedded system. Fuzzy function method is used to locate the error components in automotive embedded system. The proposed research work, the fault detection, and fault recovery are achieved through fuzzy rough set technique and interface EXFSM with DSDA approach, respectively.

Keywords Embedded faults · Fault detection · Rough sets and rough and fuzzy rough sets

1 Introduction

The task-specific system with a domain-specific application like a fault-tolerant automotive embedded system needs to have a high reliability. The reliability enhancement of such system has to emerge from a collection of reliable components whose faulty statuses are to be detected and rectified at the earlier time in the system repairable time. The automotive embedded system needs to be collective analysis in technical relation. That is any automotive system can be considered as a set of heterogeneous components or subsystems that includes mechanical, electrical, electronics, and programmable nature with devices and software components. Their association with each other depends on the context and action of the user in

P. Balachandra (✉)
Faculty of Engineering and Technology, Department of Electrical
and Computer Engineering, Mettu University, 318 Mettu, Ethiopia
e-mail: balapk1971@gmail.com

© Springer Nature Singapore Pte Ltd. 2018
D.K. Mishra et al. (eds.), *Information and Communication Technology
for Sustainable Development*, Lecture Notes in Networks and Systems 9,
https://doi.org/10.1007/978-981-10-3932-4_24

233

that environment. The assemblies of all the nature of components and their operational environments are to be decided in order to detect the faults in each of the assembly. In order to apply fault detection techniques for automobiles, the theory of rough sets can be considered since the elements in the set varies during the operation or execution of the system. The data or parameters that are passed across such a number of automotive subsystems involve a lot of transformations at the respective interfaces. The mathematical, communication, and computation techniques are to be applied in the fault detection phase of an automotive embedded system. The correctness and performance are the two aspects which pose a high challenge in this issue of fault detection within a scenario of heterogeneous component interactions. Applying a suitable set theory, organizing the interface enabling and allowing permissible transformation, and tracking the variables within their limits pose major problems in detection methodology.

The system includes electrical power supplies, electronic sensors, programmable microcontrollers, and the software embedded within the available memory. The real-time operating system function modules, the external crystal, and the temperature of the external environment are all to be included in order to determine the system reliability from a set of acceptable and unacceptable values of these component behaviors. The lower accepted and the upper permitted values of parameters of various components are grouped assets; their values are fuzzified and treated as rough sets. In classical-relation rough set theory [1], a set is that each and every element must be uniquely classified as belonging to the set or not and making the notion of the faultiness of each component. This mathematical relation helps to construct the rough set application. For any rough set property with relevant to the automotive application, required prediction of said property is determined. As per Dummett, the observational relation and sense of tolerance can't be predicted. The idea of the relation between two pairs of objects in this scenario cannot be tolerated. The internal identical relationship of the constituent objects and the external relation of indiscernibility between pairs can be determined [2]. The interrelationship between "α"- and "β"-level fuzzy rough sets is proposed, and the fuzzy approximation state space is studied to improve the mathematical computational efficiency in decision-making process achieved [3]. The earlier motor current signature analysis (MCSA) has been carried out by the application of rough sets toward a classifier in fault diagnosis analysis [4]. A rough set-based approach is proposed to handle the uncertainty of the values of the states of parameters in the power model [5]. Rough set approximation theory has an overlapping with other theories dealing with imperfect knowledge, e.g., evidence theory, fuzzy sets, and Bayesian inference [6, 7].

The fuzzy rough set and rough set technique decreases the dataset without much more loss in the information system when compared with the conventional type rough set attribute reduction techniques [8, 9]. The technological requirement which focus to apply the rough set and fuzzy rough set attribute state function reduction technique, for the fault finding detection in an embedded system for combined effect and mixed faults with diversified state attributes and their relevant values. The representation of complex class is the sum of all its members and identify all the objects belonging to that category. The intentional description of the category is used as a representation of the category based on a set of *rules* that describe the scope of the category. The different types of faults are determined according to the functionality, behavior, and the available environment of the automotive embedded system. The automotive system faults and the hardware-required faults can be realized by the hybrid technical concept using approximate reduction techniques in the real-time detection mechanism. In the proposed technique, a rough set-based data analysis methods have been applied in the design of embedded system focusing the fault tolerance software engineering concepts to an automotive system.

2 Fault Detection and Recovery in a FT System

The fault detection and recovery in a fault-tolerant system are the two main features which are to be modeled and designed through the use of correct mathematical approaches. In the proposed research work, the fault detection and fault recovery are achieved through fuzzy rough set technique and interface extended finite-state machine approach, respectively.

The fault-tolerant system consists of other phases such as fault rectification and reliability enhancement. Each and every phase of the research work focused on the core, that is, fault-tolerant behavior of the system. The limited error recovery is implemented with a controller area network (CAN)-based microcontroller out using a pair of fault-tolerant CAN subsystems. The last phase is the reliability enhancement using DSDA technique focusing predictably dependable automotive embedded system design toward safety and reliability through intercomponent trust as shown in Fig. 1 [10].

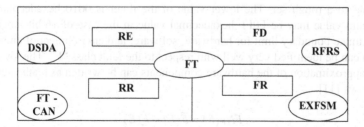

Fig. 1 Fault-tolerant phases and techniques. *FD:—Fault detection, RFRS:—Rough and fuzzy rough sets, FR:—Fault rectification, EXFSM:—Extended finite-state machine, RR:—Recovery and reliability, FT-CAN:—Fault-tolerant control area network, RE:—Reliability enhancement, DSDA: —Detection of safety and dependability aspects, FT:—Fault tolerant*

3 Embedded Faults in the System

Let us consider M be an automotive machine with a set of all three kinds of objects related to hardware fault, software fault, and the fault in the operating environment. The faults in the different objects can be assumed as an "F" representing the imprecision in knowing the faulty state condition of the objects of the machine "M". The assumed fault "F" generates from the hardware-related fault or software or environment category fault that gives the machine as faulty state or faulty condition. This gives the relation mathematically as follows:

$$F = M \times M \times M \tag{1}$$

Any three assumed position of faults make the system faulty, stating that the machine is available with one fault within the system. Let "F" be the set of faulty condition due to any one object in hardware fault or one object in software-related fault or the object oriented in the environment faults. The lowest approximation of a hardware set "H" with respect to fault "F" is a set of all hardware fault modules which can be treated as good condition and positive as such with respect to fault "F". The upper most limit of the approximation of rough set "X" with respect to "F" is the set of all kind of objects which can be in normal state and classified as "H" with respect to fault "F". The limit of the outer most region of rough set "H" with respect to fault "F" is set of all kinds of components of the modules which can be classified neither as "H" or not "H" with respect to "F". The hardware-related set "H" is crisp with respect to faultiness of the component "F" and the almost boundary outer region is empty which approaches to zero. The set "H" is rough enough or imprecise if the boundary region of "H" is nonempty set. Hence, a rough set in an embedded system is responsibly having nonempty boundary region. For example, the hardware components like memory that be with certain (normal) conditions and other permissible conditions when permitted to operate with different operational conditions in the context of data size. The system fault class of "F" is represented by F(h) where h ∈ H. For example, the hardware timer within the chip may show the value within its range, and it can be assumed that it is normal. The maximum and minimum values a hardware timer can hold are the mapping to a rough set. The lower value of the timer is 0010 hexadecimal, and maximum value may be F0FF hexadecimal value in the case of 16-bit timer. The approximate boundary limit for hardware, software, and environment of embedded system can be identified very well with respect to the fault class "F". The lower and upper approximation of the hardware components can be written as mathematically as follows [11].

$$FN_F(E) = \overline{F}(E) - \underline{F}(E) \tag{2}$$

4   Limits of Faults with Set Theory

The various faulty conditions of rectified components in an automotive embedded system can be described using approximations as in rough sets with fuzzy relations. Within the same time bound, the same type of set may also being defined as the terms of membership state functions and their cardinalities with respect to limiting conditions. The fuzzy rough membership state functions explain the conditional probability parameter of that particular situation with an object belonging to a fuzzy set given the relationship and the mathematical relation can be interpreted and correlated as a degree of freedom that the object belongs to the limit set in that point of view-related information about membership of the object quantity expressed by that corresponding relation. The rough and fuzzy rough sets can be implemented for the limits constructed using the upper and lower limits with fuzzy relation membership functions or combining effect of the fuzzy and rough set which is based on the cutting edge of fuzzy sets of the function. Relation of the fuzzy rough sets for hardware, software, and the environment-desired bonding faults is represented mathematically through the following equations as follows:

$$\mu_h^F(H) = \frac{|H \cap F(h)|}{F(h)} \tag{3}$$

$$\mu_s^F(S) = \frac{|S \cap F(s)|}{F(s)} \tag{4}$$

$$\mu_e^F(E) = \frac{|H \cap F(e)|}{F(e)} \tag{5}$$

The fuzzy rough sets within an automotive embedded system can be formed by the different classification approach of disjoint-type categories of fuzzy objects with each of three bounded segment limits shown in Fig. 2. Criteria of the hardware objects belonging to the similar category, assuming hardware characterized by the same object attributes or related properties or related features, are elaborately not distinguishable. For the hardware concept, let us consider the core co-processor and timing concept of the component are emphasized to the same hardware-oriented category, and the same is characterized by the states of attributes like large quantity of data size and related clock frequency required are indistinguishable with the hardware fuzzy components. The above 3, 4 and 5 equations show the relationship between Random Access Memory (RAM) and Read-only Memory (ROM) are elements of the hardware category of an embedded system. They are characterized by their data size and address of accessibility and become more reliability with embedded hardware co-related components. In this context, the reconfiguration-state file and the initial-state file are not distinguished. In the case of object software category, the state return aspects attributes in an embedded system and state type of memory enhanced. After this, the last nature environment-related

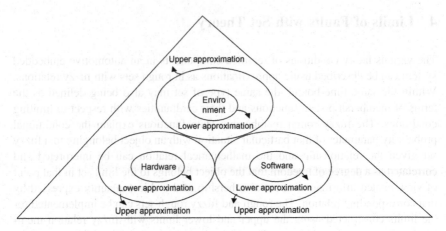

Fig. 2 Limits approximation region-bound triangle

category concept, the most used sensors, and most actuators are not distinguishable with surrounding based on the data size and minimum required threshold supply voltage input to the system. Fuzzy set and the rough sets can be formed with fuzzy aspects using the lower limit and upper limit approximations to many more attributes and necessary values in each type and every type category, let us consider the rough sets in hardware component rough sets in the upper and lower limit environment.

By applying the membership functions of individual elements in each category of permissible limits within the lower and higher values, an automotive embedded system can be considered as an aggregation of fuzzy rough set of elements in the context of their faultiness. The operational profile like the working temperature that is as per the minimum and maximum limits in the hardware specifications. The lower and higher power dissipation and the number of input/output devices are considered in the work. The H, S, and E being considered in the above are justified for the fault recovery [11].

The rough sets with embedded enhanced component "D" with hardware, software, and environment being co-emphasized as H/D, S/D, and E/D and the U/C_{11}... U/C_{1N} where $C_{11} = U - \{a_{11}\}$ can be determined. The overall fault detection capability or accuracy of the rough set approach can be calculated as Eq. (6):

$$S(h, s, e) = \alpha(C_{ij}) \, |(\delta + \mu + \sigma) \, / \, \text{Value}(C_{ij})| \qquad (6)$$

The capacity in bits (a_{11}) = 8 bits, component-used voltage (a_{12}) = 2.4 V, and power consumption in watts (a_{13}) for the corresponding values of attributes are calculated as shown in Tables 1 and 2, respectively (Fig. 3).

Table 1 The value for hardware peripherals

Peripheral component	Capacity in bits	Component-used voltage	Consumption in watts (W)	Frequency (MHz)
C_1 hard core	8	2.4	0.25	60
C_2 RAM	16	2.4	0.25	48
C_3 port	16	3.6	0.1	48
C_4 timer	8	2.4	0.1	60

Table 2 Fuzzified value of SW components as faulty

Component	Fuzzy values				
	Very low	Low	Minimum	Nobel typical	Expected
Buffer—software	0.1	0.3	0.6	0.8	1
Crystal—environment	0.1	0.2	0.6	0.8	1
Configuration—software	0.1	0.3	0.5	0.8	1
Function call—software	0.1	0.2	0.5	0.9	1
RTOS—software	0.1	0.2	0.6	0.9	1

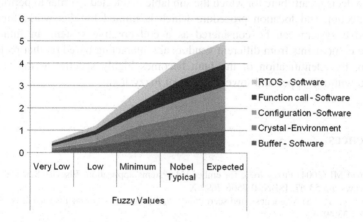

Fig. 3 Fuzzified value of SW components as faulty

The above Fig. 3 indicates that this operates within the required component which justifies the design criteria of thermo- and electric conditions based on the predefined values of the design. The numerical fuzzy-state attributes being discredited for the particular limiting interval, for the cut sets, are threshold states denoting these intervals. Further the set of attributes being formed with grouped (quantized) cuts with disjoint subsets with original attributes of the state values, which prove the expected result.

5   Conclusion

The fault detection approach through rough sets using fuzzy logic control to detect exact location and nature of faults, in terms of states in the automotive embedded system. Fuzzy function method is used to locate the error components in automotive embedded system. Any manufacturer-specific model of an automotive with a given set of specifications and interfaces of a particular subassembly can also be selected. With sufficient data for training and decision attribute through which the fault can be detected and located within the three categories of entities using fuzzy rough set technique. The clear distinction between thermal, mechanical, electrical, electronics, communication, and computing subsystems is another challenge in the design of fault-tolerant automotive embedded system. The scope of the research work is to apply the fuzzy rough set approach to fault diagnosis which enhances the correctness and performance in the detection process that indirectly enhances the fault-tolerant features.

In order to present the capability of generating fault detection or decision column with the help of two tables. One is the training table and for the second is generated toward a decision attribute for which the sub table is selected in order to perform the fault detection and location procedure which is limited for this work. Since an automotive system can be considered as a collaborative system, in future the multiple components from different vendors are interacting based on that particular scenario, the identification of the fault becomes highly specific to the branded instance with the upper and lower regions of fuzzy logic.

References

1. Ahmad MI (2004) Fuzzy logic for embedded systems application. Elsevier Science (USA), Newnws, pp 53-67. ISBN: 0-7506-7699-X
2. Varzi AC (1995) Vagueness, indiscernibility, and pragmatics comments on burns. South J Philos 33:49–62
3. De Persis C, Isidori A (2001) A geometric approach to nonlinear fault detection and isolation. IEEE Trans Autom Control AC-46 853–865
4. Johnson BW (1996) An introduction to the design and analysis of fault-tolerant systems. In: Pradhan D (ed) Fault-tolerant computer system design ch. 1. Prentice Hall Inc, pp 1–84
5. Shu X, Lai EM-K (2005) A rough set approach to instruction-level power analysis of embedded VLIW processors. pp 1–5
6. Hong JS, Lie CH (1993) Joint reliability importance of two edges in an undirected network. IEEE Trans Reliab 42(1):17–23
7. Pawlak Z (2002) Rough set theory and its applications. J Telecommun Inf Technol 7–10
8. Jensen R, Shen Q (2002) Fuzzy-rough sets for descriptive dimensionality reduction. Fuzzy Systems 2002, FUZZ-IEEE'02, Honolulu, HI, USA, pp 29–34
9. Salazar O, Soriano J (2013) Generating embedded type-1 fuzzy sets by means of convex combination. In Proceedings of the 2013 IFSA world congress NAFIPS annual meeting, Edmonton, Canada, pp 51–56

10. Pattanaik B, Subramaniam C (2013) Development of safety and dependability aspects for fault tolerant automotive embedded system. Int Rev Electr Eng (IREE) Italy 8(4):1218–1230. SNIP 0.020
11. Pattanaik B, Subramaniam C (2011) Fault detection in embedded system using rough and fuzzy rough sets "WSEAS, Greece". ACM Digital library (Reference number 659–549), pp 405–411

An Analysis on Pricing Strategies of Software 'I-Med' in Healthcare Industry

Pattnaik Manjula and Pattanaik Balachandra

Abstract Software in healthcare industry plays vital role to make their work ease and effective. Medical field has shown a tremendous development in its services from the past to the present trend depending upon the requirement of the industry. Pricing strategies for products or services encompass to improve profits. This paper focuses on strategic price for the software 'I-Med' for the healthcare industry and the various factors affecting pricing. Apart from market potential and customer perception, cause and effect study for implementing software in healthcare industry has been included in this study. Descriptive research has been adopted, and the data were collected by both primary and secondary sources. The sampling method adopted for the study is convenience sampling under non-probability sampling. The size of the sample is 150. The various statistical tools like Karl Pearsons correlation, Kruskal-Wallis test and Chi-square test are used to explain the significant difference between performance based system and responsibility taken in this study.

Keywords I-Med · Pricing strategies · Kruskal-Wallis · Non-probability sampling

1 Introduction

In this developing world, organization has to survey and find out the market needs and conditions. In order to sustain in the market place, they have to implement effective pricing strategy. This study helps to know about the effectiveness of the

P. Manjula (✉)
Faculty of Business and Economics, Department of Accounting & Finance,
Mettu University, Mettu, Ethiopia
e-mail: drmanjula23@gmail.com

P. Balachandra
Faculty of Engineering and Technology, Department of Electrical and Computer
Engineering, Mettu University, Mettu, Ethiopia
e-mail: balapk1971@gmail.com

© Springer Nature Singapore Pte Ltd. 2018
D.K. Mishra et al. (eds.), *Information and Communication Technology
for Sustainable Development*, Lecture Notes in Networks and Systems 9,
https://doi.org/10.1007/978-981-10-3932-4_25

pricing strategy. It also helps the organization to concentrate on the pricing of their product [1]. This study helps in understanding the process of strategic pricing. Squaresoft Technologies of India has developed a software package named 'I-Med' for hospital management, and it has to determine a pricing strategy for that software. Squaresoft Technologies does not have a pricing strategy for their product 'I-Med' and needs to develop an appropriate pricing strategy. This project would help Squaresoft Technologies to develop an appropriate pricing strategy. Advanced software's enables to provide numerous modern services to their customers. It helps to provide better competitive prices based on the usage [2].The approach for fixing the optimal cost based on the preliminary approach cost based on their upgradation [3]. Based on the above approach, the requirement of IT solution needed is highly expansive [4]. To acquire the software for the healthcare field, the vendor has to spend various running cost stage by stage starting from license fees. The optimal demand of health care can be stated by patients' perception about the best price [5, 6].

2 Objectives of the Study

Primary Objective: To analyze a strategic price for the software 'I-Med' for the healthcare industry.

Secondary Objective:

- To analyze the customer perception about the software from other competitive companies.
- To identify the market potential for 'I-Med' in the healthcare industry.
- To analyze the cause and effect of implementing software in healthcare industry.
- To find the various factors related to the price of the software for the healthcare industry.

3 Methods and Methodology

Explanatory research is used for this study. The study is undertaken by using convenience sampling under non-probability sampling [7]. Both primary and secondary data were used for this study. Primary data were collected by distributing questionnaires to different employees of software companies, apart from that interviews were conducted with selected persons of the companies. The various sources of secondary data are books, periodicals, journals, directories, magazines, statistical data sources, etc. The secondary source used for this study is company profile, scope, need, review of the literature. The sample size is 150. The units selected may be each person who comes across the investigator.

4 Data Analysis and Interpretation

Inference: In Table 1, it is interpreted that 52% of the respondents have 5000–10000 as their expected cost, 20% of the respondents have 10000–15000, and 28% of the respondents have more than 15 K as expected cost.

Interpretation: From Table 2 and Fig. 1, it can be interpreted that 50% of the respondents believe that there is a market potential for software 'I-Med,' and 33% of the respondents feel average about market potential for software

Interpretation: From Table 3, it can be interpreted that 80% of the respondents are ready to buy software 'I-Med' if its meet expectation, and 20% of the respondents have not ready to buy software 'I-Med' if its meet expectation.

Table 1 Total cost of the software

S. no	Particulars	No of respondents	Percentage
1	<5 K	0	0
2	5–10 K	78	52
3	10–15 K	30	20
4	>15 K	42	28
	Total	150	100

Table 2 Market potential of I-Med

S. no	Particulars	No of respondents	Percentage
1	Good	75	50
2	Average	50	33
3	Bad	25	17
	Total	150	100

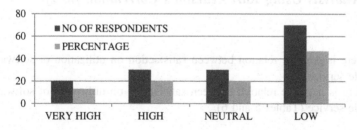

Fig. 1 Market potential of I-Med

Table 3 Purchase of I-Med

Particulars	No of respondents	Percentage
Agree	120	80
Disagree	30	20
	150	100

Table 4 Level of competition

S. no	Particulars	No of respondents	Percentage
1	Very high	20	13
2	High	30	20
3	Neutral	30	20
4	Low	70	47
5	Very low	0	0
	Total	150	100

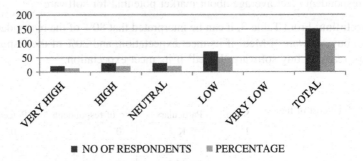

■ NO OF RESPONDENTS   ■ PERCENTAGE

Fig. 2 Level of competition

Interpretation: From Table 4 and Fig. 2, it can be interpreted that 47% of the respondents feel low, 20% of the respondents say low, and 20% of the respondents say high about level of competition of the software.

5 Statistical Tools

5.1 Analysis Using Karl Pearson's Correlation [8, 9]

Ho: There is positive relation between satisfaction on utilization of software and after-sale service.
H₁: There is negative relation between satisfaction on utilization of software and after-sale service (Tables 5 and 6).

$$r = \frac{N \sum XY - \sum X \sum Y}{\sqrt{N \sum X^2 - (\sum X)^2}\sqrt{N \sum Y^2 - (\sum Y)^2}} = 9825/10277$$

$$r = 0.956$$

Table 5 Satisfaction on utilization of software and after-sale service

Factors	Satisfaction on utilization (x)			After-sale service (y)
Highly dissatisfied	0			14
dissatisfied	20			17
Neutral	40			35
satisfied	68			63
Highly satisfied	22			21
X	X^2	Y	Y^2	XY
0	0	14	196	0
20	400	17	289	340
40	1600	35	1225	1400
68	4624	63	3969	4284
22	484	21	441	441
$\sum x = 150$	$\sum X^2 = 7108$	$\sum y = 150$	$\sum Y^2 = 6120$	$\sum XY = 6465$

Table 6 SPSS analysis Correlations

		Level of satisfaction on utilization Health Medicare Software	After the sales of the product
Level of satisfaction on utilization Health Medicare Software	Pearson Correlation	1	0.251[**]
	Sig. (two-tailed)		0.002
	N	150	150
After the sales of the product	Pearson Correlation	0.251[**]	1
	Sig. (two-tailed)	0.002	
	N	150	150

Inference: Since r is positive, there is relation between satisfaction on utilization of software and after-sale service.

5.2　Analysis Using Kruskal–Wallis Test

It is also called as H-test. This test is depends on the ranks of the sample observation. So it is also called rank-sum test. Kruskal–Wallis test employed more than two populations [10, 11].

Table 7 Various factors of performance of the software

Particulars	Adequacy	Convenience	Timeliness	Dependability
Very good	22	17	20	10
Good	60	55	75	70
Average	40	45	30	40
Bad	20	26	25	20
Very bad	8	7	0	10
Total	150	150	150	150

Order	Rank
22	10
60	18
40	14.5
20	8
8	3
17	6
55	17
45	16
26	12
7	2
20	8
75	20
30	13
25	11
0	1
10	4.5
70	19
40	14.5
20	8
10	4.5

Ho: There is a positive relationship between various factors of performance of the software.

H₁: There is a negative relationship between various factors of performance of the software (Table 7).

Formulae:

$$Z = \frac{12}{n(n+1)} \frac{R1^{\wedge}2}{N1} + \frac{R2^{\wedge}2}{N2} + \frac{R3^{\wedge}2}{N3} + \frac{R4^{\wedge}2}{N4} - 3(N+1)$$

$$Z = \frac{12}{n(n+1)} \frac{R1^{\wedge}2}{N1} + \frac{R2^{\wedge}2}{N2} + \frac{R3^{\wedge}2}{N3} + \frac{R4^{\wedge}2}{N4} - 3(N+1) = 1.2$$

$Z\alpha$ (tab) $= 7.815$, $Z\alpha$ (tab) $> Z$ cal, H_0 is accepted.

Inference: Since the tabulated value is greater than the calculated value, null hypothesis is accepted; there is a positive relationship between various factors of performance of the software.

5.3   Chi-square test—(Ψ^2)

Ho: There is no significant difference between performance-based system and responsibilities taken.

H_1: There is significant difference between performance-based system and responsibilities taken (Tables 8 and 9).

$$\text{Expected frequency} = \frac{\text{Row Total * Column Total}}{\text{Grand Total}}$$

$\psi^2 = 7.20$,

Level of significance $= 0.05$

Degrees of freedom: $= (r-1)(c-1)$

$$= (2-1)(2-1)$$

$$= 1$$

Table 8 Performance-based system and responsibilities taken

	Good	No	
Yes	60	50	110
No	10	30	40
	70	80	150

Table 9 Calculation of ψ^2 value

O	E	$(O-E)^2$	$(O-E)^2/E$
94	84.5	90.25	1.06
26	35.5	90.25	2.54
75	84.5	90.25	1.06
45	35.5	90.25	2.54

Table 10 Chi-square tests

	Value	df	Asymp. Sig. (two-sided)	Exact Sig. (two-sided)	Exact Sig. (one-sided)
Pearson Chi-square	11.308[a]	1	0.001		
Continuity correction[b]	10.037	1	0.002		
Likelihood ratio	12.349	1	0.000		
Fisher's exact test				0.001	0.001
Linear-by-linear association	11.232	1	0.001		
N of valid cases	150				

Calculated value = 7.20
Tabulated value = 3.841
Z = Z cal > Z tab
Z = 7.20 > 3.841

Hence, the alternate hypothesis [H1] is accepted (Table 10).

Inference: Since the tabulated number is more than the calculated number, H_0 is accepted, and hence, there is no significant difference between performance-based system and responsibilities taken.

6 Findings and Suggestions

6.1 Findings

These main findings are very important for this study to come out with appropriate suggestions. Following are the main findings drawn from the study,

- About 40% of the hospitals are with the service period of 10–15 years.
- More than 30% of the hospitals are involved in providing general services.
- Most of the hospitals have more than 200 beds for their patients.
- About 40% of the respondents are utilizing Health Medicare Software around for the past 10 years.
- More than 35% of the respondents say that their software license validity period is less than 2 years.
- Around 40% of the hospitals avail service agency for after-sale service for their Health Medicare Software.
- About 50% of the respondents say that their Health Medicare Software is working under Windows platform.

- About 50% of the respondents say that they have not faced any problem with their software.
- Majority of the respondents (63) have said that they are using Health Medicare Software.
- More than 70% of the respondents are satisfied with the performance of Health Medicare Software.
- Around 40% of the respondents are ready to buy a new Health Medicare Software for their hospitals.
- About 40% respondents feel that their performance rating for adequacy is good
- Around 35% respondents feel that their performance rating for convenience is good.
- More than 50% respondents feel that their performance rating for timeliness is good.
- About 50% of the respondents say that their annual support cost for the software lies between 10 and 15 K.

6.2 Suggestions

- Squaresoft Technologies can sell this product 'I-Med' with different packages to target various kinds of hospitals.
- Squaresoft Technologies can sell this product 'I-Med' Micro targeted for small size hospitals at a price of Rs. 30,000 with an annual maintenance cost of Rs. 6,000.
- Squaresoft Technologies can assure the license validity period for their product about 5 years.
- They can assure 24/7 customer care service to their product 'I-Med.'
- The product 'I-Med' Core targeted for small size hospitals at a price of Rs. 1,50,000 with an annual maintenance cost of Rs. 30,000.
- The product 'I-Med' Pro targeted for small size hospitals at a price of Rs. 3,00,000 with an annual maintenance cost of Rs. 60,000.
- 'I-Med' Micro can be sold as a plug and play product, while 'I-Med' Core and 'I-Med' Pro can be customized according to customer requirements.
- They can give some attractive price discounts in order to obtain more customers.

7 Conclusions

India is very well placed to tap the growing potential of the healthcare sector The advanced softwares which are rapidly introduced in medical field, bring an overall advancement in health sectors for better services. The future forecast growth in the medical field expected to exceed 250 million USD. Based on the increase in per capita income, it creates further investment requirement in medical field. This study is useful in estimating the demand for hospital management software in each segment of the market in India. It also helps in finding out the important factors which the hospitals consider while purchasing a hospital management software. This study will help Squaresoft Technologies to improve the sale of 'I-Med' by providing selling it at the above specified pricing strategy. The sound pricing system is hallmark of organization's success and prosperity.

References

1. Singh J (2010) Software pricing strategies University of Helsinki, department of computer science
2. Bala R, Carr S (2010) Usage-based pricing of software services under competition. J Revenue Pricing Manage 9(3):204–216
3. Ahtiala P (2006) The optimal pricing of computer software and other products with high switching costs. Int Rev Econ Financ 15(2):202–211
4. Cusumano MA (2007) The changing labyrinth of software pricing. Commun ACM 50(7):19–22
5. Gruber G (2008) Pricing strategies in online and offline retailing. Institute of management information systems, Vienna University
6. Health Care Industry Report New Horizons, Voyage to Value (2014) By Robert wood Johnson Foundation
7. Hui ECM, Wong JTY, Wong KT (2010) Marketing time and pricing strategies
8. Kothari CR (1985) Research methodology. Vishwa Prakhasam Pvt Ltd, New Delhi
9. Research Methods for Business-Uma Sekaran (2013)
10. Kotler P (2013) Marketing management, 8th edn. Prentice Hall, London
11. Mc Donald JH (2014) Handbook of biological statistics, 3rd edn. Spark House Publication, Baltimore, MD, pp 157–164

FPGA-Based Partial Crypto-Reconfiguration of Nodes for WSN

Prajakta Patrikar and Gayatri Phade

Abstract FPGA has a large number of applications in various fields. Wireless sensor network is a vast field, and implementation of FPGA for WSN can minimize a number of complexities. Secure communication between the networks must be taken care of as a lot of networking takes place. Dynamic partial reconfiguration of the specific part minimizes area utilization and power consumption. In this paper, cryptographic algorithm Blowfish has been used. Partial reconfiguration is performed considering the variable key lengths. Low power node (L) has less key size, moderate power node (M) has moderate key size, and high power node (H) has large key size. It is also seen how Blowfish is better as compared to AES cipher. Experimental evaluation is given by considering two nodes on FPGA (ALTERA DE2) kit. Programming is done in VHDL.

Keywords Reconfiguration · Cryptography · Blowfish · FPGA

1 Introduction

As complex wireless sensor networks are increasing, security has become a great issue. Secured communication between two or more nodes is essential as the data to be transmitted should not be leaked or manipulated by various types of attacks or unknown users. Along with handling the issue of security, care must be taken about the reconfiguration of various parts in a network. Also, reconfiguration time must be as small as possible so that fast operations can be achieved in much less time. For secure interaction between two networks consisting of nodes, the node must reconfigure itself in real time. Cryptographic algorithms are used for secure data transfer between the networks. Partial reconfiguration of an FPGA is used to modify

P. Patrikar (✉) · G. Phade
Sandip Institute of Technology and Research Center, Nashik, India
e-mail: prajaktapatrikar@gmail.com

G. Phade
e-mail: gayatri.phade@sitrc.org

© Springer Nature Singapore Pte Ltd. 2018
D.K. Mishra et al. (eds.), *Information and Communication Technology for Sustainable Development*, Lecture Notes in Networks and Systems 9, https://doi.org/10.1007/978-981-10-3932-4_26

certain elements of the network nodes. Using FPGA with WSN allows us to implement diverse cryptographic algorithms on the same device without modifying the network or node [1]. There are various types of cryptographic ciphers such as RSA, DES, AES, Blowfish, and TWOFISH. Also, we can use hash functions, for example, SHA1, SHA2, SHA3, for encryption purposes. Each of them has some advantages as well as disadvantages. In this paper, we are using BLOWFISH algorithm for reconfiguration of nodes with FPGA as its base. The basic methodology used in this work is given as follows:

- development of Blowfish crypto core,
- development of Blowfish De-crypto core,
- reconfiguring Blowfish for various key sizes, and
- integration in the system.

The rest of the paper is organized as follows: Sect. 2 describes other contributions in the similar work. Section 3 gives the comparison between AES and Blowfish. Section 4 describes the encryption and decryption of blowfish algorithm. Section 5 describes the reconfiguration of the system. Section 6 gives the experimental evaluation. Section 7 shows the results, and Sect. 8 concludes the paper.

2 Other Contributions

The paper [2] provides the concept of software-defined networking (SDN) and gives the relation between the need for reconfiguration and development of SDN. SDN provides reconfiguration and control to the devices. SDN can play a major role in the future of electronic and photonic reconfiguration.

Cryptanalysis for blowfish has not been found yet though many scientists tried their hands but were only successful up to 3 or 4 rounds. The work in [3] gives information about how blowfish is efficient in terms of speed and how speed can be increased with the help of space and power using the method of pipelining.

The paper [4] shows that blowfish gives better performance as compared to DES, 3DES, and AES. AES requires more processing power and hence gives poor performance as compared to other algorithms.

Managing dynamically reconfigurable systems is a big challenge; as a result, MDC (Multi-Dataflow Composer) tool is developed by the authors in the paper [5]. Also, power consumption is a great issue in RVC (Reconfigurable Video Coding), and the work showcases the power management done for coarse-grained reconfiguration at the data flow level. The result combines structural and dynamic strategies and shows that approximately 70% of the consumed power was saved.

Reconfigurable computing (RC) has gain a lot of importance year by year due to its use in enumerable applications. The paper [6] gives information about the extensive use of RC and has proven an alternative for refabrication of the devices.

In order to check the strengths and weaknesses of RC, its application in complex networks is still needed.

3   Comparison of AES and Blowfish

Blowfish is much faster as compared to AES. The comparison is made on the basis of speed, block size, and key size [7]. Table 1 shows that the key size required by both the algorithms is same, but block size required by both the algorithms is different. Block size required by Blowfish algorithm is 64 bits less than that required by AES algorithm which is 128 bits (Table 2).

4   Blowfish

This section gives information about Blowfish cipher, its encryption, and decryption process. Blowfish (symmetric block cipher) was designed in 1993 by Bruce Schneier as an alternative to DES (Data Encryption Standard). It has a variable key length varying from 32 bits to 448 bits. Blowfish is unpatented and can be freely used by users. It uses 16 rounds and has a feistel network. The feistel network uses P-boxes (permutation boxes) and S-boxes (substitution boxes) as elements and has a feistel function F.

Figure 1 shows the feistel structure of the blowfish algorithm. XORing is used for linear mixing. Each P array entry is used for every round, and at last, XORing takes place between every half of the data block and one of the two remaining P entries which are unused. The 32-bit input is divided into 8 bits as shown in Fig. 2 which forms input to the four S-boxes. Further addition is performed at the output (2^{32} addition), and XORing is performed. Decryption of Blowfish cipher is exactly same as its encryption but in the reverse order. That is, P_{17} and P_{18} blocks are XORed first, and then, the remaining P entries are taken in the reverse order.

Table 1 Comparison of AES and Blowfish [7]	Algorithm	Key size (Bits)	Block size (Bits)
	AES	128	128
	BLOWFISH	128	64

Table 2 Power node selection as per input	Input mode	Power node selected (L/M/H)
	00	L
	01	M
	10/11	H

Fig. 1 Blowfish algorithm [3]

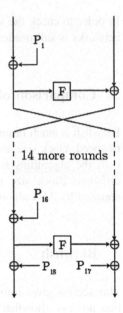

Fig. 2 Feistel function of blowfish [3]

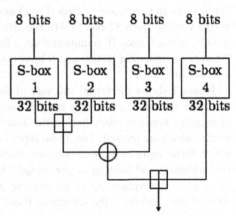

5 Reconfiguration

A configuration consists of one reconfigurable module for every reconfigurable partition. One full BIT file and one partial BIT file are generated by each configuration for each reconfigurable module. In partial reconfiguration, partial BIT file is generated. FPGA technology allows programming and reprogramming without refabricating the entire design [8]. In partial reconfiguration, separate configuration file is generated for each algorithm. This file is stored in a flash memory. Partial reconfiguration is divided into two types:

- **Static partial reconfiguration**: It stops the system while any part is partially modified. That is, the whole system remains static except for the part being modified.
- **Dynamic partial reconfiguration**: It does not stop the whole system while any part is partially modified. That is, the whole system keeps on running simultaneously while any part is being modified.

Figure 3 shows basic premise of partial reconfiguration. The function to be modified in Reconfig Block "A" requires partial BIT files denoted as A1.bit, A2.bit, A3.bit, A4.bit, as shown in Fig. 3. The gray part of FPGA shows *static logic,* and the Reconfig Block "A" part shows *reconfigurable logic*.

In this work, we are modifying the cores for low-power WSN. So that the data can be securely sent from transmitter to receiver in low power, and the overall energy efficiency of the network can be improved. We are using a low-power Blowfish cipher with variable key length for reconfiguration. Based on the device power, the key size is varied. Low power node has less key size, moderate power node has moderate key size, and high power node has large key size. The encryption strength for low power node is weak, that for moderate power node is normal, and that for high power node is strong. The number of rounds required by each power node is different:

- $L = 5$ rounds work;
- $M = 10$ rounds work; and
- $H = 16$ rounds work.

These power nodes can be selected depending upon the input. If the input mode is 00, then low mode will be selected; if the input mode is 01, then medium mode will be selected; and if the input mode is 10 or 11, then high mode will be selected. This selection is same for encryption as well as decryption.

Fig. 3 Basic premise of partial reconfiguration [8]

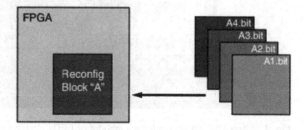

6 Experimental Evaluation

FPGA (ALTERA DE2) kit is used as shown in Figs. 4 and 5. For application purpose, two nodes are considered: node T_X for transmission and node R_X for reception. If the input mode for node T_X is same as that for node R_X, say $T_X = Rx = 00$, then even amount of LEDs glow at the output as shown in Fig. 4, denoting that the data is transmitted and received successfully. If the input mode for T_X is different than that of node R_X, say $T_x = 00$ & $R_X = 01$, then odd amount of LEDs glow at the output as shown in Fig. 5, denoting that the data which is transmitted is not received successfully.

This means:

- Input mode for T_X equal to input mode for R_X—data transmitted securely.
- Input mode for T_X not equal to input mode for R_X—data not transmitted securely (Table 3).

Figure 6 shows the input to FPGA as 00110011. The input mode for T_x as well as R_x is 01, and the output obtained is same as that of the input that is 00110011. This means the data which is transmitted is received successfully.

Fig. 4 Altera DE2 kit showing even amount of LEDs glowing at the output

Fig. 5 Altera DE2 kit showing odd amount of LEDs glowing at the output

Table 3 Data transmission

T$_x$	R$_x$	LEDs glowing in even/odd amount	Data transmitted
00	00	Even	Yes
00	01	Odd	No

Messages		
/e_blowfish_fpga/input	00110011	00110011
/e_blowfish_fpga/mode_tx	01	01
/e_blowfish_fpga/mode_rx	01	01
/e_blowfish_fpga/output_tx	11011100	11011100
/e_blowfish_fpga/output_rx	00110011	00110011
/e_blowfish_fpga/in_tx	00000000000000000000000000	00000000000000000000000000000110011
/e_blowfish_fpga/out_tx	00000011110000111100000011	00000011110000111100000011011100
/e_blowfish_fpga/out_rx	00000000000000000000000000	00000000000000000000000000000110011

Fig. 6 Transmission of data from transmitter to receiver

7 Results

I. Comparison of AES and Blowfish is done on the basis of speed block size and key size, and it is found that Blowfish cipher is better as compared to AES.

II. Partial reconfiguration is achieved considering the variable key lengths. The key size is varied depending upon the device power; accordingly, power nodes are selected (L, M, H) depending upon the input mode (refer Table 2).

III. The transmitter and the receiver nodes are considered on a single FPGA kit (Altera DE2), and if the input mode for transmitter and receiver is same, then even LEDs glow at the output showing successful transmission of data.

8 Conclusion and Future Work

Secure communication between two nodes is achieved with the help of cryptographic algorithm Blowfish using partial reconfiguration for wireless sensor network with FPGA as the base.

Wireless sensor network consists of a number of nodes. Each node will require one FPGA for secure communication with the other node. Thus by using multiple FPGAs, we can show secure interaction between "n" number of nodes in a complex wireless sensor network.

References

1. Cardona LA, Lorente B, Ferrer C (2014) Partial crypto reconfiguration of nodes based on FPGA for WSN, ISBN: 978-1-4799-3530-7, Pagination: 468-471, ICCST 2014 IEEE, pp 0–4
2. Zilberman N, Watts P, Rotsos C, Moore A (2015) Reconfigurable network systems and software-defined networking. Proc IEEE 103(7):102–1124
3. Cody B, Madigan J, MacDonald S, Hsu K (2007) High speed SOC design for blowfish cryptographic algorithm. In: 2007 IEEE international conference on very large scale integration. pp 284–287. ISSN: 2324-8432
4. Karim A, Tamimi A (2008) Performance analysis of data encryption algorithms. pp 1–13
5. Palumbo F, Sau C, Raffo L (2015) Coarse-grained reconfiguration: dataflow-based power management. IET Comput Digit Tech 9(1):36–48
6. Tessier R, Pocek K, DeHon A (2015) Reconfigurable computing architectures. Proc IEEE 103 (3):332–354. ISSN: 0018-9219
7. Thakur J, Kumar N (2011) DES, AES and blowfish: symmetric key cryptography algorithms simulation based performance analysis. IJETAE 1(2). ISSN: 2250-2459
8. Vivado Design Suite User Guide (2015) Partial reconfiguration, UG909 (v2015.2) 24 June 2015

Performance Scrutiny of Thinning Algorithms on Printed Gujarati Characters and Handwritten Numerals

Sanket B. Suthar, Rahul S. Goradia, Bijal N. Dalwadi, Sagar M. Patel and Sandip Patel

Abstract We analyze the behavior of most common thinning algorithms on printed Gujarati text and handwritten numerals. We are focusing mostly on two types of algorithms: The first is serial thinning and second is parallel thinning. Thinning is more crucial when we focusing on structural feature-based character recognition. Thinned character reduced complication of the shape of the character. This analysis focuses on the actual result we get after applying serial and parallel thinning algorithms. Total five algorithms are used for experiments and applied on small, medium, big size of character data and on skewed character data. The results are useful where we designing classifiers for Gujarati text.

Keywords Gujarati text · Handwritten numerals · Thinning
OCR · Algorithms

S.B. Suthar (✉) · S.M. Patel · S. Patel
Department of Information Technology, Charotar University of Science and Technology,
Changa 388421, India
e-mail: sanketsuthar.it@charusat.ac.in

S.M. Patel
e-mail: sagarpatel.it@charusat.ac.in

S. Patel
e-mail: sandippatel.it@charusat.ac.in

R.S. Goradia
Department of Electronics and Communication, G. H. Patel College of Engineering
and Technology, V.V. Nagar, Anand 388120, India
e-mail: rahulgoradia@gcet.ac.in

B.N. Dalwadi
Department of Information Technology, Birla Vishvakarma Mahavidyalaya,
V.V. Nagar, Anand 388120, India
e-mail: bijal.dalwadi@bvmengineering.ac.in

© Springer Nature Singapore Pte Ltd. 2018
D.K. Mishra et al. (eds.), *Information and Communication Technology
for Sustainable Development*, Lecture Notes in Networks and Systems 9,
https://doi.org/10.1007/978-981-10-3932-4_27

1 Introduction

Thinning is preprocessing phase of OCR, and it is an essential step for many structural feature extraction methods. Applying thinning on binary Gujarati characters which converts binary image to single pixel wide line. Major benefits of thinning reduce space and shapes that can be identified more easily.

Here, focusing on Gujarati characters and handwritten numerals, we can say characters are perfectly thinned if it is one pixel wide, character is not broken, and character's curves should no more distracted and contain minimum noise at joints.

Thinning and skeleton is slightly different in terms of application. Skeleton is used for pattern reorganization, image coding, and quantitative metallography where thinning is used for circuit diagrams, engineering diagrams, finger print reorganization, biomedical diagnosis, optical character reorganization, handwritten, and printed characters. Performance of thinning algorithms is highly problem dependent. Within the same domain, it gives different result for different patterns.

These limitations motivated us for the further analysis to find best algorithm among all for Gujarati text. Among all thinning algorithms, we use five basic algorithms and focuses on number of factors that affect the thinning operation like type of documents (machine-/laser-printed), skew characters, and font size.

We also focus on connectivity, shape, and position of the junction point after the results we get. Input symbols considered for printed characters and handwritten numerals are from various font size, thickness, and skew present.

2 Review of Literature

2.1 Segmentation

The second component of typical OCR system is segmentation after optical scanning. Segmentations remove vowel(s)/modifier from basic character. Extracted character further proceeds for experiments. Segmentation is the process in which characters from the text are segmented by connected component, and so in Gujarati language, all connected components from the original character are removed. We collected almost "100" samples of each Gujarati character and all 10 numerals from 0 to 9. Thus, we have database of "18518" characters and numerals collected from different sources such as magazines, newspapers, and books (Ramayana, Bhagavad Gita).

2.2 Preprocessing

Scanned image always has some noise which depends on the scanning machine and also on which techniques you used for scanning. Results not always guaranteed one pixel thick, and some characters are broken. Further, it gives very poor recognition rates. Thinning converts characters in one pixel thick and so most of the problems are solved. We also take 10 and −10 skewed characters in experiments and check how efficiently algorithm works.

2.3 Hilditch Sequential Thinning Algorithm

Naccache and Shinghal presented an algorithm of Hilditch for thinning edges. This algorithm is applied by moving a 3 by 3 window on the binary character and applies some passes. A pixel is marked for deletion if it satisfies all rules applied to it. This basic procedure is applied iteratively until no further points are deleted [1].

2.4 ZS Parallel Thinning Algorithm

| Original | ZS algorithm |

The ZS algorithm described in [2] uses 3 × 3 masks and is a two subiteration algorithm. The criterion used in first subiteration is to remove the southeast pixel which satisfies a set of 4 conditions.

2.5 LW Parallel Thinning Algorithm

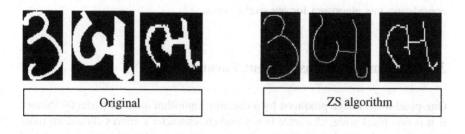

| Original | ZS algorithm |

We applied LW parallel thinning algorithm on same characters on which ZS applied. Here, we can clearly observe that results of LW thinning algorithm are slightly different from the result of ZS. Thus, we can say it gives different results for different patterns and shapes.

The Lu and Wang [3] proposed an algorithm referred to as LW algorithm to tackle diagonal line problem of ZS algorithm. In this algorithm, the first condition of ZS algorithm $2 \leq BP \leq 6$ is modified as $3 \leq BP \leq 6$. This makes the LW algorithm able to preserve 8-connectivity. Thus, the slant lines are retained, but they are of two pixel wide.

2.6 ZW Parallel Thinning Algorithm

For thinning, bwmorph divides the image pixels into two subfields in a checkerboard fashion. Each iteration has two subiteration. In the first subiteration, pixels in the first subfield are deleted if and only if they satisfy the conditions designed to eliminate pixels while preserving topology. In the second subiteration, pixels in the second subfield are deleted if and only if they satisfy a slightly different set of three conditions.

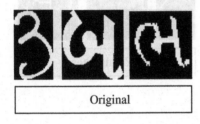

Original

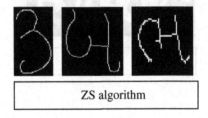

ZS algorithm

2.7 Stentiford Algorithm

In 1983, Stentiford uses concept of mask and introduced a new approach for skeletonization algorithms using a mask concept. Four masks are used to scroll the image in an orderly manner in the form: M1, M2, M3, M4 [4]. We are also considering this algorithm for our excitements.

3 Performance Measurement Parameters

One-pixel-thick image produced by a thinning algorithm must be perfectly thinned if it is one pixel wide, character is not broken, character's curves should no more distracted and contain minimum noise at joints, etc. By observing results of different parameters such that we preserve three characteristics of our input images and these are **"Thinning Quality"**, **"Sensitivity Noise"**, and **"Quality of Shape"**. Thus, to preserve quality of thinning we consider parameters [14].

Thinning rate (TR) measures how much character. Noise sensitivity (NS) measures the extra pixels remaining after thinning or not. In number of component (NOC), we write the function which counts total number of components present in image which is inbuilt function provided by MATLAB (R2009a).

con_st = bwconncomp(thin_img);
Components = con_st.NumObjects;

To identify the shape distortion in thinned image, the very important factor is to calculate number of endpoints in character. In Gujarati script for each character, a number of endpoints are different for the accurate character and distorted character. From that factor, we can find the image is deformed or not. We marked the pixel as an endpoint if they have less than 2 neighbor pixels.

Here, we converted results count number of endpoints (CNE) in between 0 and 1. We can clearly say that the how much quality of shape it preserves for algorithm we are going to use. Number of endpoints tends to 1 gives no spurious edges. Thus, it preserves the shape. The equation we are used is shown below.

The connectivity number used to observe the character is broken or not. If broken, then thinning algorithm not worked well on the characters.

4 Experiments and Results

We have collected the printed (laser and machine) and handwritten Gujarati characters from different sources such as magazines, newspapers, and books. 100 samples of each character are collected for laser- and machine-printed, and 300 samples of each numeral are collected for handwritten numerals.

4.1 Dataset Collection

- Machine-printed text: 5693 characters taken from books, 5000 from newspaper, and 910 skewed characters (+10, −10 and 15°)
- Laser-printed text: 4825 character taken from books and newspaper and 500 skewed characters (+10, −10 and 15°)
- Gujarati handwritten numerals: 3000 character taken from books and newspaper and 300 skewed characters (+10, −10, and 15°).

4.2   Experiment 1

The set of images considered for testing involves Gujarati "Handwritten Numerals," set of various sizes, skewed, thick, and also handwritten numerals. Total 3300 numbers of symbols are there in our dataset (Table 1).

4.3   Experiment 2

The set of images considered for testing involves Gujarati "Small Size Characters," set of various sizes, skewed, thick, and also handwritten numerals (Table 2).

4.4   Experiment 3

The set of images considered for testing involves Gujarati "Medium Size Characters," set of various sizes, skewed, thick, and also handwritten numerals (Table 3).

Table 1 Analysis of handwritten numerals

Thinning algorithm	TR	NOC	NS	EA	Overall result
ZS	0.985845	1	0.998153	0.862570	0.84879
ZW	0.999562	1	0.996393	0.810615	0.80733
LW	0.985086	1	0.997906	0.824639	0.810639
Stentiford	0.952240	1	0.991341	0.360471	0.340283
Hilditch	0.990746	1	0.997066	0.802355	0.792598

Table 2 Analysis of small size dataset

Thinning algorithm	TR	NOC	NS	EA	Overall result
ZS	0.980719	1	0.992928	0.784320	0.763758
ZW	0.997690	1	0.986087	0.793016	0.780176
LW	0.977593	1	0.991460	0.793016	0.768626
Stentiford	0.990074	1	0.993386	0.790976	0.777945
Hilditch	0.985531	1	0.989753	0.792140	0.772679

Table 3 Analysis of medium size dataset

Thinning algorithm	TR	NOC	NS	EA	Overall result
ZS	0.988679	1	0.996971	0.864930	0.852548
ZW	0.999143	1	0.993673	0.835861	0.829861
LW	0.987579	1	0.996329	0.849339	0.83571
Stentiford	0.993280	1	0.995801	0.858531	0.849181
Hilditch	0.999628	1	0.994584	0.821218	0.816466

4.5 Experiment 4

The set of images considered for testing involves Gujarati "Big Size Characters," set of various sizes, skewed, thick, and also handwritten numerals (Table 4).

4.6 Experiment 5

The set of images considered for testing involves Gujarati characters with 10° skewed (Table 5).

4.7 Experiment 6

The set of images considered for testing involves Gujarati characters with −10° skewed (Table 6).

Table 4 Analysis of big size character dataset

Thinning algorithm	TR	NOC	NS	EA	Overall result
ZS	0.991324	1	0.998878	0.839904	0.831683
ZW	0.999205	1	0.996137	0.830570	0.826704
LW	0.990765	1	0.997724	0.846543	0.836816
Stentiford	0.982108	1	0.993053	0.853579	0.832483
Hilditch	0.995056	1	0.997537	0.841145	0.834925

Table 5 Analysis of 10° skewed dataset

Thinning algorithm	TR	NOC	NS	EA	Overall result
ZS	0.983464	0.999043	0.993537	0.9923	0.968656
ZW	0.998145	0.999043	0.987181	0.9932	0.977713
LW	0.977949	0.999042	0.991937	0.9929	0.962254
Stentiford	0.975354	0.994351	0.989301	0.6532	0.626724
Hilditch	0.986841	0.999043	0.987943	0.9426	0.918101

Table 6 Analysis of −10° skewed dataset

Thinning algorithm	TR	NOC	NS	EA	Overall result
ZS	0.976561	0.999043	0.990013	0.9953	0.961343
ZW	0.996250	0.999043	0.977089	0.9942	0.966853
LW	0.972798	0.999042	0.986873	0.9969	0.956135
Stentiford	0.974628	0.994351	0.977098	0.6332	0.599594
Hilditch	0.996226	0.999043	0.976654	0.9485	0.921977

Fig. 1 Average performance
of all five algorithms on
individual test case

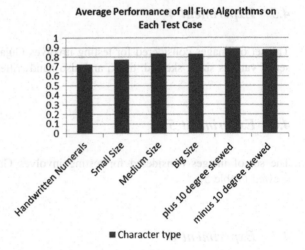

Average Performance of all Five Algorithms on
Each Test Case

■ Character type

Figure 1 shows that till now we discussed and compared individual performance of five thinning algorithms on different datasets which are very from each other in terms of size, handwritten, printed, etc., but after applied all five together on individual dataset, we observed that by calculating the average of the column overall result from each table shows that it gives excellent performance for "10° skewed dataset."

5 Conclusion

From this performance analysis, it is concluded that for whole dataset ZW gives very good result among all the thinning algorithms as our result shows. On skewed dataset, all algorithms give result not good as on normal dataset and give no guarantee having single component after thinning. The dataset contains variety in terms of big size characters "LW" works well as compared to other four algorithms, and as Handwritten Numerals are concerned "ZS" gives results well then other four because the average result measured after applying "ZS" is 0.84879 which is highest.

References

1. Hilditch CJ (1969) Linear skeletons from square cupboards. In: Meltzer B, Michie D (eds) Machine Intelligence, vol 4. Elsevier, New York: Amar, pp 403–420
2. Zhang TY, Suen CY (1984) A fast parallel algorithm for thinning digital patterns. Commun ACM 27(6):236–239
3. Lu HE, Wang PSP (1985) An improved fast parallel algorithm for thinning digital patterns. Proceedings of the IEEE conference on computer vision and pattern recognition, pp 364–367

4. Stentifod FWM, Mortimer RG (1983) Some new heuristics for thinning binary handprinted characters for OCR. IEEE Trans Syst Man Cybern SMC-13(1)

5. Wang PSP, Hui L, Fleming Jr T (1986) Further improved fast parallel thinning algorithm for digital patterns. In: Wang PSP (ed) Computer vision, image processing and communication systems and applications, pp 37–40

6. Holt CM, Stewert A, Client M, Perrot RH (1987) An improved parallel thinning algorithm. CACM 30(2):156–160

7. Chin RT, Wan HK, Stover DL, Iverson RD (1987) A one pass thinning algorithm and its parallel implementation. Comput Vision, Graph Image Process 40(1):30–40

8. Pal S, Bhattacharyaa P (1989) A preserving one pass parallel thinning algorithm. Indian Institute of Management, Calcutta, Working paper series No 123(89)

9. Guo Z, Hall RW (1989) Parallel thinning with two subitration algorithm. CACM, 32(3): 359–373

10. Chen CH, DeCurtins JL (1993) Word recognition in a segmentation-free approach to OCR. Proceedings of the second international conference on document analysis and recognition, pp 573–576. IEEE (1993)

11. Kardos Peter (2011) Sufficient conditions for order-independency in sequential thinning. Acta Cybernetica 20:87–100

12. Khalid S (2010) A universal algorithm for image skeletonization and a review of thinning techniques. Int J Apply Math Comput Sci 20(2):317–335

13. Kalles D, Morries DT (1993) A novel fast and reliable thinning algorithm. Image Vis Comput 11(9):588–603

14. Tarabek P (2008) Performance measurements of thinning algorithms. J Info Control Manag Syst 6(2)

15. Lam L, Lee SW, Suen CY (1992) Thinning methodologies-a comprehensive survey. IEEE Trans Pattern Anal Mach Intell 14(9)

An Exploratory Analysis of Foreign Tourist Visits for Indian State Attractions Using Multinomial Logistic Regression Technique

Hari Bhaskar Sankaranarayanan

Abstract India attracts good amount foreign tourists for its rich heritage and cultural destinations. Several states in India spend efforts to promote and market their tourist destinations. Ministry of Tourism publishes the number of foreign tourist visitors on an aggregated basis per state. The tourist destination in states such as Tamil Nadu, Maharashtra, Uttar Pradesh and Delhi is highly skewed in positively attracting foreign visitors. The paper analyzes the pattern and understands the relevance of such skew on attractions using data from travel review sites such as tripadvisor. The dataset is fused with visitor statistics from government data sources. This paper also explores an approach to predict the degree of the number of visitors per attraction in a particular state using multinomial logistic regression technique. The experiment results are presented and discussed for further studies.

Keywords Tourism · India · Multinomial logistic regression
Machine learning

1 Introduction

Indian destinations are famous for its rich cultural and heritage value among foreign visitors. In the year 2014, tourism in India has grown at the rate of 10.2% year on year with 7.68 million visitors [1]. Tourism generates 20.2 billion US dollar revenues per year and accounts for 6.7% of Gross Domestic Product (GDP). Ministry of Tourism promotes and markets the destinations to attract foreign and domestic visitors alike. Travel ecosystem in India is booming with the advent of new airports, air traffic growth, infrastructure modernization, tour operators offering packages, online travel agents (OTAs) providing convenience by offering the end-to-end touring experience end. India has 32 heritage sites listed by UNESCO [2]. The data show that the top 10 states contribute to high visitor rates such as 88% of foreign

H.B. Sankaranarayanan (✉)
Amadeus Software Labs, Bangalore, India
e-mail: hari.sankaranarayanan@amadeus.com

© Springer Nature Singapore Pte Ltd. 2018
D.K. Mishra et al. (eds.), *Information and Communication Technology
for Sustainable Development*, Lecture Notes in Networks and Systems 9,
https://doi.org/10.1007/978-981-10-3932-4_28

tourists. India is still placed at a 52nd position out of 141 in terms of tourism competitiveness worldwide [3]. Online reviews act as a key planning tool for determining the attractions to visit by foreign tourists. Tripadvisor is widely used by tourists across the world to gather word of mouth recommendations and influencing trip destinations for hotel and activity bookings. Recent study highlights that tripadvisor reviews influenced 352 million trips which account for 472 billion US dollars to travel spending for worldwide trips [4]. In this paper, we will analyze the dataset available from the Ministry of Tourism along with the data from tripadvisor on Indian attractions and fuse them to study the visit patterns, and predict the degree foreign tourists on a particular attraction using multinomial logistic (MNL) regression techniques. The paper is organized into following sections. Section 2 discusses the related work in this area, Sect. 3 discusses the dataset, Sect. 4 discusses the analysis approach, Sect. 5 presents the results of experiments and discussion around them, Sect. 6 highlights the limitations and future work and Sect. 7 provides the conclusion.

2 Related Work

Measuring tourist attraction visitations based on probability approach is discussed in a research study conducted by National Park Service for Arizona state attractions [4]. Multinomial logistic regression models are widely used for traveler choice and demand prediction such as micro-econometric models for destination choice [5]. A study on using destination image characteristics to predict visitor intention is conducted for New York State Hudson river-based communities; it reveals the choice of visitors on characteristics such as cultural offerings, local character and infrastructure [6]. The tourism demand forecasting and modeling approaches are comprehensively discussed in a survey paper, and it highlights that there is no one particular method that can be relied on for predicting tourism demand [7, 8]. Thus, the motivation of this research is to understand the uniqueness of Indian state attractions and how we can develop a model for predicting the number of visitors to an attraction.

3 Information About Datasets

The dataset collected is from the year 2014 unless specified. The dataset is gathered for 10 states and 293 attractions listed in tripadvisor. Table 1 highlights the sample set foreign tourist visits to respective states [1].

The next dataset contains information about tripadvisor information about every state attraction. Table 2 highlights an indicative dataset for every state.

Table 3 depicts the states with good and bad access to tourist attractions. The criteria used are the proximity to international airports such as Chennai, Delhi,

Table 1 Top 10 states of India by foreign tourist visitors

State	Number of foreign visitors (2014)
Tamil Nadu	46,57,630
Maharashtra	43,89,098
Uttar Pradesh	29,09,735
Delhi	23,19,046
Rajasthan	15,25,574
West Bengal	13,75,740
Kerala	9,23,366
Bihar	8,29,508
Karnataka	5,61,870
Haryana	5,47,367

Table 2 Tripadvisor data for state attractions (sample dataset)

State	Attraction name	Attraction type	Award of excellence	Number of reviews	Number of tours
Tamil Nadu	Monuments of Mahabalipuram	Ancient ruins	Certificate of excellence	784	3
Maharashtra	Gateway of India	Speciality museums	Certificate of excellence	5,098	16
Uttar Pradesh	Taj Mahal	Architectural buildings	2015 traveler choice winner	15,986	94
Delhi	Qutub Minar	Architectural buildings	2015 traveler choice winner	5,986	25
Rajasthan	Hawa Mahal— palace of wind	Architectural buildings	Certificate of excellence	3,938	23

Bangalore, Kolkata and Mumbai, tour operator options and connectivity through major highways.

The next dataset is the number of visitors to a particular attraction. The list is not comprehensive considering the diversity of attractions. The data source available here is the monuments which are maintained by Archaeological Survey of India (ASI). Table 4 provides the visitor details for the monuments shortlisted from tripadvisor data and matched to ASI dataset. This dataset will also form as a training set for predicting the visitors for other monuments.

The datasets are applied with discrete values for the number of visitors (Table 5), the number of reviews (Table 6), state visitor share (Table 7) and heritage value of attractions (Table 8).

Table 3 Access characteristics of attractions (sample dataset)

State	Attraction locations	Airport	Road/tours connectivity	Access rating
Tamil Nadu	Trichy, Madurai, Kanyakumari, Thanjavur	Chennai international airport	Good highways	Good
Maharashtra	Mumbai, Aurangabad	Mumbai international airport	Good highways	Good
Uttar Pradesh	Varanasi, Agra, Lucknow	Delhi international airport	Presence of large tour operators	Good
Delhi	Delhi	Delhi international airport	Metro rail, tour operators	Good
Bihar	Saranath, Nalanda	Delhi international airport	No tour operators from tripadvisor list	Bad

Table 4 Visitor count for ASI maintained monuments (sample dataset)

Attraction name	Number of foreign visitors
Taj Mahal, Agra	6,48,511
Agra fort, Agra	3,43,483
Qutub Minar, Delhi	2,76,043
Humayun's tomb, Delhi	2,56,421
Fatehpur Sikri, Agra	2,31,099
Itimad—ud Daulah, Agra	66,186

Table 5 Classification for number of foreign visitors

Classification	Number of visitors
High	>2,00,000
Medium	75,000–2,00,000
Low	Up to 75,000

Table 6 Classification for number of tripadvisor reviews

Classification	Number of reviews
High	>1000
Medium	500–1000
Low	<500

Table 7 Classification for state's foreign visitor share

Classification	State's foreign visitor share (%)
High	>10
Medium	5–10
Low	Less than 5

Table 8 Classification for heritage value based on type of attraction

Heritage value	Type of attraction
Yes	Historic sites, ancient ruins, architecture buildings, speciality museums, selected sacred and religious sites

4 Analysis Approach

A parsimonious model can be built at by combining the above features and analyzing them using MNL regression technique. Figure 1 highlights the phases of analysis.

4.1 Phase 1: Identification of Features

The objective of this phase is to list the dependent features that can influence the overall visitor number to an attraction. For the purpose of this experiment, tripadvisor data, attraction access rating, state visit share, heritage value of an attraction based on the type of attraction are identified as dependent features.

4.2 Phase 2: Data Fusion

From the above dataset listed in tables, the data are fused into a comma separated values for further processing. The keys are matched based on attraction name and states.

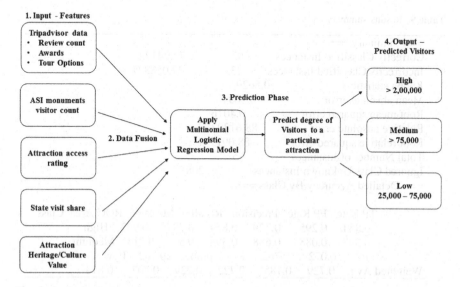

Fig. 1 Analysis phases

4.3   Phase 3: Prediction

Multinomial logistic regression is chosen for this analysis since it fits into the prediction of a feature based on dependent feature which is statistically related to each other. Also, this technique provides classification on logit models that can provide a degree of classification of visitors. The prediction phase is done in Weka tool for machine learning.

4.4   Phase 4: Output

The output is generated as a form of "High," "Medium," "Low" prediction classes which denotes the number of visitors to a particular attraction. The results of the prediction with the dataset listed in Sect. 3 will be discussed in the next section.

5   Results and Discussion

The results of the prediction with 72.91% accuracy are presented in Table 9.

Table 10 presents the coefficient values.

The positive coefficient on tour operators, airport access, medium and low reviews and heritage value leads to "High" classification of an attraction. The high

Table 9 Results summary

=== Summary ===		
Correctly Classified Instances	62	72.9412 %
Incorrectly Classified Instances	23	27.0588 %
Kappa statistic	0.5629	
Mean absolute error	0.3691	
Root mean squared error	0.4016	
Relative absolute error	88.6129 %	
Root relative squared error	88.0917 %	
Total Number of Instances	85	
Ignored Class Unknown Instances	208	
=== Detailed Accuracy By Class ===		

	TP Rate	FP Rate	Precision	Recall	F-Measure	ROC Area	Class
	0.854	0.295	0.729	0.854	0.787	0.902	High
	0.393	0.088	0.688	0.393	0.5	0.715	Medium
	1	0.072	0.762	1	0.865	0.762	Low
Weighted Avg.	0.729	0.185	0.722	0.729	0.707	0.814	

Table 10 Regression coefficient for prediction classes

Class: High		Class: Medium		Class: Low	
Variable	Coefficient	Variable	Coefficient	Variable	Coefficient
Tours	0.3229	Tours	-0.3448	Tours	-44.3906
Award	-0.1213	Award	0.4968	Award	-0.1882
Reviews = Low	0.8627	Reviews = Low	-0.3329	Reviews = Low	-42.1538
Reviews = High	-1.1831	Reviews = High	0.318	Reviews = High	45.7866
Reviews = Medium	0.6459	Reviews = Medium	-0.0126	Reviews = Medium	-11.0244
State = Bihar	-0.6021	State = Bihar	-0.3176	State = Bihar	5.4422
State = Delhi	-1.0544	State = Delhi	0.8387	State = Delhi	-2.7471
State = Haryana	15.6277	State = Haryana	-0.9205	State = Haryana	11.2744
State = Karnataka	-0.462	State = Karnataka	0.0624	State = Karnataka	-45.6926
State = Kerala	-2.3479	State = Kerala	-0.0147	State = Kerala	-19.2861
State = Maharashtra	0.6994	State = Maharashtra	-0.3887	State = Maharashtra	48.2954
State = Rajasthan	-1.1188	State = Rajasthan	0.1208	State = Rajasthan	-30.9492
State = Tamil Nadu	-1.108	State = Tamil Nadu	-0.2418	State = Tamil Nadu	17.766
State = Uttar Pradesh	-0.0735	State = Uttar Pradesh	-0.1545	State = Uttar Pradesh	-14.3084
State = West Bengal	-2.363	State = West Bengal	0.9	State = West Bengal	-16.2322
Airport access	0.6021	Airport access	0.3176	Airport access	-5.4422
Heritage value	1.976	Heritage value	-0.1791	Heritage value	-72.3714
Visit share = Low	0.9987	Visit share = Low	0.0373	Visit share = Low	-23.5857
Visit share = Medium	-0.6329	Visit share = Medium	0.3745	Visit share = Medium	-10.199
Visit share = High	-0.5708	Visit share = High	-0.3566	Visit share = High	34.6571

amount of reviews, awards, airport access and medium visit shares leads the attraction toward "Medium" classification. Low classification is a resultant of low heritage value and airport access. From the results, it is evident that heritage value, tour availability and airport access play a vital role along with review and visit share values for predicting the visitor footfall for an attraction.

6   Limitations and Future Work

The current research work has the following limitations such as there may be the possibility of multiple other features such as traveler profile, preferences, expenditure, safety, availability of facilities such as hotels that can be used to train the model. The model of prediction is simplistic as it predicts only the degree of classification of the number of visits. The dataset is available only for ticketed monuments and the visitor count for the common point of interests (e.g., Gateway of India) are not available since they are not ticketed. The assumption is that foreign tourists will place high weights for heritage and cultural attraction compared to rest of attraction categories. The current model is parsimonious in nature; hence, the accuracy can be improved with more features and improved models can be built with empirical and combinatorial approaches. This research work can be extended in multiple ways such as adding more features such as cleanliness index, promotion and safety aspects to predict the visitor count in much precise manner. The model can be applied for developing overall tourism growth and prediction model for various attractions. The same model can be extended to predict expenditure of tourist in an attraction and further insights can be developed to add tours, events, transport and other facilities for improving affinity toward an attraction. This research can be extended to domestic visitors by including features such as demographics, tourism purposes such as leisure, pilgrimage, visiting friends and relatives. We can further enrich the dataset with geolocation information from mobile devices and social media information.

7   Conclusion

The initial experiments show promising results in terms of estimating the individual attraction visitor count based on parsimonious model and MNL regression algorithm. This research is a stepping stone toward building models around travel propensity model for foreign tourists visiting India. The research can be used to predict and act on insights to promote tourist destinations with high heritage value. It is an opportunity for tour operators to develop innovative packages and make the end-to-end travel process as an enjoyable one for foreign tourists. Government and travel agencies can come up with novel mechanisms such as mobile apps to track visitors and promote the heritage sites which are getting low footfall of tourists using this prediction data.

References

1. Government of India, Ministry of Tourism. http://www.tourism.nic.in/
2. UNESCO World Heritage List. http://whc.unesco.org/en/list
3. World Economic Forum, Travel and Tourism Competitiveness Report 2015. http://reports. weforum.org/travel-and-tourism-competitiveness-report-2015/economy-rankings/
4. Tripadvisor, Key Findings: The Global Economic Contribution of TripAdvisor study. https:// www.tripadvisor.com/TripAdvisorInsights/n2687/key-findings-global-economic-contribution-tripadvisor-study
5. National Park Services, Measuring Probabilities in Attraction Visitors. https://www.nps.gov/ tourism/ResearchTrendsandDatainfo/measuringprobabilitiesinattractionvisitors.pdf
6. Eymann A, Ronning G (1997) Microeconometric models of tourists' destination choice. Regional science and urban economics
7. Schuster RM, et al (2009) Using destination image to predict visitors' intention to revisit three Hudson river valley New York communities
8. Song H, Li G (2008) Tourism demand modelling and forecasting—a review of recent research. Tourism management

Performance Exploration of Different Dispersion Compensation Schemes with Binary and Duo Binary Modulation Formats Over Fiber-Optic Communication

Hiroshama Nain, Urvashi Jadon and Vivekanand Mishra

Abstract Fiber-optic mediation offers high speed for data mediation and also offered huge bandwidth for use. Due to its adaptable merits and insignificant mediation losses, it becomes one of the most vital used communication techniques. Although Fiber-optics mediation has a plenty of merits, 'Dispersion' is the chief restrictive feature. Dispersion is the core issue responsible for impediment progress of optical mediation systems in terms of bit rate and long-haul mediation. Dispersion originate distortion and expansion in the pulses, thus recital of the fiber-optic communication. This inquiry paper gratified that inspection with 20 Gbps mediation network arrangement has been prepared with binary non-return to zero and duo binary modulation designs in symmetric dispersion compensation schemes. The coordination is imitated and evaluated by means of pre-, post-, and symmetric dispersion compensation outlines. All organizations are linked in expressions of bit error rate (BER), Q-factor, eye opening, and jitter. The outcome is investigated using optical software tool, i.e., OPTSIM.

Keywords Dispersion · Dispersion compensation · Dispersion compensating fiber (DCF) · Quality factor · Bit error rate · Jitter · Eye diagram
Eye opening · Duo binary · Erbium-doped fiber amplifier (EDFA)
Photo detector (PD) and optical analysis tool (OPTSIM)

H. Nain (✉) · U. Jadon
Department of Electronics and Communication Engineering,
ITM University, Gwalior, Madhya Pradesh, India
e-mail: er.hiroshama.nain@gmail.com

U. Jadon
e-mail: urvashijadon27@gmail.com

V. Mishra
Department of Electronics and Communication Engineering,
SVNIT, Surat, Gujarat, India
e-mail: vive2009@gmail.com

© Springer Nature Singapore Pte Ltd. 2018   281
D.K. Mishra et al. (eds.), *Information and Communication Technology
for Sustainable Development*, Lecture Notes in Networks and Systems 9,
https://doi.org/10.1007/978-981-10-3932-4_29

1 Introduction

Optical fiber mediation is a procedure of carrying information from transmitting and receiving side of the network by driving a light pulse through a fiber. In Fiber-optics mediation, refractive index is mediate in the $|E|^2$, here E expresses electric field, i.e., the power density inward the fiber [1]. Nonlinearity and dispersion in fiber optics have developed an expanse of theoretical exploration and of unlimited status in the optical fiber-based mediation network. Single-mode fiber is used in many investigations for huge data rate mediation with minimum loss, but dispersion is an imperative loss that deteriorates complete scheme presentation [2]. Dispersion of the conducted optical light pulses origins alteration for both digital and analog diffusions along fiber optic. Dispersion slanted the outline of the pulse as a light wave broadcasting through an optical fiber. This kind of effect is well known as inter symbol interference (ISI). Thus, a growing integer of errors may be met on the digital fiber-optic network as the ISI becomes extra prominent. Dispersion is classified on the basis of characteristic, which are linear characteristics and non-linear characteristics. On the basis of linear characteristic, dispersion is divided into three forms which are modal dispersion, waveguide dispersion, and material dispersion, whereas on the bases of nonlinear characteristics, dispersion is of two types: polarization mode dispersion and chromatic dispersion. Modal dispersion arises in optical fiber which devouring manifold modes mediation. It ascends since modes survey dissimilar paths when mediated from the fiber and accordingly attain at the next end of the optical fiber at diverse time. Chromatic dispersion is instigated through delay variance between the group velocities of the dissimilar wavelength comprising the basis range. Polarization mode dispersion is correlated to the differential group delay, time alteration in the group interruption among two orthogonal polarized modes, which origins signal diffusion in digital networks, and alterations in analog network [3]. Duo binary modulation scheme is an active technique in providing eminent speed for fiber-optic mediation networks by rise dispersion tolerance property of optical fiber, to advance spectral proficiency and to diminish the compassion to nonlinearity. Here, the idiom 'duo' refers to the repetition the ability of bit of a traditional binary system. The main merit of this modulation format is to maximize the tolerance property of the optical network, for the effects generated by the chromatic dispersion. To overcome the possessions by ISI, this is proficient by accumulation of a binary data series to a single bit postponed form of itself, which can be found by passing the binary data series over the delay-and-add filter [4]. Duo binary encoder consists of EXOR and ADD as shown in Fig. 1.

Fig. 1 Duo binary encoder

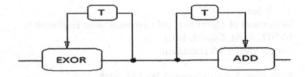

2 Dispersion Compensating Fiber

Dispersion compensation is a most commonly used scheme for growing the mediation expanse in optical network functioning at huge data rates. There are numerous procedures for dispersion compensation, but from all of them, dispersion compensating fiber is highly acclaimed for long-haul optical mediation networks [5]. Dispersion compensating fiber is an informal and proficient approach for recompensing dispersion. Optical fibers residing dispersion recompense possessions has negative dispersal fluctuating from −65 to −95 ps/nm/km and with the assistance of recompense the positive dispersal of mediation in single-mode fibers. Due to the nonlinear characteristics of proliferation, the presentation of the network is contingent on the level of power at the contribution of multiple sorts of fiber optics, on the place of the DCF and on the quantity of dispersal. Slighter dimension of dispersion compensating fiber is superior for operative optical mediation network [6]. The cataloging of altered dispersion compensating schemes is shown in Fig. 2.

This scheme basically shows whether dispersion compensating fiber is positioned before and after to the single-mode optical fiber. For an unmitigated mediation network, a dispersion compensating fiber should have minimum insertion losses, polarization mode dispersion, and nonlinear effects [7]. By locating a single-mode optical fiber which contains positive dispersion property just before dispersion compensating fiber with negative dispersion property, the net amount of dispersion in the optical mediation network becomes zero.

$$D_{SMF} * L_{SMF} = -D_{DCF} * L_{DCF} \tag{1}$$

Here D refers to dispersion limit, L refers to the dimension (length) of the optical fiber, SMF refers to the single-mode optical fiber, and DCF refers to the dispersion compensating fiber. As it shown in Fig. 2 that there are three forms used for compensating a dispersion in an optical fiber network by using dispersion compensating fiber, the three required arrangements are—pre-compensation, post-compensation, and symmetric compensation. In pre-compensation technique, the dispersion compensating fiber is spotted just before the single-mode optical

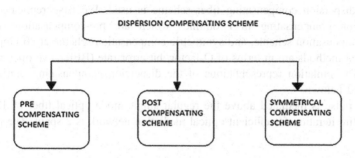

Fig. 2 Classification of dispersion compensating schemes

fiber for recompensing the effects occured by dispersion over a fiber-optic network. Similarly, in post-compensation technique, the dispersion compensating fiber is sited after the single-mode optical fiber for indemnity of dispersion on to a fiber-optic network. But in case of symmetric compensation procedure, dispersion compensating fiber is located before and after the single-mode optical fiber for minimizing the dispersion over a fiber-optic network [8].

3 Experimental Configuration

The consequence is inspected using optical software tool, i.e., RSOFT's OPTSIM. To accomplish huge capability, the average single-mode optical fiber should be elevated to recompense the dispersion. The simulation illustration of the system is split into three mediation sections: one is transmitting section, another one is the optical channel, and the last one is receiving section. In transmitting section for generating a signal, we use a PN sequence generator which basically generates a binary logical sequence and here the date rate and bit rate are made to order. A binary and duo binary non-return to zero modulation layout is used and for producing electrical pulse, non-return zero electrical driver is used which converts the logical sequence into electrical pulse. For duo binary modulation format, electrical pulse is then led to a band-pass Bessel electrical filter for riddling. In the same segment, a CW laser is used with multiple values of input power level. An amplitude dual arm Mach zander modulator and sin square Mach zander modulator are used for modulating a signal for different modulation formats and converting it into optical signal for optical channel. An optical channel section consists of four modules: optical fiber, amplifier with fixed power, optical combiner, and optical splitter. In the receiving section, a preamplifier is used to amplify a signal before detection through a sensitivity receiver. A band-pass raised cosine optical filter is used for filtering process before sensitivity receiver, and its main overlooks on central frequency. At receiving section, sensitivity receiver having PIN configuration, which is used to convert the optical signal into electrical one for the electrical scope which provides the eye analysis according to the input signal. Organization is functioned at data rate of 20 Gbps. In this technique, for diminishing dispersion effect, dispersion compensating fiber scheme is used. We have replicated three dispersion compensating fiber outlines which are pre-compensation scheme, post-compensation scheme, and symmetric compensation scheme at 20 Gbps. The overhead methods are in terms of Q-factor, bit error rate (BER), eye opening, and jitter. The imitation representations of the dispersion compensating outlines are specified below:

The pulse is mediated above the regular single-mode optical fiber of 120 km. The entire iteration for efficient optical mediation network is 2 spam. The module

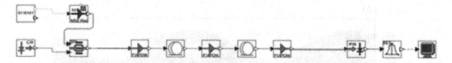

Fig. 3 Simulation illustration of pre-compensating scheme

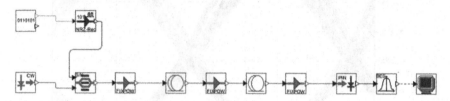

Fig. 4 Simulation illustration of post-compensating scheme

(a)

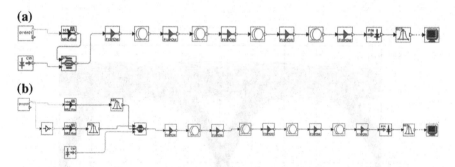

(b)

Fig. 5 Simulation illustration of **a** Symmetric compensation scheme using binary NRZ modulation format and **b** Symmetric compensation scheme using duo binary modulation format

has been imitated for three forms of dispersion compensation outlines which are pre-compensation, post-compensation, and symmetrical compensation. In pre-compensation methodology, as given in Fig. 3 to recompense for the dispersion, dispersion compensating optical fiber of 25 km is aligned afore the single-mode fiber optic of 120 km in dimension to minimize the losses occur by dispersal. In post-recompense methodology, as presented in Fig. 4 to recompense for the dispersion, dispersion compensating fiber of 25 km is used after single-mode fiber optic of 120 km in dimension for reducing dispersal. In symmetrical compensation outline, as formed in Fig. 5, two dispersion compensating fibers each of 25 km are located before and after the two single-mode optical fibers of 120 km in dimension each.

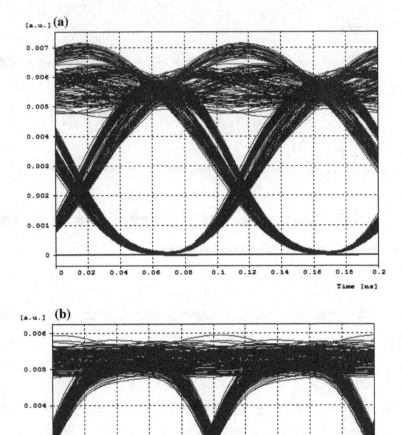

Fig. 6 **a** Eye illustration of the pre-compensation outline on 20 Gbps, **b** Eye illustration of the post-compensation outline on 20 Gbps

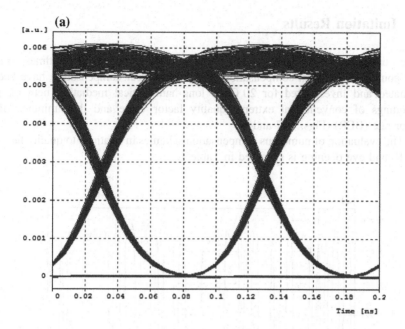

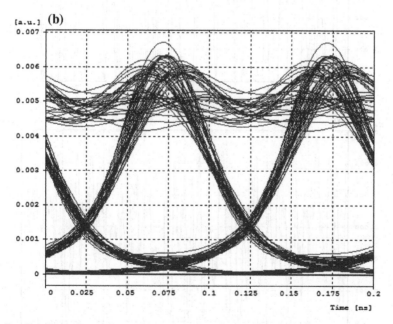

Fig. 7 **a** Eye illustration of the symmetrical compensation with binary NRZ modulation format on 20 Gbps with 120 and 24 km dimension of the optical fiber and **b** Eye diagram of the symmetrical compensation scheme with duo binary modulation format on 20 Gbps with 120 and 24 km dimension of the optical fiber

4 Imitation Results

The three numerous categories of dispersion recompense outlines, i.e., pre-compensation, post-compensation, and symmetrical compensation have been imitated and investigated for 20 Gbps long-haul optical mediation network in standings of conventional extreme quality factor value and then minutest Bit error rate (BER) (Figs. 6, 7 and 8).

The evaluation of numerous compensation schemes in relations to quality factor, BER, and eye opening is specified in Table 1.

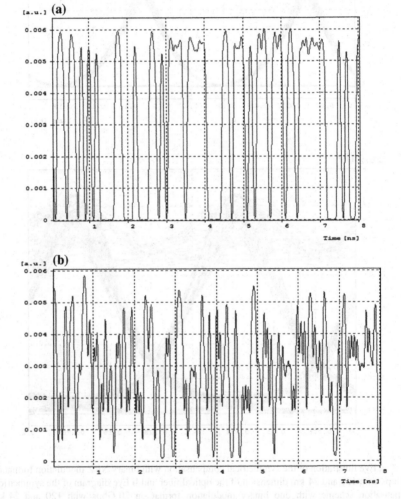

Fig. 8 a Electrical spectrum of the binary modulation layout for symmetrical compensation scheme and **b** Electrical spectrum of duo binary modulation layout for symmetrical compensation scheme

Table 1 Comparison of disparate sort of dispersion compensating structures

Compensating strategy	Q-factor (dB)	Bit error rate (BER)	Eye opening (a.µ.)
Pre-compensation	26.020428	$1e^{-040}$	0.00474951
Post-compensation	27.168122	$1e^{-040}$	0.00467761
Symmetric compensation	31.082041	$1e^{-040}$	0.00501172

Table 2 Differentiation among unprecedented modulation format used in symmetrical compensating technique

Modulation FORMAT	Eye opening (a.µ.)	Bit error rate (BER)	Jitter (ns)
Symmetric compensation using binary	0.00501172	$1e^{-040}$	0.0250043
Symmetric compensation using duo binary	$1.61522e^{-005}$	0.0227501	0.0238412

5 Conclusion

By mean of excessive data rate, i.e., 20 Gbps used the dispersion prompted expansion of the diminutive pulses promulgating in the optical fiber origins crosstalk among the alongside time slits, prominent inaccuracies when the mediation expanse rises outside the dispersion extent of the optical fiber to requite dispersion, dispersion compensating fiber were worn (Table 2). Three diverse compensation outlines which are pre-, post-, and symmetric compensation are imitated. Furthermore, an assessment among two modulation formats which are binary and duo binary is prepared above a schematic of symmetrical compensation scheme. These outlines are equated in expressions of Q-factor, BER, eye opening, and jitter. In subsequent imitation, we have determined that symmetric compensation is the preeminent dispersion compensation arrangement than pre- and post-compensation, as well as we invented symmetric compensation scheme operative exertion in duo binary modulation format.

Acknowledgements Author would like to express their gratitude to RSOFT enterprise assemblage for making available the OPTSIM imitation software for optical mediation network. Author would also like to show appreciation to the assessors for their appreciated recommendations which fetch the script in extant profile.

References

1. Keiser G (2014) Optical fiber communication system, 4th edn, McGraw-Hill
2. Govind PA (2014) Fiber-optic communication systems, 3rd edn, Wiley
3. Kumar S (2013) Performance analysis of dispersion compensation in long haul optical fiber with DCF, 6(6):19–23, e-ISSN: 2278-2834, p-ISSN: 2278-8735
4. Patel GH (2013) Dispersion compensation in 40 Gbps WDM network using dispersion compensating fiber, 02(02), ISSN: 0975-6779

5. Khanaa V, Mohanta K, Saravanan T (2013) Performance analysis of a two channel optical WDM system using binary and duo binary modulation formats, vol 6, p-ISSN: 0974-6846
6. Agrawal R, Mishra V (2012) Comparison of pre, post and symmetrical dispersion compensation scheme with 10 Gbps NRZ link for SCM system. In: International conference on electrical engineering and computer science (ICEECS)
7. Shrivastava A, Saxena M (2014) Analysis of optical communication system for compensation of dispersion by comparing using fiber Bragg grating
8. Ravi Teja N, Aneesh Babu M, Prasad TRS, Ravi T (2012) Different types of dispersion in an optical fiber. Int J Sci Res 2(12) ISSN: 2250-3153

Cognitive Radio Networks: State of Research Domain in Next-Generation Wireless Networks—An Analytical Analysis

M. Anusha and Srikanth Vemuru

Abstract Cognitive radios (CRs) are the nearly novel innovation in which issues such as under-utilization of spectrum and shortage of spectrum is comprehended in light of the progressive thoughts. Intellectual radio permits gathering of clients to recognize and entrance for accessible spectrum assets for utilizing ideally. Late learning demonstrates that the vast majority of the selected spectrum was under-utilized. Then again, the expanding number of remote sight and sound applications prompts a spectrum shortage. CRs are proposed as a capable innovation for tackling the awkwardness among spectrum shortage and spectrum under-utilization. In CRs, spectrum detecting was completed by keeping that in mind at the end of the goal to find the idle spectrum sections. This manuscript demonstrates that quality & capacities of CRs procedures and makes that all the extra intense more than the other focused radio. Anxiety is specified on appliance zones, where CR procedures are executed and demonstrated away to have high ground than the accessible intellect and adjusting radios.

Keywords Cognitive radio · Wireless networks · SDR · MAC Intelligence · Spectrum

1 Introduction

The point of the CR innovation is to give the most extreme proficiency for the spectrum by enhancing its use by utilizing dynamic spectrum admission strategies. The way for empowering the great spectrum proficiency for giving the ability of impart the remote diverts to authorized clients in the majority proficient way and that objective will be accomplished by utilizing the spectrum administration systems are progressive and productive. CRs are remote gadgets that can be programmable by sensing their surroundings and that can adjust their strategies to

M. Anusha (✉) · S. Vemuru
Department of Computer Science and Engineering, K L University,
Vaddeswaram, Andhra Pradesh, India

© Springer Nature Singapore Pte Ltd. 2018
D.K. Mishra et al. (eds.), *Information and Communication Technology for Sustainable Development*, Lecture Notes in Networks and Systems 9, https://doi.org/10.1007/978-981-10-3932-4_30

access the channels powerfully; transmission waveform, utilization of spectrum, and systems administration conventions are required for excellent system and for execution of applications [2]. The fascinating component of CR innovation is one of the handset devices that could naturally under-utilize the spectrum. Quickness of a radio is the point at which it uses the accessible administration from locally available remote PC organizes, and collaborates with systems of favored conventions, with no perplexity in discovering the fitting remote system for sound video or information [3]. Moreover, frequencies determination and use minimize/keep away from impedance with existing radio frameworks.

2   Cognitive Radios (CRs)

To get over this confinement, another idea known as CR has been enhanced the premise of software-defined radio. CR included with the cerebrum of knowledge for its own has conferred by another ability of making decision. CR gives an interesting answer for the methodology for spectrum under-utilization issue. CR could sense the encompassing surroundings and relying upon the data along with the requirements the circumstance and adjusts its physical layer by reconfiguring. It could modify the design, so that it could deal with difficult circumstances. Besides, it could adjust to the new circumstance [4].

$$CR's = SDR + Intellect + Reconfigurablility$$

CR concocted another method for taking care of spectrum under-utilization issue. It could sense the encompassing atmosphere by attempting to discover the electromagnetic spectrum that was not used to its ideal limit. Subsequent to decision such scope of frequencies, CR needs for utilizing appropriately for permitting the unauthorized users for utilizing vacant spaces. The principles of CR from its forerunners are as follows: It was characterized through programming along with completely reconfigurable; every rationale in the CR could be actualized by programming; and contingent upon the data, CRs could sense and modify its conduct and which should be possible during physical adjustments along with that additionally through programming directions.

Present numerous meanings and explanations of CR have until now been created by both in the scholastic establishments and different gauge organizations. In the Layman expression, CR might be characterized as [3] "a psychological radio was a remote correspondence framework that could insightfully use some accessible data about the actions of active channel circumstance, codebooks, and communication for different hubs that can share the spectrum." Fundamentally, psychological radio can be characterized as "a radio is cognitive," "Meditate, therefore est." [5]. Numerous specialists plus open authorities concur that updating the product radio's control procedures can increase the charge of programming radios; however, difference over the level of "comprehension" needs the brought about contradiction above the exactness of the meaning of CR. A portion of unmistakably definition about the CR is talked about in the subtle element. In 1999, Joseph Mitola III instituted the expression "psychological radio," by characterizing the CR as [6] "a radio which utilizes the model is based on thinking by accomplishing a predetermined level of fitness in spectrum-associated spaces."

Soon after Simon Haykin characterized CR as [7] "a smart remote correspondence framework knows about its encompassing surroundings, utilizes the approach of comprehension by working to achieve from nature, and adjust its interior states to factual varieties in the making so as to approach RF boosts-related changes in certain working restrictions such as transmit force, regulation methodology, and bearer recurrence continuously, considering two necessary objectives:

a. Extremely reliable correspondences at which, whatever, and wherever required.
b. Expert usage of the spectrum radio.

FCC concentrates on the function of transmitter characterized CR as [8] "a radio could alter its transmission limits in view of connection with nature on which it can work." NTIA, USA [9], licensed spectrum administrative, can characterize CR as follows: "A radio or framework can detect the functionality of electromagnetic surroundings and can independently and powerfully modify its radio working limits for adjusting framework operation, for example, alleviate obstruction, access optional business sector, encourage interoperability, and augment throughput." The worldwide spectrum administrative group, ITUWp8A, characterizes psychological radios can spotlight on the capacity as follows [2]: "A radio or else framework can faculties the known about the operational surroundings and could progressively by self-governing change its radios on working limits appropriately." USA IEEE characterizes the CRs as [10] "a radio recurrence transceiver can be intended to keenly by recognizing that a specific section of the spectrum radio is right now being used, and to bounce into and out of, as vital the incidentally unused spectrum in a flash, with no impedance of transmissions with other approved clients."

Also, SDR environment [5] built up two gatherings on CR and characterized as: "A radio which have the attention to modify its surroundings [2] and in light of those progressions adjusts its working attributes somehow to enhance its execution or to minimize a misfortune in execution." In any case, the SDR forms special attention group [5] by creating intellectual radios for utilizing applications to accompanying the definition.

"A versatile, multi-dimensionally mindful, self-ruling radio's framework gains that from its encounters for spectrum, reasons choose the future activities to address the issues of clients." Virginia Tech CR Working Group characterizes that CRs as adaptive radios [3]:

"CRs are those which have the following abilities:

a. knowledge about its surroundings and their abilities,
b. object-driven self-ruling process,
c. Learn and understand the activities that affect their objectives, and
d. Recollect and connect the precedent activities, execution along with surroundings."

Table 1 Matrix of the CR

Definer	Adjusts intelligently	Self-directed	Sensing surroundings	Transmitter	Receiver	Awareness on surroundings	Determining objective	Studying on surroundings	Conscious on ability	Discuss waveform	No interference
FCC	◊	◊	◊	◊							
Haykin	◊	◊	◊		◊	◊	◊	◊			◊
USA IEEE	◊	◊	◊	◊	◊	◊					
ITU-R	◊	◊	◊	◊	◊	◊					
Mitola III	◊	◊	◊	◊	◊	◊	◊	◊	◊	◊	
NTIA US	◊	◊	◊	◊	◊	◊	◊				
SDRF CRWG	◊	◊	◊	◊	◊	◊	◊				
SDRF SIG	◊	◊		◊	◊	◊	◊		◊		
VT CRWG	◊	◊	◊	◊		◊	◊	◊	◊		

The functionality of all the above-mentioned descriptions is condensed in Table 1. Several broad abilities between the greater parts of the descriptions are as follows:

Perception—Specifically otherwise in a roundabout way, framework gains data about its working surroundings.

Flexibility—Fit for changing its waveform.

Insight—Equipped for applying data to accomplish target objective.

Joining these normal elements, we could characterize CR as "completely programmable, remote gadgets that could sense their surroundings by progressively adjusting their channel access strategy, communication waveform, organizing conventions, and spectrum utilization required for best system and execution of application."

3 Strengths of CR

To streamline asset use, cutting edge systems require brilliant gadgets like CR to have the capacity to show their area, systems, clients, and bigger environment. Figure 1 demonstrates the different CR abilities, which make it additional normal from different radios. In light of checked spectrum of those limits, CR could adjust for the proper recurrence groups, edges, and conventions [6].

Mitola [7] spoke to the real capacities to adjust the communication limits in altering situations during a cognitive sequence.

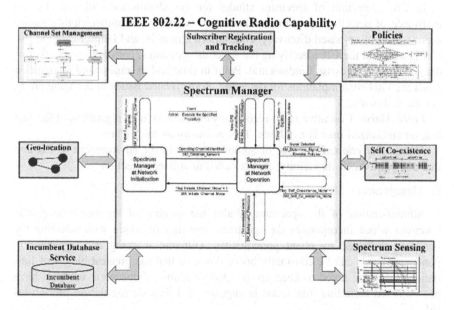

Fig. 1 Capabilities of CR [3]

The six-stage discernment sequence appeared in Fig. 1 as informed below:

Watch   Get information on the working surroundings on the assistance for detecting along with flagging instruments. Calculate the watched data along with the importance. Depending on the calculated assessment, CR decides its choices for enhanced asset improvement.

Choose   Takes the choice, by considering the perspective of the option that assesses additional positively with the other different alternatives.

Act   CR actualizes on the choices that are considered for enhancement of the assets. Achieved modifications are reflected on the obstruction report exhibited on the CR in the exterior world as speak.

Learn   In the ongoing procedures, CR uses its perceptions and choices for enhancing its future operations, making novel displaying states and choices by joining the component of knowledge.

A CR can dependably sense wide transfer speed and identify the accessible vacant spectrum for utilizing unfilled openings for the correspondence except those are essential by the authorized users. The air interface for CR depends on 4 primary methods [8]:

a. Sensing,
b. Management,
c. Sharing, and
d. Mobility.

a. *Sensing*

In CRs, detection of spectrum alludes for the identification of gaps by the assistance of detecting spectrum methods, for example, transmitter/vitality recognition, obstruction-based discovery, coordinated channels, and helpful location [9]. Not just should the CR identifying the spectrum openings, constant observation of the spectrum is likewise fundamental. Point in time and exactness and recognition reach are vital contemplations for detecting. Some related issues in detecting are as per the following:

False Alarm: Cognitive radio recognizes the primary user regardless of the fact that no authorized user is available. This is known as false alarm.

Missed Detection: Existence of authorized user in the environment of CR and that does not watch its vicinity; this is called as missed detection.

b. Management

Administration of the spectrum alludes the gaining of the finest accessible spectrum which incorporates the spectrum investigation along with selecting the bands as indicated by client prerequisites. Different working and transmission parameter needs can be constantly broke down so that the greatest blends of limitations might be tuned to keep up the QoS. Various enhancement methods have been utilized including man-made brainpower and delicate registering procedures [6].

c. Sharing

Once if the CR knows the transmitting recurrence, it educates its collector regarding the spectrum band picked. So that a typical conveying channel could be built up. Furthermore, a reasonable spectrum planning strategy can be given; it could be respected to be like nonspecific MAC issues in the existing frameworks [6].

d. Mobility

Versatility of the spectrum or handover alludes to the modification of working recurrence on band. Versatility happens on. CR modifies its recurrence endless supply of authorized users sign. For acquiring elevated QoS, CR wants to change to an additional recurrence, keeping up consistent correspondence prerequisites amid the move to enhanced spectrum [6].

4   MAC Layers and Architecture of CR

4.1   DLL (Data Link Layer)

The essential capacity of CR's data link layer is spectrum sharing, which could be found in Fig. 2. DLL is called for sharing the spectrum, since the issue identifies with a radio with the entrance of the spectrum to ordinarily worries with the sub-layer MAC. Essential contrast among the nonexclusive MAC along with MAC

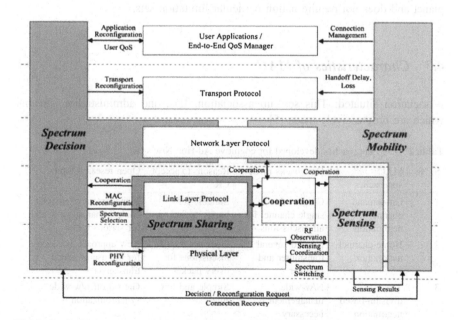

Fig. 2 Basic CR-layered architecture as conventional OSI approach [11]

in CR is that conjunction in the middle of authorized and unauthorized clients and active determination of the recurrence for transmitting in the scope of accessible spectrum and transceiver handshakes where there are two or more intellectual radio bands should concede to common channel whereupon to convey [3]. By and large, productive MAC, and blunder the control along with revision of the primary capacity of the connection layer.

Blunder Control: FEC (forward error correction) and ARQ (Automatic Repeat Request) are the primary mistake control plans of the remote systems. In any case, ARQ retransmission-based instrument causes additional vitality utilization and diminishes transfer speed utilization [6]. Along these lines, FEC plan is a decent option for asset-compelled intellectual radio system.

Medium Access Control: MAC convention plans to furnish system with intends to get to the medium in a reasonable and effective way, which is a testing target taking in perspective the constrained assets, thick system sending, and the applications particular QoS prerequisites (Table 2).

4.2 MAC Sub-layer

The IEEE 802.22 MAC sub-layer gives instruments to adaptable, bolster intellectual abilities and proficient information transmission for both dependable security of occupant administrations in the television band and self-conjunction among 802.22 frameworks [1]. The IEEE 802.22 MAC sub-layer is pertinent to any district on the planet and does not require nation particular limitation sets.

4.3 Characteristics of MAC

Association situated: This sets up association, IDs, and administration streams which are progressively made [5].

Table 2 MAC approaches developed for cognitive Ad-Hoc N/w's [6]

S. No	MAC approaches	Drawbacks in CR	Reasons to accept CR	Open research topics
1	On-demand channel negotiation	Conflict due to single channel for all discussions	Suitable for bursty traffic	Management of various control channels essential for serious traffic
2	Home channel negotiation	Needed several transmitter and receiver	Does not require discussions for every packet	New approach necessary for the use of single transceiver
3	Time division-based negotiation	N/w-wide management is necessary	Simple and few rules	Get rid off n/w-wide synchronization

Table 3 Quality of service obtained by MAC sub-layers

S. No	QoS	Applications
1	UGS	VoIP, T1/E1
2	rtPS	MPEG video streaming
3	nrtPS	FTP
4	BE	E-mail
5	Contention	BW request, etc.

Quality of service: Different sorts of Quality of service administrations are bolstered as shown in Table 3. ARQ bolstered, uni-cast, multi-cast and show administrations are upheld [5].

4.4 Cognitive Radio Functionality

Self-motivated and adaptable reserving the silent phases are permits to the framework for adjusting QoS necessities for the clients which needs to calm down the system for bolster detecting of spectrum. In calm phases, spectrum from 1 image to 1 super outline. Subscribers can alarm the BS, the vicinity of client in various methods. Devoted-Urgent-Co-presence-Situation (UCS) communication needs MAC communication. Base station could ask for one or more endorsers of move to an extra direct during various ways of utilizing frame control header (FCH) or devoted MAC communication.

4.4.1 L-L-C Sub-layer

L-L-C sub-layer in 802.22 has no particular dissimilarity for intelligent assignments.

5 Applications of CR

With the up coming innovation for CR, numerous conceivable outcomes are being investigated for utilizing the functionalities and capacities for different appliances. At present, there are numerous software-defined radio strategies that could be improved for utilizing CR elements. Taking after are few much of the time supported uses of subjective radio.

a. Indoor sensing applications,
b. Multimedia applications,
c. Multi-class heterogeneous sensing applications,

d. Real-time surveillance applications,
e. Improving spectrum use and proficiency,
f. Improving connection dependability,
g. Less cost radios,
h. Collaborative techniques, and
i. Automated radio asset administration.

6 Conclusion

Experiencing this manuscript, CR has proposed a capable innovation for comprehending the irregularity among spectrum lack and under-utilization. Different CR capacities such as detecting the spectrum, channel set administration, versatility, sharing, geo-location, self-concurrence supporter enlistment, and following and occupant databases make the additional customary from the different radios by checking the spectrum limitations. CR can adjust the suitable recurrence groups, edges along with conventions. Usual CRs are known as IEEE 802.22 Working Group which could be useful for provincial and distant regions as IEEE 802.22 has a scope of around 100 km. Because of the better general execution, CRs are relied upon to shape a spine for the cutting edge spectrums halfway or entirely, which might to the considerable degree relieve the spectrum lack issue. Additionally, amid the crisis circumstances getting to the confined unlicensed spectrum for transmission will be the most elite application capable element for CR.

References

1. Bannered JS, Karmakar K (2012) A comparative study in cognitive radio implementation issues. Int J Comput Appl 45(15)
2. Anusha M, Vemuru S et al (2015) Transmission protocols in cognitive radio mesh networks. Int J Electric Comput Eng (IJECE), 5(4)
3. Mody AN, Chouinard G, IEEE 802.22 wireless regional area networks enabling rural broadband wireless access using cognitive radio technology, IEEE 802.22–10/0073r03, June 2010
4. Stevenson CR et al IEEE 802.22: The first cognitive radio wireless regional area network standard, IEEE standards in communications and networking IEEE Communications Magazine, Jan 2009
5. Hester L, Ridley AD (2008) Cognitive radio networks: not your father's wireless network, The Telecommunications Review, A NOBLIS Publication, pp 44–54
6. Byun S, Balasingham I, Liang X Dynamic spectrum allocation in wireless cognitive sensor networks: improving fairness and energy efficiency. Proceedings of IEEE VTC, pp. 1–5, Sep 2008
7. Anusha M, Vemuru S (2015) Cognitive radio networks: a survey. Int J Emerg Trends Technol Comput Sci (IJETTCS) 4(6)
8. Azad AKM, Kamruzzaman J, A framework for collaborative multi class heterogeneous wireless sensor networks. In: Proceedings of IEEE ICC, pp 4396–4401, May 2008

9. Anusha M, Vemuru S (2015) Enhancement of wireless mesh network using cognitive radio's. Eur J Appl Sci 7(3)
10. Anusha M, Vemuru S (2016) An efficient MAC protocol for reducing channel interference and access delay in cognitive radio wireless mesh networks. Int J Commun Anten Propag (IRECAP) 6(1)
11. Singh JSP, Singh J, Kang AS (2013) Cognitive radio: state of research domain in next generation wireless networks—a critical analysis. Int J Comput Appl 74(10)

Smart Bike Through Server Using GPS Technology

Rajneesh Tanwar and Ashwani Chaudhary

Abstract Population is increasing day by day and everyone uses bike for personal work. Cases of bike accidents and bike thefts are also increasing simultaneously. Connection of all bikes with a centralized server by inserting a programmable chip in bike starting mechanism which will control the speed of bike, authenticate the user by key and allow ON/OFF of engine, stores the vehicle identification number (VIN) and also tell the location of bike by using Global Positioning System (GPS). User can ask for details of bike location by contacting server and can also make bike engine stop in case of theft. If chip is removed from the bike, connection between battery and server will be lost and bike will not be able to start in any case. By using GPS, speed of bike is also managed automatically i.e., rider will not able to exceed speed of bike more than limit of road. This will decrease the ratio of accidents on road; speed will be maintained and lost or theft bike can be caught easy.

Keywords Server · Programmable chip · VIN · GPS · Battery

1 Introduction

Now a day, in all over the world, personal vehicles are counted as only two, first is four-wheel vehicle i.e., cars and second is two- wheel vehicles i.e., bike or bicycle. In today's world, security for car is concerned and many advanced techniques are invented which are possibly securing car on the basis of speed thrills, accidents, car theft, auto control, etc. But today in modern era, still two wheelers are used in many parts of world as a necessary means of personal transport but till now there is no

R. Tanwar (✉)
Department of Information Technology, Amity University, Noida, India
e-mail: rajneeshtanwar15@gmail.com

A. Chaudhary
C.B.P. Government Engineering College, New Delhi, India
e-mail: ashwanicha@gmail.com

© Springer Nature Singapore Pte Ltd. 2018
D.K. Mishra et al. (eds.), *Information and Communication Technology for Sustainable Development*, Lecture Notes in Networks and Systems 9,
https://doi.org/10.1007/978-981-10-3932-4_31

security concern is considered for these vehicles. Likewise cars, bikes, or bicycles are also manufactured in such a way that these are also able to describe its security to user. GPS (Global Positioning System) [1] is one of the biggest and widely used technologies in whole world whenever concern for the location tracing or tracking comes [2]. With the use of GPS technology, vehicle can easily be tracked and user or owner of vehicle can easily able to trace the location of vehicle and have advantage to lock vehicle if micro controller to control locking system is attached [3]. With the help of GPS technology, if accident happened on the road, then the information is passed to server telling details about the location of accident. Accidental situation can also get managed and easy curing arrangement can be done [4]. GPS carry out its working in very smart way. Each of the satellites in a circle permits a beneficiary to distinguish no less than four of the operational satellites. The satellites convey microwave signs to a collector where the implicit PC utilizes these signs to work out your exact separation from each of the four satellites and after that triangulates your careful position on the planet to the closest few meters taking into account these separations. Signals from only three satellites are expected to complete this triangulation procedure; the count of your position on earth in view of your separation from three satellites [5].

2 Survey of Related Work

Up to now, World is applying all theft protection techniques on car [6] which is basically expensive. Till now, many car are present in the market which are connected to the server and opening or driving authentication is all done by the server. But till now, no work is done for bike, as in bike, authentication part is very difficult. Speed control system is also present in today's world but for only cars not for other vehicle [7].

My approach is to provide all authentications and speed controlling techniques in bike so that bike owners are able to detect and track their bike and also feel comfortable if other is driving as controlling system of speed is done by server which is not in hand of user.

3 Methodology

Methodology is proposed for tracing and speed controlling over vehicle like bike, a chip is mounted between key lock and battery connection. If this chip gets detached from bike, then connection between key and battery will get lose. This proposed system will trace the bike location with the help of mounted chip and also control the speed of bike i.e., GPS will instruct the maximum speed on road, and chip will control the speed of bike according to the output from the GPS system.

In the proposed system, first, CHIP which is going to be used for managing all authentication and connection to server plus GPS location tracking system should be fitted very securely in bike as in rainy season it can cause problem and thief can easy get access to that which will led to loss of security to the bike. The location of chip should be connected to the battery as it will allow to start and controlling speed of bike according to the rules or GPS system.

Figure 1 describes the role of mounted chip. The mounted chip performs following function and also stores some values

- Always contains bike's Engine Number and Chassis Number for unique identification and authentication.
- Always connected to battery which will help chip to continue its server and GPS functioning in OFF of bike i.e., engine is OFF.
- Always connected to GPS for analyzing the speed of the bike and also tell the location of bike in both driving and steady state.
- Always connected to server which will analyze the output from GPS and give command to chip according to the output, also control the ON and OFF mechanism of bike, and lastly, this will also authenticate the right rider or owner for details about bike.

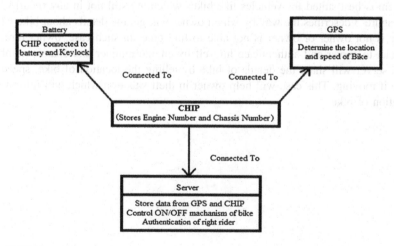

Fig. 1 CHIP

At server site, the server will determine the location of bike from GPS and also trace the speed of bike according to output of the GPS. It will automatically give command to chip for managing the speed of vehicle. Server will also authenticate the user for getting the information about the bike. This authentication is on the basis of engine number and chassis number which is stored in chip and maintained by server. A toll-free number is provided for user to get contacted with server.

3.1   Flow of Whole System

Figure 2 describes the whole working of the system in which there is 3 chances for entering the key for open the lock of vehicle. If number of attempt becomes more than three then the locking system will be blocked by server through chip. This will again get activated by contacting server and authenticating engine number and chassis number. This will again let the user to go for three more attempts.

This system will also show how the GPS, Chip, and server are connected to each other and coordinating their information in moving plus steady condition of bike. In this system, control in the speed of bike is also shown which is done by chip. GPS will tell the current speed and maximum speed; server will match both and give automatic command to chip according to the output from GPS system.

The proposed system will need the chip having GPS module and connected to server and battery too for controlling the speed and tracing the location of bike. This system is best suited for vehicles like bikes which are still not in any security.

Figure 3 describes the way by which owner will get the details about bike when rider is not owner or owner is not able to find bike. In such case, owner directly contact to server and authenticate himself by engine number and chassis number, then server will share the details of bike by telling the location of bike, speed of bike if moving. This case will help owner in theft situation which will inform the location of bike.

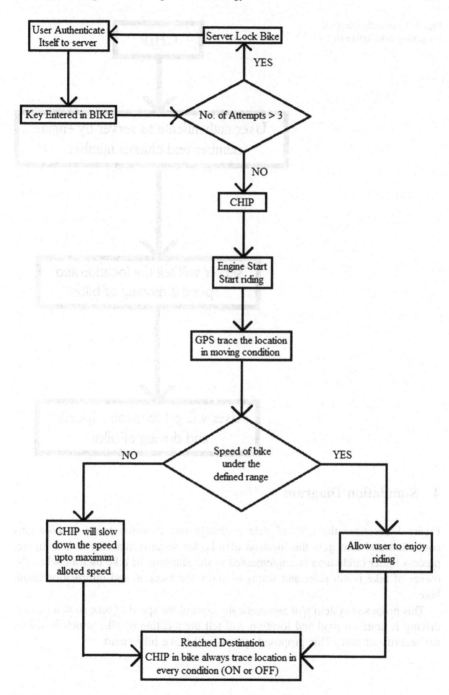

Fig. 2 Flow of system

Fig. 3 User authentication
for getting bike information

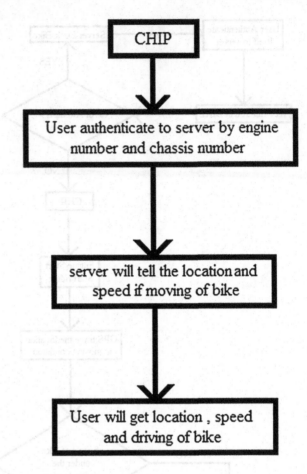

Fig. 3 User authentication for getting bike information

4   Simulation Diagram

Figure 4 describes the flow of data exchange and how the speed of bike gets controlled and also gets the location of bike for security reasons. In this figure, process under circle area is implemented in the situation of bike theft or when the owner of bike is not rider and wants to know the location and information about bike.

This proposed system will automatically control the speed of bike so that no rash driving is done on road and location will tell the location of bike which increases the security of bike. This proposed system will make bike smart.

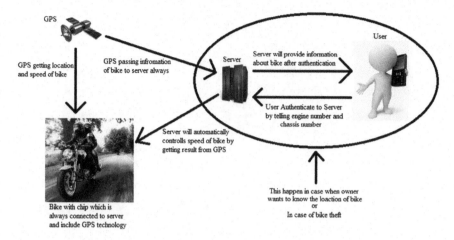

Fig. 4 Simulation flow diagram for proposed system

5 Conclusion

Implementation of such proposed system will help to know the location of the bike and also speed of the bike will get controlled which also result in less accident overroad. In many cases, when owner is not driver and worrying about his/her vehicle, this system will able to give details about the vehicle location and the speed of bike. Such system is very helpful, especially for youth protection and caring. In theft cases, this system will work in very good manner and is able to determine the location of bike.

References

1. http://www.mio.com/technology-gps-work.htm
2. Obuhuma JI, Moturi CA (2012) Paper on use of GPS with road mapping for traffic analysis. Int J Sci Technol Res
3. Thakor N, Vyas T, Shah D (2013) Automatic vehicle accident detection system based on ARM &GPS. Int J Res Technol Stud l(1), ISSN: Applied
4. Xu Z (2011) V-V location based broadcast communication, automated crash notification via the wireless web: system design and validation 19(6):1048–1059
5. Ramani R, Valarmathy S, SuthanthiraVanitha N Dr, Selvaraju S, Thiruppathi M, Thangam R (2013) Vehicle tracking and locking system based on GSM and GPS. I J Intell Syst Appl 09:86–93
6. Car Security and Tracking System with Position, Route, and Speed Calculation (2011) EDP topic VG06
7. Mohol S, Pavanikar A, Dhage G (2014) GPS vehicle tracking system. Int J Emerg Eng Res Technol 2(7), ISSN 2349–4395 and ISSN 2349–4409

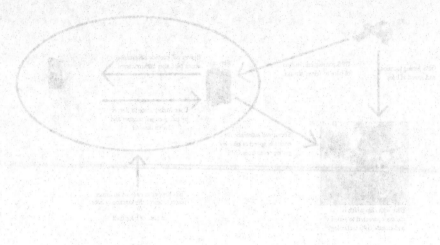

Fig. 4 Simulation flow diagram for proposed system

5 Conclusion

Implementation of such proposed system will help to know the location of the bike and also speed of the bike will get controlled which also result in less accident byroad. In many cases, when owner is not driver and worrying about his/her vehicle, this system will help to give details about the vehicle location and the speed of bike. Such system is very helpful especially for youth protection and caring. In their cases, this system will work in very good manner and is able to determine the location of bike.

References

1. http://www.autocontechnology.app/working

2. Chilukuri R, Midathala V (2012) Paper on use of GPS with road mapping for traffic analysis and Soc (Consol Res

3. Tseik S, Voo T, Sheh D (2011) Automatic vehicle accident detection system based on ARM. KLUES-Int J Res Technol Stud 1(7) ISSN: Applied

4. Xu Z (2011) V-V Location based broadcast Communication, decoupled crash notification via the vehicular web, system design and validation. PhD J 1045–1095

5. Ramani R, Valarmathy S, Suthanthira Vanitha N, Selvaraju S, Thiruppathi M, Thangam R (2013) Mobile tracking and accident system based on GSM and GPS-1 J Intell Syst Appl (Issue-02)

6. Car Security and Tracking System with Ignition Control and Speed Control (2013) IDP topic VCOn

7. Mohan S, Ravindra, Dalage O (2014) GPS vehicle tracking system. Int J Emerg Eng Res Technol 2(2), ISSN 2349–4395 and ISSN 2349–4100

Preventing Node Replication Attack in Mobile Wireless Sensor Networks

Aarti Singh and Kavita Gupta

Abstract Node replication is a challenging security issue in sensor networks. Attacker compromises one node, creates copy of that node, and captures the sensed information forwarded to the base station. This work proposes a centralized technique for preventing the MWSN from replicated attack. The proposed technique detects the attacker at very early stage, thereby receiving sensed data at the base station in secure manner. Experimental results indicate the feasibility of the proposed scheme that it is suitable for MWSN in real-time environment.

Keywords Node replication · MWSN · Security

1 Introduction

Mobile wireless sensor networks (MWSN) are collection of small size sensor nodes that are having sensing capability and are densely deployed in hostile environments for sensing task. This sensed information is forwarded to the base station or sink that can be either static or mobile. MWSN find applicability in military applications, health monitoring, tracking the movement of animals, environment mapping, and control [1–3].

In static sensor networks, nodes are stationary, so all the nodes along with sink node have their fixed location, and thus, their geographic position can easily be identified [4]. In MWSN, nodes are of mobile nature, so their location detection is not an easy task. Traditional location detection schemes are not useful in MWSN. So in these types of networks, attackers compromise a node and create its replica. This replica is then used with cryptographic details of the original node to

A. Singh (✉)
Guru Nanak Girls College, Yamuna Nagar, Haryana, India
e-mail: singh2208@gmail.com

K. Gupta
Maharishi Markandeshwar University, Mullana, India
e-mail: kavita_mittal25@yahoo.co.in

© Springer Nature Singapore Pte Ltd. 2018
D.K. Mishra et al. (eds.), *Information and Communication Technology for Sustainable Development*, Lecture Notes in Networks and Systems 9,
https://doi.org/10.1007/978-981-10-3932-4_32

communicate with other peer nodes in the network. These replicated nodes are known as pretenders as they use identity of an existing node to communicate with peer nodes in the network.

Since pretender node has stolen all the credentials of compromised node, distinguishing the pretender node from real node is not an easy task. Thus, pretender detection is a major challenge [5, 6] in MWSN. The literature highlighted [7] that there are various types of attacks associated with node identity such as replication of nodes, selective forwarding, and sinkhole attack [8]. Creating multiple copies of a node or using stolen identity during data aggregation process is a major challenge faced in MWSN [9].

This work focuses on node replication attack in MWSN through asymmetric key cryptographic mechanism managed centrally. This work uniquely contributes toward prevention and detection of pretender attack in MWSN.

2   Related Work

Mobile wireless sensor networks (MWSN) [10] consist of n sensor nodes that are deployed in hostile environment. These sensor nodes sense data or information and forward it to the base station that can be static [11] or dynamic. It acts like a gateway for external network using routing protocols suitable for MWSN [12–14]. This work assumes a homogeneous network where all nodes have similar properties such as memory, energy, and computational capabilities. They move freely within the bounded network region with different speeds, and hence, it is not easy to predict their speed, position, or IDs of colliding nodes.

A pretender is defined as a malicious node which steals the identity of other node for communication in the network. It steals identity information [15, 16] of a node for communicating with sink and the other nodes. However, it is assumed that the base station is secure and pretender cannot generate new IDs by using base station's credentials. The proposed work focuses on sybil attack, so the literature related to this attack has been explored in this section.

Radio-based detection scheme [17–19] is used for authenticating sensor nodes for pretender detection that is based on signal strength or other physical characteristics.

Parno et al. [4] presented network-based detection scheme deploying claimer witness framework. It is suitable for static networks by storing their location information with respect to some nodes in the network that can detect and report duplicate nodes when multiple locations are used by one identity.

A centralized scheme is proposed by [20] where speed of nodes is calculated by the base station using their location information received from their neighbors. ID-based cryptography scheme is proposed by [4, 21]. Here, public keys are embedded in nodes before deployment. Thus, nodes are recognized by their public key so that other nodes can verify the authenticity of message. According to [22], ID-based authentication is not efficient for message authentication.

From the above literature review, it is clear that security of sensor nodes is of the utmost importance [23, 24]. Compromising a node leads to shift of network control toward hacker. Thus, there is dire need of mechanism for detecting and preventing replication attack in MWSN. The next section presents proposed mechanism.

3 Proposed Work

This work proposes an asymmetric key cryptography mechanism for preventing node replication in MWSN. Every node in the network is required to be registered with the base station for taking part in communication. The base station assigns a unique asymmetric key pair to that node of the form: $< public(pbk), private(ptk) >$. The base station maintains a log containing key assigned to every node in the network along with time of key generation. In MWSN, participating nodes are mobile in nature and are possessed with fixed amount of energy. The amount of energy consumed in movement of node and in communication is known, and thus, lifetime of a node can be calculated if its energy level is known. Using this concept, the base station allocates key to a node for its lifetime, i.e., a fixed period of time, after which key gets deactivated. If this concept is not used, key pair of a compromised node may be used by some replicated nodes, even after its lifetime, creating a severe hole in security.

Figure 1 provides high-level view of proposed mechanism; here, whenever a node is registered with based station, a tuple containing its public_key and private_key is generated. Base station assigns a node id to that node and also shares its own public key, i.e., BS_public_key with that node to be used in all future

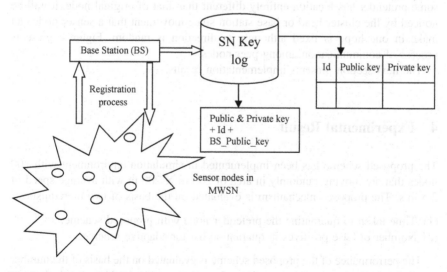

Fig. 1 Overview of proposed frame work

Fig. 2 Communication
among peer nodes

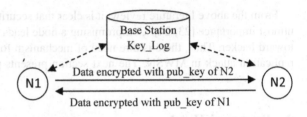

communications. Public key of all registered nodes is shared among cluster members for making communication possible.

The base station uses Eq. (1) to determine the validity of keys. Here, E is the initial energy level of node, e be the energy consumption rate of node per unit time (ms), and t be the remaining lifetime of the node. Remaining lifetime parameter helps identify pretender nodes using identity of dead nodes. Table 1 shows the remaining time of sensor nodes.

$$t = E/e \ldots \tag{1}$$

Now, whenever two nodes say N1 and N2 want to communicate with each other, N1 encrypts data with public_key of N2 and sends the data, and N2 decrypts the message with its own private key. While sending reply, N2 encrypts data using public_key of N1 and sends it to N1, and N1 decrypts the message using its own private key. Since nodes in MWSN are mobile and they move from one cluster to other while moving, thus nodes need to access public keys of other nodes in current cluster from their cluster head which in turn gets this information from the base station. In this process, every node needs to authenticate itself either to cluster head or base station, and chances of pretender getting exposed at early stages increase. If some pretender has location entirely different than that of original node, it will be noticed by the cluster head or base station since movement that a sensor node can make in one hope is fixed although its direction is random. Figure 2 presents process of communication among peer nodes.

The next section presents implementation results.

4 Experimental Result

The proposed scheme has been implemented in simulation environment with 200 nodes that are moving randomly in area of 200 m X 200 m with average speed of 2.5 m/s. The proposed mechanism is evaluated on the basis of two prerequisites:

(1) Time taken to quarantine the pretender node with proposed scheme;
(2) Number of false positives in quarantine list for adaptive scheme.

The performance of the proposed scheme is evaluated on the basis of the number of neighbors, i.e., node density and number of pretenders in the network. In poor

Fig. 3 Node density w.r.t. time

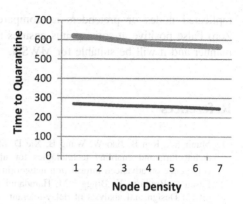

Node Density

Fig. 4 Number of pretenders w.r.t. time (proposed versus existed)

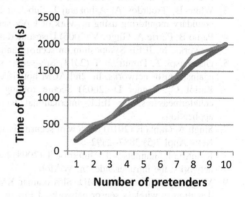

Number of pretenders

case, we assume that node density varies from 1 to 7 and number of pretenders from 2 to 25. Density of nodes can be increased by expending the communication range of the nodes.

Figures 3 and 4 present pretender detection time w.r.t. the number of pretenders and node density respectively. Bold line indicates the performance of proposed scheme which is clear than that of traditional scheme.

5 Conclusion

The node replication is a severe attack in MWSN. Many preventive security mechanisms have been proposed for static network where both nodes and base station are fixed. But this work proposes a node replication detection and prevention scheme for mobile wireless sensor networks. This is a centralized scheme where major tasks are handed by base station and cluster head. This scheme detects the pretender or replicated nodes at very early stage leading to reduction in number of replicated nodes or pretenders as compared to traditional preventing scheme. Zero False positive at some points ensures that this scheme works in self-healing manner and it will be suitable for MWSN.

References

1. Munir SA, Ren B, Jiao W, Wang B, Xie D, Ma J (2007) Mobile wireless sensor network: architecture and enabling technologies for ubiquitous computing. In: 21st international conference on advanced information networking and applications
2. Ehsan S, Bradford K, Brugger M, Hamdaoui B, Kovchegov Y, Johnson D, Louhaichi M (2012) Design and analysis of delay-tolerant sensor networks for monitoring and tracking free-roaming animals. IEEE Trans Wireless Comm 11(3):1220–1227
3. White B, Tsourdos A, Ashokaraj I, Subchan S, Zbikowski R (2008) Contaminant cloud boundary monitoring using network of UAV sensors. IEEE Sensors J l. 8(10):1681–1692
4. Parno B, Perrig A, Gligor V (2005) Distributed detection of node replication attacks in sensor networks. In: IEEE Symposium on security and privacy
5. Giannetsos T, Dimitriou T (2013) Spy-sense: spyware tool for executing stealthy exploits against sensor networks. In: 2nd ACM hotwisec
6. Karlof C, Wagner D (2003) Secure routing in wireless sensor networks: attacks and countermeasures. In: IEEE international workshop on sensor network protocols and applications
7. Singh A, Gupta K (2016) Review of security issues in mobile wireless sensor networks. J Adv Netw Appl 7(5):2887–2892
8. Krontiris I, Giannetsos T, Dimitriou T (2008) Launching a sinkhole attack in wireless sensor networks; the intruder side. In: WiMob
9. Manjunatha TN, Sushma MD, Shivakumar KM (2013) Security concepts and sybil attack detection in wireless sensor networks. J Emerg Trends Technol Comput Sci 2(2):383–390
10. Singh A, Gupta K (2016) Optimal cluster head election algorithm for mobile wireless sensor networks. In: Proceedings of ICTCS'16
11. Zhu WT, Zhou J, Deng RH, Bao F (2012) Detecting node replication attacks in wireless sensor networks: a survey. J Netw Comput Appl 35:1022–1034
12. Arboleda LM, Nasser N (2006) Cluster-based routing protocol for mobile sensor networks. In: Proceedings of the 3rd ACM conference on quality of service in heterogeneous wired/wireless networks
13. Yu M, Li JH, Levy R (2006) Mobility resistant clustering in multi-hop wireless networks. J Netw 1(1):12–19
14. Khan AR, Ali S, Mustafa S, Othman M (2012) Impact of mobility models on clustering based routing protocols in mobile WSNs. In: 10th IEEE conference on frontiers of information technology (FIT)
15. Yu CM, Lu CS, Kuo SY (2008) Mobile sensor network resilient against node replication attacks. In: 5th IEEE Conference on Sensor, Mesh and Ad Hoc Communication and Networks (SECON)
16. Alrashed EA, Karaata MH (2014) Imposter detection in mobile wireless sensor networks. Int J Comput Commun Eng 3(6)
17. Hall J, Barbeau M, Kranakis E (2003) Detection of transient in radio frequency fingerprinting using signal phase. In: Proceedings of IASTED international conference on wireless and optical communications (WOC)
18. Bhuse V, Gupta A (2006) Anomaly intrusion detection in wireless sensor networks. J. High Speed Netw 15(1):33–51

19. Hussain S, Rahman MS (2009) Using received signal strength indicator to detect node replacement and replication attacks in wireless sensor networks. In: Proceedings of SPIE
20. Ho JW, Wright M, Das SK (2009) Fast detection of replica node attacks in mobile sensor networks using sequential analysis. In: INFOCOM
21. Zhu WT, Zhou J, Deng RH, Bao F (2011) Detecting node replication attacks in mobile sensor networks: theory and approaches. Secur Commun Netw, 5:496–507
22. Liu J, Baek J, Zhou J, Yang Y, Wong J-W (2010) Efficient online/offline identity-based signature for wireless sensor network. Int J Info Secur (9), 287–296
23. Krontiris I, Giannetsos T, Dimitriou T (2008) LIDeA: a distributed lightweight intrusion detection architecture for sensor networks. In: SECURECOMM
24. Yu CM, Lu CS, Kuo SY (2009) Efficient and distributed detection of node replication attacks in mobile sensor networks. In: Vehicular technology conference (VTC)

Logistic Regression with Stochastic Gradient Ascent to Estimate Click Through Rate

Jenish Dhanani and Keyur Rana

Abstract Majority of Web users utilize search engines to locate Web site links. Based upon the search queries provided by the users, search engines display *sponsored advertisements* together with actual Web site link results to procreate monetary benefits. However, users may click the concerned sponsored advertisements that generate revenue for the search engines based upon a predefined pricing model. Furthermore, by analyzing previous information of users, advertisements, and queries; search engines estimate *click-through rate* (CTR) for predicting users' clicks. CTR is a ratio of clicks to number of impressions associated with a particular advertisement. In this paper, we propose a model, based on CTR, to estimate probabilities of clicks using logistic regression that determines parameters using *stochastic gradient ascent* method (SGA). Moreover, this paper also summarizes the comparative analysis of SGA and *batch gradient ascent* (BGA) methods, in terms of accuracy and learning time.

Keywords CTR · Internet advertising · Logistic regression · Stochastic gradient ascent · Batch gradient ascent

1 Introduction

To increase the sales of products and services, an advertisement has become an important entity of branding. Majority of advertisement media incorporates newspaper, magazine, television, radio, banner board, and Internet [1]. In recent years, Internet has emerged as a major source of information that is rapidly increasing Internet users, by providing versatile services that include communication, data transfer, social networking, and product selling. Moreover, by

J. Dhanani (✉) · K. Rana
Sarvajanik College of Engineering and Technology, Surat, India
e-mail: jenishdhanani26@gmail.com

K. Rana
e-mail: keyur.rana@scet.ac.in

© Springer Nature Singapore Pte Ltd. 2018
D.K. Mishra et al. (eds.), *Information and Communication Technology for Sustainable Development*, Lecture Notes in Networks and Systems 9, https://doi.org/10.1007/978-981-10-3932-4_33

considering growing Internet popularity, advertisers have identified it as a mammoth market to advertise their products and services. Also, report published by the Interactive Advertising Bureau (IAB-2015) shows that US advertising market comprises of $49 billion [2]. Henceforth, targeting individual user or group of users, by providing personalized advertisements considering different parameters such as user's demographic information, behavior data, search history, and queries have established Internet advertising as a widely accepted platform for advertisement campaigning [3, 4].

Significant consumers of Internet advertising include *users* and *advertisers*: who want to advertise their products or services over the Internet [5], *publishers and search engines*: who display advertisements on their Web sites [6, 7], and *ad networks*: who play a role of the mediator between publishers and advertisers [8], by collecting advertisements from advertisers and exhibiting the same over publisher's Web sites such as media net [9] and charging commissions from both the advertisers and publishers. *Cost per impression (CPM), cost per click (CPC), and cost per action (CPA)* are the three popular pricing models for Internet advertising [2, 4, 5, 10].

Selection of an advertisement should be such that it should result in some action, such as a click from the user. There are three approaches to select an advertisement, namely, sponsored search advertising [1, 2, 4, 5, 11, 12], contextual advertising [4, 5, 8, 13, 14, 15], and behavior-targeted advertising [16, 17]. Sponsored search is advertisements that result in the search query [18]. Majority of search engines have adopted sponsored search advertising. In search engines, when user fires a query, it results in Web links (also known as the organic search result) and sponsored search links, which is a small text-based or multimedia-based advertisements [11, 19]. The contextual and behavior-targeted advertising are beyond the scope of this paper.

Advertisement selection plays an important role in the Internet advertising. If an advertisement is relevant to the user, then an advertisement has a higher probability of being clicked by the user. The probability of an advertisement being clicked by the user can be estimated from click through rate (CTR). CTR is a qualitative characteristic of an advertisement [20]. It is the ratio of number of clicks to the impressions of an advertisement [1, 10, 16]. To display the advertisement and to predict the clicks, CTR is widely used in sponsored search. Search engine displays the advertisement that has a higher CTR. To make advertising more personalize, search engine needs to handle and process the data of billions of user and millions of advertisements. In this paper, we have proposed a model to estimate CTR using logistic regression where parameters are estimated by stochastic gradient ascent (SGA) method. SGA is time efficient as compared to batch gradient ascent (BGA).

Section 2 surveys the works related to CTR prediction using different approaches. Sect. 3.2 discusses the data preprocessing applied to KDD data set. Section 3.3 outlines logistic regression with SGA and BGA. Sect. 4 presents and analyzes the experimental results. Finally, in Sect. 5, we conclude the paper.

2 Related Work

The major focus has been given on the Internet advertising and role of CTR into the Internet advertising research. The search engine can maximize the revenue by showing relevant advertisement based on user's query and information. An advertisement has a higher probability of being clicked by the user if it is relevant to user's query. The click by the user brings revenue for the search engine. The click is predicted from CTR. In the literature, several learning methods have been discussed to estimate CTR, namely, logistic regression [1], naive Bayes, Bayesian learning [21], multivariable linear regression [22], boosted tree [11], Poisson regression [17], and multiple criteria linear programming [12, 23].

Logistic regression is a well-known binary classifier that we have used to classify that, whether the user has clicked on the advertisement or not. Logistic regression is applied broadly in the field of statistical machine learning and data mining such as classification of documents [24], credit scoring [25, 26], and text classification. In logistic regression, parameter estimation plays an important role in building a model because the performance of a model is dependent on the values of estimated parameters. In the literature, we found two methods for parameter estimation viz. batch gradient ascent [27, 28] and stochastic gradient ascent [27, 29]. To estimate CTR, a system should process the data of user and advertisement. Such data is usually very large in volume. Time efficient and accurate machine learning algorithm should be used to estimate CTR from the large volume of data. In logistic regression, stochastic gradient ascent can be used for parameter estimation from a large volume of data [30, 31].

3 Proposed Work

This section describes our proposed model as depicted in Fig. 1. The figure is divided into two phases: (1) *training phase*: It consists of data preprocessing and building a model from the training data set and (2) *testing phase*: The built model is applied to the testing data set to measure the performance. Various modules of the system are described below.

3.1 The Data Set

For this study, we have collected data of CTR from the track 2 of KDD cup 2012 Web site in text format [32]. This data set includes one training file and different associated files. Training and associated files include different features such as *click, impression, user_id, query_id, advertisement_id, title_id, description_id, age of the user, gender,* and *keywords of advertisement*.

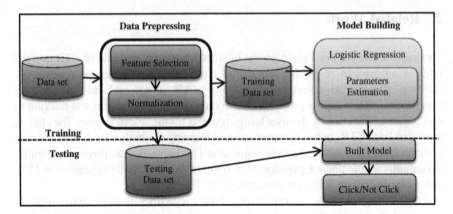

Fig. 1 Proposed model of click prediction

3.2 Data Preprocessing

Data preprocessing is one of the important steps for CTR prediction. The collected data is noisy, incomplete, and inconsistent which cannot be directly given as input to the learning algorithm. The data preprocessing consists of techniques such as feature selection and normalization. The features are selected manually because not all the features available in the data set are required for our approach. The next step is to normalize the value of all features in between 0 to 10 because the value range of each feature is different. Min-max [13] method has been used to normalize all features.

3.3 Logistic Regression

In this work, Logistic regression is used to estimate CTR and predict click from the CTR. Logistic regression is based on the logit function that is natural log of an odds ratio. The predicted value of CTR is always in the range of 0–1. For given m data points such as $\{(X^{(1)}, Y^{(1)}), (X^{(2)}, Y^{(2)}), (X^{(3)}, Y^{(3)}), \ldots, (X^{(m)}, Y^{(m)})\}$ and k features such as $X_0, X_1, X_2, X_1, \ldots, X_k$ of data set where bias term $X_0 = 1$, the hypothesis function of logistic regression is $f(z)$ or $h_\beta(X)$

$$f(z) = \frac{1}{1 + e^{-z}} \tag{1}$$

where $z = \beta_0 + \beta_1 X_1 + \beta_2 X_2 + \cdots + \beta_k X_k$ and $\beta_0, \beta_1, \beta_2, \beta_3 \ldots \beta_k$ are the parameters to be estimated $h_\beta(X)$ or $p(Y = 1|X; \beta)$ is the probability of advertisement being clicked $(Y = 1)$ for the given features $(X_0, X_1, X_2, X_1, \ldots, X_k)$. $p(Y = 0|X; \beta)$ or $1 - h_\beta(X)$ is the probability of advertisement being not clicked $(Y = 0)$ for given features $(X_0, X_1, X_2, X_1, \ldots, X_k)$. If the result of $h_\beta(X)$ is above 0.5, then it is considered that "click" is predicted else "not click" is predicted.

Maximum likelihood estimation (MLE) method is used to fit a model for a given data points. The aim of MLE is to maximize the value of average objective function $J(\beta)$ for given data points. It is formulated in Eq. (2) for m data points.

$$J(\beta) = \frac{1}{m} \sum_{i=1}^{i=m} \left(Y\left(\log(h_\beta\left(X^{(i)}\right)\right) + (1 - Y)\left(\log(1 - h_\beta\left(X^{(i)}\right)\right)\right) \quad (2)$$

To estimate the parameters in logistic regression, we have explored batch gradient ascent (BGA) and stochastic gradient ascent (SGA) methods in this paper.

3.3.1 Stochastic Gradient Ascent (SGA)

SGA takes the initial random value of parameters $(\beta_0, \beta_1, \beta_2, \beta_3 \ldots \beta_k)$. To estimate the new value of parameters, the product of learning rate (α) and gradient of the objective function is added to the old value of all parameters as shown in Eq. (3).

$$\beta_j^{new} = \beta_j^{old} + \alpha\left(\left(Y^{(i)} - h_\beta\left(X^{(i)}\right)\right)X_j^{(i)}\right) \quad (3)$$

where $Y^{(i)}$ is actual output, and $h_\beta(X^{(i)})$ is predicted output of the ith data point.

Once, all parameters are updated through all data points that consider as one iteration and move further for next iteration. The difference between BGA method and SGA method is its gradient of the objective function is calculated from all data points as shown in Eq. (4).

$$\beta_j^{new} = \beta_j^{old} + \alpha\left(\frac{1}{m} \sum_{i=1}^{i=m} (Y^{(i)} - (h_\beta(X^{(i)}))X_k^i\right) \quad (4)$$

where $Y^{(i)}$ is actual output, $h_\beta(X^{(i)})$ is predicted output of the ith data point, and m is the number of data points.

4 Experimental Results

In this paper, we have used track 2 KDD data set to estimate CTR [32]. Features used to estimate the CTR are *click, impression, query, age of the user, gender of the user*, and *keywords of advertisement*. In this experiment, one million data points have been used to click prediction. The data set is divided into two chunks: Training data set that contains 67% record drawn randomly and testing data set that contains 33% record drawn randomly. We have applied logistic regression with the baseline model batch gradient ascent. We implemented logistic regression in which

Table 1 Comparison of objective function and time for different learning rate of BGA and SGA for one million data points

Learning rate (α)	BGA			SGA		
	Value of obj. function	Time (s)	Accuracy (%)	Value of Obj. function	Time (s)	Accuracy (%)
0.001	−0.219	1481	94.86	−0.196	300	95.03
0.005	−0.215	1452	94.91	−0.195	293	95.03
0.01	−0.213	1455	94.91	−0.197	294	95.03
0.05	−0.201	1456	95.03	−0.331	294	95.03
0.1	−0.197	1462	95.03	−0.554	295	95.03
0.15	−0.195	1460	95.03	−0.815	295	95.03
0.2	−0.195	1457	95.03	−1.491	294	95.01
0.25	−0.195	1455	95.03	−1.749	294	94.87
0.3	−0.194	1451	95.03	−1.663	294	94.95

parameters are estimated through batch gradient ascent and stochastic gradient ascent methods in java.

We compare the objective function value and time to build a model using stochastic gradient ascent and batch gradient ascent as shown in Table 1. We analyze the result with different learning rate (α).

By analyzing these experimental results, it is observed that SGA algorithm requires approximately 400% less time as compared to BGA algorithm. The main cause of time difference is because BGA algorithm requires passing through all data points in order to update the parameters, whereas the SGA algorithm updates parameters at every data points. From the comparison of the objective function value of SGA and BGA at different learning rate, it is observed that for the increasing values of learning rate, the value of objective function in BGA algorithm increases as the number of iterations increases. Since, we search for the maximum value of the objective function, it can be empirically seen that BGA results toward the optimal solution. For the smaller values of learning rate, BGA needs more number of iterations as compared to the higher values of learning rate to reach maximum value of objective function. In SGA, the value of objective function varies a lot for the higher values of learning rate, i.e., $\alpha \geq 0.1$. These experiments also highlight comparison of accuracy. Though SGA takes lesser time as compared to BGA, the accuracy of SGA and BGA is almost similar.

5 Conclusions

In this paper, we have discussed the prediction of binary class 'click' and 'not click'; by computing CTR that maximizes the revenue of a search engine for sponsored search. We had proposed a model to predict clicks using logistic regression with stochastic gradient ascent, a parameter estimation method. Furthermore, this paper examined that stochastic gradient ascent method takes less time, in comparison with batch gradient ascent method for building classifier model. Moreover, we analyzed that difference of both the methods in terms of accuracy is negligible.

Future work may include devising a new classifier model, considering more features such as social data of users, for training the model.

References

1. Kumar R, Naik SM, Naik VD, Shirall S, Sunil VG, Husain M (2015) Predicting clicks: CTR estimation of advertisements using logistic regression classifier. In: IEEE International Advance Computing Conference. IEEE, pp 1134–1138
2. IAB (2014) Interactive Advertising Bureau (IAB): internet advertising revenue report 2014 full year. Survey report
3. Langheinrich M, Nakamura A, Abe N, Kamba T, Koseki Y (1999) Unintrusive customization techniques for web advertising. Comput Netw 31:1259–1272
4. Shatnawi M, Mohamed N (2012) Statistical techniques for online personalized advertising. In: 27th annual ACM symposium on applied computing. ACM, New York, pp 680–687
5. Chowdhury NM (2007) A survey of search advertising. Project report, Information retrieval at University of Waterloo
6. Dave V, Guha S, Zhang Y (2012) Measuring and fingerprinting click-spam in ad networks. In: ACM SIGCOMM computer communication review. ACM, New York, pp 175–186
7. Pearce P, Dave V, Grier C, Levchenko K, Guha S, McCoy D, Voelker GM (2014) Characterizing large-scale click fraud in zeroaccess. In: ACM SIGSAC conference on computer and communications security. ACM, New York, pp 141–152
8. Broder A, Fontoura M, Josifovski V, Riedel L (2007) A semantic approach to contextual advertising. In: 30th annual international ACM SIGIR conference on research and development in information retrieval. ACM, New York, pp 559–566
9. Media net digital advertising. http://www.media.net/
10. Moon Y, Kwon C (2011) Online advertisement service pricing and an option contract. Electron Commer Res Appl 10:38–48
11. Trofimov I, Kornetova A, Topinskiy V (2012) Using boosted trees for click-through rate prediction for sponsored search. In: 6th international workshop on data mining for online advertising and internet economy. ACM, New York
12. Wang F, Suphamitmongkol W, Wang B (2013) Advertisement click-through rate prediction using multiple criteria linear programming regression model. Procedia Comput. Sci. 17:803–811
13. Fan T, Chang C (2011) Blogger-centric contextual advertising. Expert Syst Appl 38:1777–1788
14. Chen Y, Chen Z, Chiu Y, Chang C (2015) An annotation approach to contextual advertising for online ads. J Electron Commerce Res 16:123–137
15. Liu P, Zhang R (2014) Automatic keywords generation for contextual advertising. In: 23rd International conference on world wide web companion. ACM, New York, pp 345–346
16. Jaworska J, Sydow M (2008) Behavioural targeting in on-line advertising: an empirical study. In: Bailey J, Maier D, Schewe KD, Thalheim B, Wang XS (eds) Web information systems engineering—WISE 2008, vol 5175., LNCS Springer, Berlin Heidelberg, pp 62–76

17. Chen Y, Pavlov D, Canny JF, Ave H (2009) Large-scale behavioral targeting categories and subject descriptors. In: 15th ACM SIGKDD international conference on knowledge discovery and data mining. ACM, New York, pp 209–218
18. Fain DC, Pedersen JO (2006) Sponsored Search: A Brief History. Bulletin of the American Society for Information Science and Technology 32:12–13
19. Schroedl S, Kesari A, Neumeyer L (2010) Personalized ad placement in web search. In: 4th annual international workshop on data mining and audience intelligence for online advertising. ACM, New York
20. Gupta R, Khirbat G, Singh S (2014) A novel method to calculate click through rate for sponsored search. Comput Res Repos
21. Graepel T, Candela JQ, Borchert T, Herbrich R (2010) Web-scale bayesian click-through rate prediction for sponsored search advertising in microsoft's bing search engine. In: 27th international conference on machine learning. Omnipress, pp 13–20
22. Gao Z, Gao Q (2013) Ad-centric model discovery for predicting ads's click-through rate. Procedia Comput Sci 19:155–162
23. Wang F, Zhang P, Shang Y, Shi Y (2013) The application of multiple criteria linear programming in advertisement clicking events prediction. Procedia Comput Sci 18:1720–1729
24. Brzezinski JR, Knafl GJ (1999) Logistic regression modeling for context-based classification. In: 10th workshop on database and expert systems applications. IEEE, pp 755–759
25. Yeh I, Lien C (2009) the comparisons of data mining techniques for the predictive accuracy of probability of default of credit card clients. Expert Syst Appl 36:2473–2480
26. Fitzpatrick T, Mues C (2016) an empirical comparison of classification algorithms for mortgage default prediction: evidence from a aistressed mortgage market. Eur J Oper Res 249:427–439
27. Ng A (2000) Supervised learning models. machine learning (CS229). Lecture notes
28. Barber D (2006) Bayesian reasoning and machine learning. Cambridge University Press
29. Elkan C (2014) Maximum likelihood, logistic regression, and stochastic gradient training. principles of artificial intelligence: learning (CSE 250B). Lecture notes
30. Bottou L (2010) Large-scale machine learning with stochastic gradient descent. In: 19th international conference on computational statistics. Springer, Berlin, pp 177–186
31. Bottou L (2012) Stochastic gradient descent tricks. In: Montavon G, Orr GB, Müller KR (eds) Neural networks: tricks of the trade, LNCS, vol 7000. Springer, Berlin, pp 421–436
32. KDD cup (2012). http://www.kddcup2012.org/c/kddcup2012-track2
33. Guha S, Cheng B, Francis P (2010) Challenges in measuring online advertising systems. In: 10th ACM SIGCOMM conference on internet measurement. ACM, New York, pp 81–87
34. Cheng H, Cantú-Paz E (2010) Personalized click prediction in sponsored search. In: 3rd ACM international conference on web search and data mining. ACM, New York, pp 351–360
35. Chen Y, Yan TW (2012) Position-normalized click prediction in search advertising. In: 18th ACM SIGKDD international conference on knowledge discovery and data mining. ACM, New York, pp 795–803
36. Dave KS, Varma V (2010) Pattern based keyword extraction for contextual advertising. In: 19th ACM International conference on information and knowledge management. ACM, New York, pp 1885–1888
37. Chen J, Stallaert J (2014) An economic analysis of online advertising using behavioral targeting. MIS Q 38:429–449
38. Chen J, Stallaert J (2010) An economic analysis of online advertising using behavioral targeting. Forthcom MIS Q
39. Mcdonald AM, Cranor LF (2009) An empirical study of how people perceive online behavioral advertising. Technical report, Behavioral advertising at CMU CyLab
40. Yan J, Liu N, Wang G, Zhang W (2009) How much can behavioral targeting help online advertising? In: 18th international conference on world wide web. ACM, New York pp 261–270

Cloud-Based Pollution Scheming System Using Raspberry Pi

Vaishnavi Kulkarni and Prashant Salunke

Abstract Pollution is increasing with time and measures to control the same needs to be taken or not now but later, the whole range of living beings have to suffer due to ignored issue of increasing pollution. Amalgamation of trending technologies with the control measures can be the best choice. This paper aims at development of a centralized cloud-based system with a self-determining machine using Raspberry Pi that can monitor and control pollution in the location of heavy traffics or in industries, and collected data can be stored on cloud server that can be accessed through Web browser from anywhere using provided log-in credentials. Due to real-time monitoring of environment, quick actions can be taken for the presence of pollutants around and this can help in controlling pollution with preventive measures. System will sense and detect the contents/pollutants in the environment that will be controlled using a controller and control measures, which will also have sensing capabilities; this data will be communicated with cloud server using Raspberry Pi interface and with log-in credentials. This complete data can be accessed using Web browser anywhere in the world with Internet connection.

Keywords Air pollution · Environmental pollution · Sensors
Cloud computing · Raspberry Pi · Internet of Things · Internet of Vehicles

1 Introduction

The air that living beings breathe has a very crisp composition; 99% from this composition contains nitrogen, oxygen, water vapor, and inert gases. Air pollution is nothing but the addition of foreign things to this air composition. A common and

V. Kulkarni (✉) · P. Salunke
Sandip Institute of Technology and Research Center, Nashik, India
e-mail: vaishnavikulkarninsk@gmail.com

P. Salunke
e-mail: prashant.salunke@sitrc.org

© Springer Nature Singapore Pte Ltd. 2018   327
D.K. Mishra et al. (eds.), *Information and Communication Technology
for Sustainable Development*, Lecture Notes in Networks and Systems 9,
https://doi.org/10.1007/978-981-10-3932-4_34

very harmful type of air pollution is caused when we release particles knowingly or unknowingly into the air from burning fuels may it be any fuel.

Only burning fuels is not the reason behind air pollution; it is also caused due to the pervasive spread of hazardous and unsafe gases, such as sulfur dioxide (it comes in some other combined forms too), carbon dioxide in large content, carbon monoxide, nitrogen oxides (it has the highest distribution in the composition of the air), and other harmful chemicals. These gases are highly reactive in nature and take part in chemical reactions by combining with other elements in the nature. These gases form acid rains and smog-like things due to their presence in atmosphere.

The other form of air pollution could be greenhouse gases, such as dioxides in the form of sulfur and carbon. These gases are the reason behind global warming; it is caused due to the greenhouse effect. The greenhouse effect is well explained by EPA; according to them, greenhouse effect is caused when the greenhouse gases as mentioned above absorb the infrared rays those are released from our planet due to many activities here; this traps the heat on the earth and stops its way to escape into the sky, and hence, out of atmosphere on the earth seems naturally warm; this warmth is increasing every year now due to the release of number of gases in the atmosphere.

Air pollution is a cause of many deaths on earth, and even ambient atmosphere kills many here. According to some statistics, around 2 million people die due to excessive air pollution. This is the immediate result of the air pollution; but along with this, it also acts as a slow poison and affects human health in very bad way; it causes simply from headache to the fatal chronic diseases and cancers. This just explains the health effects in a long term, but we cannot avoid the incidents like Bhopal gas tragedy that took place in the year 1984.

After implementation of this project, it will get easier to monitor and control environment using cloud server, where all data will be stored that can be analyzed and seen in real time through web browser with the interface that will be provided to the cloud server; in case of any emergency alert will be generated on the screen displaying contents of gases (optional) along with alarm at actual place under monetization and control measures can be taken.

There can be many environment pollution control systems, but without the flexibility of Web access or the Internet, that is one of the assets of digitization, we can leverage this opportunity and are coming with the valuable implementation of this project that would provide help to the society and environment, flexibility of the data access, easy control measures, quick actions, and effortless data analysis from anywhere in the world.

Above was the technical scope and novelty, along with that this implementation can be very helpful for the Ministry of Environment, Forest and Climate change under Government of India to monitor pollution at many different areas simultaneously and analyze the data monitored to take quick and necessary actions.

2 Overview

There has been number of regulations and schemes put forth by government to control environmental pollution. Many attempts went unsuccessful while few created major impacts to improve quality of life, by inculcating values and habits for environment protection. The environmental data such as temperature, humidity, and air quality play an important role in the well-being of the environment [1].

The principal emanation standards were presented in India in the year 1991 for petrol and for diesel vehicles; it was presented exactly next year of the release of petrol vehicles [2]. The principles, taking into account European regulations were initially presented in 2000. Dynamically, stern standards have been taken off from that point forward. Every new vehicle made-up after the release of these standards' usage must be consistent with the regulations presented and mandated. Right from October month the year 2010, Bharat stage III standards have been advocated all over the India. In 13, noteworthy urban communities, Bharat stage IV discharge standards are set up subsequent to April 2010.

Vehicle pollution monitoring has been initiated at many places as a step toward pollution control measures, sometimes along with wellness monitoring or accident detection [3]. This application is in similar faces with the industrial pollution monitoring, but at more advanced level as the critical industries such as chemical industries and atomic plants have mandatory processes that are practiced on regular basis for company, people, and environment health checkup. It is necessary to maintain the data for annual review, and it is in fact asset of the company. Hence, the monitored data analysis can be done in number of ways from manual to automate. Cloud server storage of data is one of the methods.

Distributed computing alludes to a computational model where errands are executed by a few interconnected PCs offering more prominent capacities as far as preparing force and capacity than stand-alone arrangements. Scientists have connected the standards of distributed computing to the fields of mechanical technology and in addition portable registering [4]. This has brought about the development of themes, for example, cloud applies autonomy and the portable cloud.

Cloud computing may it be Glasgow Pi Cloud, AWB, eNlight, Openstack, etc., along with the smart robot consisting of Raspberry Pi as proposed below can play a major role in improvement of the environmental conditions on earth [5]. Cloud computing, i.e., SaaS, IaaS makes a vital role in the pollution control and global warming due to its many features initiated due to virtualization technology [6]. We can say that Virtualization is an environmental asset and can benefit human and nature widely [8].

After the literature review, it was noted that there is no such system that could help environmental inspectors to check for pollution in specific areas. An approach that could help inspectors monitor pollution effectively from anywhere in the world and could help them in taking proper decisions in order to take control measures or preventive and corrective actions against it. For example, traffic jams, ill-managed vehicles, flow of traffics, etc., could be the reason for pollution in specific area. Due to centralized pollution scheming systems, a track of keeping fluctuations in the pollution levels could be easily managed and reasons and patterns in pollution could be identified with action plan to manage the same.

3 Proposed System

Proposed system is being represented with the diagram below Fig. 1, learning about the problems faced by the world due to pollution caused by industrial wastes and uncontrolled processes, the system proposed as shown in Fig. 1 can be helpful to monitor measure and control the pollution significantly. There are many systems developed for monitoring pollution but here the novelty of this project is centralized cloud system that can enable anywhere access to the data monitored.

Briefing about the system design broadly, it can be divided into three parts such as sensors, active system, and cloud platform. Further, we can delegate the system mobility with location tracking to gain exact position of pollution augment. Let us see each component of the block diagram in detail.

3.1 Sensors

Primary industrial pollutants are nitrogen dioxide (NO_2) and carbon monoxide (CO) [7]. To monitor these pollutants, the sensors we are using are MICS-2710 for NO_2 and MICS-5525 for CO.

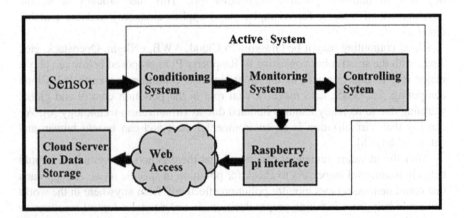

Fig. 1 Block diagram

3.2 Active System

It comprises of signal conditioning system for sensors that can be applied to the monitoring system that have display of screen, which will demonstrate the count of pollutant concentration, through which quick view of the pollutants in air at that place can be monitored. Signal conditioning comprises of combinations of resistors and capacitors, and its output is given to the ADC. The important component, we can say heart of the complete system is Raspberry Pi.

Raspberry Pi: It is called as a single-board computer; it has provision of Ethernet/Wi-fi, USB mouse and keyboard connection, and SD card. Question arises that why Raspberry Pi? When we have variety of processors and controllers available. Reference [5] shows that the Raspberry Pi venture can possibly alter computer science instruction and is as of now collecting backing from industry and the training division, and in addition boundless media scope. The venture and the subsequent framework empower computing science understudies to utilize case ICT and to design complete frameworks. The machines are as of now accessible in model A and B, including little differences in RAM size and the number of input/output ports in the RPi. Device, in any case, offers a noteworthy properties of cloud-based servers, for example, restricted capacity and fringe ability—but at a littler scale. The utilization of an instructive processing stage to build a cloud domain proposes a favorable blend of education and exploration.

3.3 Cloud Platform

Raspberry Pi can be connected to the Internet using Ethernet of Wi-fi module. Cloud platform can be accessed in the presence of Internet, and it provides ease of access from anywhere with additional many features such as data security flexibility, low cost, ease of access, user-friendly, platform-independent. There are number of cloud services available; we are using a ThingSpeak an open data platform for the Internet of Things. Using ThingSpeak, we can collect data, analyze it, and can act upon it.

Maximum of the process and system flow has been already explained in the block diagram, but the work flow will give the details on hardware and software partition in the proposed system. A dataflow, as shown in Fig. 2, consists of the stepwise detailing of the proposed system. Process flow of the proposed system is divided into hardware and software elements, where hardware element defines the details and work flow of the hardware that is used in the system and the software element will provide the details and workflow of the software part used in the system. Although this is at initial stage of the implementation, this could provide a brief overview of the proposed system. Many alterations are expected in every system as the project is proceeded and challenges are faced.

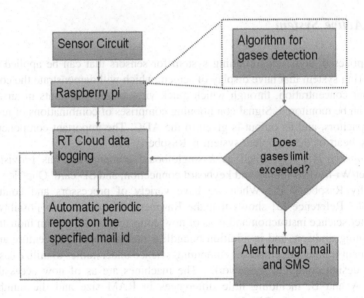

Fig. 2 Data flow

Cloud access in the form of user interface providing details can be provided with the log-in credentials, and with availability of these details, cloud server on which all real-time data in the form of monitoring of pollution is stored can be accessed through Web browser; testing was done for the proposed system, and readings in the form of graphs were taken where we can get into detail by clicking any instance. Figure 3 shows the graph with readings of temperature with respect to time. Similar readings were taken for other parameters, and the testing will be continued till the final product with desired specs is obtained.

Implementation of this system will lead to the useful way out for all the environmental inspectors, and they would be able to easily monitor the streets and regions across the globe from anywhere on the globe. This also initiates a thought for smart city concept, wherein environmental sensors or the modules like one proposed in the system could be deployed across a cluster and the data can be collected through it and monitored through Web. Clusters could then be a part of more big clusters this way, data could be available quick as a wink and everything would be connected and everything would be smart. For example, the recent even–odd rule implemented in Delhi, during the active days of this rule, proposed system could help in monitoring the reduction in pollution. Traffics those are major cause of pollution sometimes can also be managed to avoid pollution in the area by diverting it through other ways and very interesting thing is there won't be any human intervention everything would work smartly and automatically as everything would be connected, a boon of Internet of Things (IoT). These modules could also

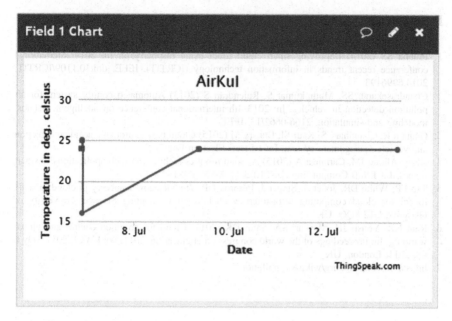

Fig. 3 Temperature monitoring graph

be deployed on the moving vehicles with GPS coordinates to increase efficiency or to increase scope of monitoring; it is referred to as Internet of Vehicles.

4 Conclusion

This entire project primarily concentrates on the idea of identifying the level of pollution and providing anywhere access to monitor it. We can notice a significant growth of air pollution from the last decade; this is resulting in many environmental issues. There will be immense industries that do not consider the measures for controlling pollution; however, many of the companies are considering the fact of pollution control which has as of now brought about a new face to manage the environmental issues. Along these lines, this framework will be profoundly advantageous is checking this issue. It will be easy to utilize this framework in the present system of the industries, its ease of data access from anywhere and log of data collected makes it perfect for the analysis of environment by environmentalists in the industries. This will provide a good solution for the analysis of environment by environmentalists.

References

1. Prabha SS (2014) Improving smart cloud robot using raspberry pi. In: 2014 international conference recent trends in information technology (ICRTIT). IEEE. doi:10.1109/ICRTIT. 2014.6996193
2. Chandrasekaran SS, Muthukumar S, Rajendran S (2013) Automated control system for air pollution detection in vehicles. In: 2013 4th international conference on intelligent systems, modeling and simulation, 2166-0662/13. IEEE
3. Gautam R, Chaudhary S, Kaur SI, Bhusry M (2015) Cloud based automatic accident detection and vehicle management system. Int J Electr Electron Eng 7(2)
4. Reyes Álamo JM, Carranza A (2015) A cloud robotics architecture with applications to smart homes. Int J Soft Comput Eng (IJSCE) 5(1). ISSN: 2231–2307
5. Tso FP, White DR, Jouet S, Singer J, Pezaros DP. The Glasgow raspberry pi cloud: a scale model for cloud computing infrastructures. School of Computing Science, University of Glasgow, G12 8QQ, UK
6. Raut NB, Yousif JH, Maskari SA, Saini DK (2011) Cloud for pollution control and global warming. In: Proceedings of the world congress on engineering, 2011, vol I WCE 2011, July 6 –8, 2011, London, UK
7. https://en.wikipedia.org/wiki/Air_pollution

A Deadlock Detection Technique Using Multi Agent Environment

Rashmi Priya and R. Belwal

Abstract In order to develop a Mobile Agent-based system for deadlock detection in Internet-based systems, Agent A and Agent C were used in the deadlock detection process. In order to achieve the deadlock detection process, a mathematical model has been designed. The implementation of this prototype has been carried using Java platform technology. The result achieved through the experiments satisfactorily matches the analysis of result through mathematical model in knowledge-based multi-agent system.

Keywords Agent · Distributed computing · Deadlock · Resolution

1 Introduction

As both servers and clients keep moving and interacting, there is a call for distributed applications. This is the reason why an improvised technique for traditional approach of distributed computing is needed. Movements of agent are limited by imposing traditional distributed solutions into mobile agent systems as traditional solutions limit the movement of agents. There are many assumptions on which traditional distributed algorithms depend. Some of those are data location, communication mechanisms, or network organization i.e., agent host location, number of nodes, and connections between hosts.

Mobile agents are becoming more popular in areas of application development. Because of movement of both clients and servers is free through the network. Some of the fundamental assumptions of deadlock detection do not hold true.

R. Priya (✉)
TMU, Moradabad, India
e-mail: rashmi.slg@gmail.com

R. Belwal
AIT, Haldwani, India
e-mail: r_belwal@rediffmail.com

© Springer Nature Singapore Pte Ltd. 2018
D.K. Mishra et al. (eds.), *Information and Communication Technology for Sustainable Development*, Lecture Notes in Networks and Systems 9, https://doi.org/10.1007/978-981-10-3932-4_35

Hence, new techniques and solutions and techniques are to be developed for mobile agents to solve their distributed coordination problems.

It is providing competitive environment to traditional distributed computing techniques.

1.1 Distributed Deadlock Detection ("Traditional Approach")

Deadlock detection properties are similar in case of Single processor Systems and distributed systems. The deadlocks are harder to detect, avoid, prevent, and resolve because of the non-availability of centralized information and resource control. These deadlocks are harder to detect. It becomes the responsibility of server to detect and resolve deadlocks in these cases of distributed systems. As the transactions are distributed and a number of servers are interconnected across multiple servers, detection process of deadlock becomes a cumbersome problem. In such situation, detection of deadlock becomes complex problem as number of servers and transactions are interconnected and distributed across network through multiple servers. Under such scenario, distributed deadlock detection solutions can be divided into five categories: centralized, path-pushing, edge-chasing, diffusing computation, and global-state detection. A number of solutions have been proposed to detect and resolve deadlocks in distributed systems. A number of solutions are achieved by optimizing number of messages or frequency of detection.

Distributed deadlock detection is implemented simply using a centralized server to maintain the global wait-for graph. A local copy of the wait-for graph is added which is analyzed on a periodical basis for each server in the network. In this technique, periodically, each server in the network adds to the central server the local copy of the wait-for graph, which is analyzed for deadlocks. As the single server can fail to handle many transactions, this method does not follow fault tolerance.

2 Mobile Agent Systems

This section presents how the properties of mobile agent systems influence traditional distributed problem-solving techniques. Traditional distributed techniques such as distributed shared memory or distributed barriers can be adapted to work in mobile agent systems, but impose restrictions or inefficiencies that are often unacceptable to mobile agent applications. Special attention is paid to distributed deadlock detection, but the problems and their solutions are applicable to a wider range of problems. Traditional distributed deadlock detection solutions commonly

Table 1 Mobile agent system functions

Property
Agent transport
Agent communication
Agent tracking
Agent management
Agent naming

have location and fault assumptions. The following two sections compare the properties of mobile agent systems and traditional distributed computing environments that cause these assumptions to fail. In particular, agent movement and failure recovery are discussed.

A large number of mobile agents are available in the systems. They are used in various day-to-day requirements. Many such environments are available free of cost or are available on purchase. There are advantages and disadvantages related to each environment. Moreover, every environment has some unique properties.

Any default standard does not exist at present for mobile agent system. There is not much competition from the market for mobile agent development as such the mobile agent developer faces difficult situation.

The mobile agent's fundamental properties of a mobile agent system should provide support for the basic properties of an agent.

Additionally, a mobile agent environment must support agent properties such as mobility, learning, and flexibility. Depending on the selected agent attributes, different mobile agent system descriptions are produced (Table 1).

Description

Description
Also called agent migration
Transfer of agent state and type
Supported methods for agent-to-agent communication
Types include asynchronous and synchronous
Localization of agents in a distributed system i.e., finding an agent
Creation/Termination, suspension/resumption
Provide globally unique agent names and locations
Control structure
Region support
Also called constrained execution
Different from security
Types include: hierarchy, partially supervised, set of agents that can be "found"
Allows the mobile agent system to cooperate with other mobile agent systems
Language support security

3 Deadlock Shadow Agent Technique

The foundation and assumptions of this algorithm are presented first, followed by the detection, and resolution phases of the algorithm. Finally, a deadlock detection process is described in the section below.

The following sections deals with topic of the thesis as proposed i.e., it deals with the solution to distributed deadlock detection.

The solution follows the topology to be static in the beginning of the working of algorithm. A Mobile ad hoc Networks (MANET) type topology update protocol must execute in the background to keep routing tables synchronized with the actual topology. This movement is accomplished via node- or resource-based routing and the agent requests to be routed to a particular environment or resource. It is assumed that the host environment to route the agent on a best effort basis. This technique is similar to the routing scheme used in Internet Protocol (IP) routers. The mobile agents of all types are independent from the network topology. The agents can move independent of nodes with no detailing of number of nodes and their connections. They can move independently throughout the network. This property allows host environments to communicate the details of Agent C. The assumption is taken that the Agent C can only lock and unlock the resources. It is assumed that the resource should be at the same environment physically. The agents manipulate the resource requests to the counterparts which detect the deadlock.

This solution assumes that agents must inform the host environment whenever any resource is blocked by them. Moreover, here, the properties of a mobile agent solution are presented. These properties are responsible for deadlock detection.

These assumptions provide the environment for the distributed deadlock detection in a mobile agent system.

The state of agent is communicated to deadlock detection agents. The additional requests desired by the blocked agents are not granted. The blocked agents during the deadlock detection process are available in the agent system. The agents are assigned identifier preceding the blocking if any done by Agent C. The agent identifiers can be assigned even when a resource is locked.

4 Algorithm Overview

"The agent-adapted mechanisms are used by mobile agents that are based on those found in traditional edge-pushing global wait-for graph algorithms [20]." The assimilation of host environments including Agent A, Agent B, and Agent C combined together helps to detect and resolve any deadlocks that are present in the case of mobile agent system. The advantages of the present technique in a mobile agent environment are dealt in the next section.

The combined elements and their tasks that are part of this distributed deadlock detection solution are described in the following table. Following sections describe the phases of the algorithm, which *are* initiation, detection, and resolution.

It is noted that deadlock detection probes come into existence from traditional solutions. The segmental partition of global wait-for graph also receives inspiration from traditional solutions.

There are three types of agents contained in the mobile agent system: Agent C, Agent B, and Agent A. In the given solution, Agent B maintains a part of the wait-for graph in place of their target Agent C. During the deadlock detection process, the Agent B progressively checks for deadlocks in the global wait-for graph. It tries to form deadlock detection agents to construct the global graph. Detector agents are small reactive agents and are used in place of probes in traditional edge-pushing solutions (Table 2).

Table 2 Deadlock detection elements	Element name description
	Agent A
	Agents in the system that actively consume resources and perform tasks
	Only agent that can lock resources
	Represent the entities which implement algorithms and perform tasks
	For deadlock detection and resolution, there is no active role
	Agent B
	It is formed by host environments
	It helps to maintain the resources locked by Agent C
	It helps to follow the network
	It initiates the deadlock detection phase
	It helps to inform the target when the blocking of agent occurs
	It analyzes the data collected by Agent B
	It starts deadlock resolution when needed
	It helps to detect and recover when faults occur in deadlock detection process
	A representation of global wait-for graph occurs
	Agent C
	It is an agent created by shadow agents
	It informs the host environment when the target agent is blocked
	It is a type of mobile agent
	It helps in the construction of global wait-for graph
	It helps in opposing deadlock situation
	Host environment
	It helps in hosting of the mobile agents
	APIS are provided for consumer and shadow agents
	APIS for resources move, block, unblock, etc.
	It helps in the coordination of agents
	It progresses through the network
	It manages the work through token
	It is taken by Agent C

5 Deadlock Detection

Upon creation, Agent B starts the deadlock detection cycle by initiating detection agents. These are created with predecessor Agent B group of locked resources and the residing environments of their location. Once initialized, the birth of a dedicated detection agent permits agent B to simultaneously look for deadlocks and reciprocate to other Agent B to move to the resources that are locked by Agent C, which is the target.

At each resource, the movement of Agent D occurs by noting the network location of each resource whenever it gets locked. The routing of detector agents is taken forward. They call the host environment to find out whether some other agents are blocked on the same resource.

The processing is executed at the same time for each and every agent whenever blocked agents are found. The Agent D locates the related Agent B and finds out for the deadlock detection information that is present on a resource which is held by a remote agent (Tables 3 and 4).

6 Deadlock Resolution

The resolution of deadlock is discussed in this section whenever the replica of an agent detects a deadlock or when a cycle is formed in the global wait-for graph which needs to be dissolved. The figure presented as under shows how the deadlock detection information gathered in above section is used to break a deadlock. The copy of an agent must identify the cycle to resolve the deadlock in a successful way and in order to determine steps to break the deadlock.

The information collected by the Agent D is enough to form the deadlock cycle.

It helps in the determination of resource that can be unlocked. The following table represents the graph represented by the deadlock detection Table 5.

Here, we see that the resolution phase is now started. Each shadow determines the deadlock cycle. It also determines lowest priority resource in the cycle. In this

Table 3 Deadlock detection information

Name of agent	This denotes to the unique name to identify the agents
Resources which are blocked	This provides the information that the agent is blocked
Primary locks	This indicates the lock in resource information format

Table 4 Resource information

Resource owner	The agent name that has locked this resource
Resource name	The name of the resource with uniqueness
Environment name	The name of environment containing the resource
Priority	The priority of the resource

Table 5 Deadlock information

Agent name	Blocked on resource	Primary locks
A1	R2 (R2, A3, E2, 2)	R1 (R1, A1, E1, 1), R4 (R4, A1, E4, 4)
A3	R1 (R1, A1, E1, 1)	R2 (R2, A3, E2, 2)

case, the resources involved in the cycle are determined to be R1 and R2. Even though both shadow agents find that R2 has the lowest.

Blocked on resource R2 (R2, A3, E2, r2) priority and must be unlocked, only S3 owns R2 and is responsible for breaking the deadlock. The following events occur to break the deadlock and allow processing to continue:

1. S3 instructs D3 to travel to E2 and lock resource R2 and notify A1 when unlocked.
2. D3 moves to E2.
3. D3 unlocks E2.
4. D3 leaves E2 to notify its shadow of success.
5. E2 notifies A1 that the resource it wanted is now free.
6. A1 locks R2 and continues processing.
7. D3 arrives at E1 and notifies its shadow that R2 is no longer locked.
8. A3 performs recovery from having its resource unlocked.

When D1 and D3 return to their shadow agents (S1 and S3), their constructed tables are analyzed for possible deadlocks before being added to the shadow's local detection information. Each shadow scans for its target agent in the returned table to recognize a deadlock situation. In this example, each shadow agent detects a deadlock, since their target agents appear in the returned tables on the first iteration.

7 Measurement Discussion

For example, the value of Mo in a fault-free ring network of three agents (configuration b) should be: Mo = 3 × 3 × 4 or Mo = 36. The value presented shows a median value of 25 migrations in this configuration. The insertion of random delays into the simulation and their associated influence on event sequencing explains the difference between the measured and theoretical value for Mo. If an agent is sufficiently delayed, other agents can migrate and perform multiple lock checks before the delayed agent arrives. This allows the delayed agent to gather more information in a single trip than normally would be possible, speeding the detection of the deadlock. The effect of sequencing changes can also be seen in the number of repetitions, which is also below the theoretical maximum values.

This means the value presented for the failure-prone configuration of three agents in a deadlock is 33 that the additional work created by detection agent failures was offset by the savings achieved through event-sequencing changes. As the number of faults increases with 13 total faults and long delays, this number is below the theoretical maximum of 36; the number of migrations will approach and then exceed the theoretical maximum.

For example, the impact of faults on the total number of migrations is difficult to measure as it depends on other factors such as network topology and at which point of the detection process the failure occurs.

The number of migrations required to detect a deadlock in configuration 'c' (a ring topology) is 54 while the equivalent mesh topology requires 32 messages. As a result, the comparison adding more nodes into this configuration, but keeping the number of agents constant, causes a proportional increase in both of these values between topologies is not a meaningful measurement.

The number of repetitions (NR) is used to compare deadlock detection efficiency between topologies. NR is only impacted by the number of agents in the system; therefore, if the number of agents is kept constant, the number of required repetitions should stay constant between topologies.

The difference over the ten executions while agents in the ring topology required about three repetitions and agents in the mesh topology required four repetitions to detect the deadlock in these values can be compared between fault-free and failure-prone scenarios to determine the impact of failures on the number of repetitions required to detect a distributed deadlock can be attributed to sequencing differences (and their savings); therefore, it can be determined that ring configurations are more sensitive to event sequence changes than mesh configurations.

When the detector agent returns and if the suggested mobile agent algorithm assumes that network topology is static and always organized in a ring, a significant reduction in the number of migrations can be realized using this assumption, and a simplistic deadlock detection technique can be created where a detector agent performs a walk of the entire network and gathers deadlock information to its starting point, the data is checked for deadlocks.

Unrestricted agent movement and fault tolerance in most real-world IP networks in this scheme follow the number of migrations equals the number of nodes in the network causes similar increases in the number of migrations is similar to the mesh topology simulation; therefore, detector agents can often migrate directly to the desired host environment. Additionally, it has been observed and documented that most deadlocks occur between two transactions. As such resource locked by a remote mobile agent roughly maps to a transaction in traditional distributed deadlock detection the measurements for systems with two agents are of particular relevance. Due to the common nature of deadlocks between few agents or transactions, the number of messages needed to detect and resolve a deadlock between many agents is not a significant concern or disadvantage.

The measurements gathered in mobile agent environments demand network organization (topology) flexibility and fault tolerance. Incorporating these properties into a mobile agent solution impacts the amount of processing required for the algorithm during simulation of the solution suggested by this thesis confirms that these properties require additional processing and messaging. Although due to the parallel nature of mobile agent systems, there is no conclusive evidence that the

Table 6 Simulation environment configurations

Configuration	Number of agents	Number of agent
A1	2	2
B1	3	3
C1	4	4
D1	20	20

Table 7 Mesh connected and fault-free measurements

Configuration	Med message vol	Median number repetitions
A1	8 during detection	A1:2
C1	32 during detection	A2:3

Table 8 Mesh-connected and failure-prone measurements

Configurations	Median message Vol	Median Rep	Total faults	Missed deadlocks
A1	2 during reso	A1:2	1	0
B1	15 during reso	A1:3	2	0
C1	2 during reso	A3:5	2	0
D1	1174 during reso	Amed:20	58	0

additional processing significantly impacts deadlock detection performance or efficiency. Additionally, the lack of similar mobile agent solutions makes comparison and evaluation difficult and inconclusive (Tables 6, 7 and 8).

8 Results and Discussion

The mobile agents which are mentioned in this paper are created using Java Agent Development framework. Through mathematical analysis, a formula is derived with respect to movement of agents which helps to identify the deadlock conditions. The result obtained experimentally is in approximation to the theoretical derivations. The measurements gathered in mobile agent environments demand network organization (topology) flexibility and fault tolerance. Incorporating these properties into a mobile agent solution impacts the amount of processing required for the algorithm during simulation of the solution suggested by this thesis confirms that these properties require additional processing and messaging. Although due to the parallel nature of mobile agent systems, there is no conclusive evidence that the additional processing significantly impacts deadlock detection performance or efficiency. Additionally, the lack of similar mobile agent solutions makes comparison and evaluation difficult and inconclusive.

Google AdWords: A Window into the Google Display Network

Vekariya Subhadra, Kapadiya Urvashi, Fruitwala Pranav
and Vyas Tarjni

Abstract In the twenty-first century, the advertisers are slowly migrating from the traditional advertising platform to digital advertising platform. Google AdWords provides two platforms to market your products and reaches the customers. Google Search Network and Google Display Network are the two options in Google AdWords to publish your advertisement on the Google. Google Search Network covers the Google search platform for advertising, and the advertisers can advertise with the Google search results. The advertisers who wish to reach diverse customers can use the Google Display Network platform. This platform reaches 90% of the people around the globe as the advertisements are not only displayed in the search results but also displayed in all the partner sites of Google such as YouTube, blogger, and also the sites registered to the Google Ad-Sense. This paper discusses the in-depths of using the Google Display Network platform for advertising.

Keywords AdWords · Google Display Network · Mobile ads

1 Introduction

The Google Display Network [1] includes mobile sites and apps and is a collection of specific Google Web sites and partner Web sites which includes YouTube, Google Finance, Blogger, and Gmail. There are different ways to show ads on

V. Subhadra (✉) · K. Urvashi · F. Pranav · V. Tarjni
Institute of Technology, Nirma University, Ahmedabad 382481, India
e-mail: 15mcen30@nirmauni.ac.in

K. Urvashi
e-mail: 15mcen11@nirmauni.ac.in

F. Pranav
e-mail: 14mcen06@nirmauni.ac.in

V. Tarjni
e-mail: tarjni.vyas@nirmauni.ac.in

© Springer Nature Singapore Pte Ltd. 2018   345
D.K. Mishra et al. (eds.), *Information and Communication Technology
for Sustainable Development*, Lecture Notes in Networks and Systems 9,
https://doi.org/10.1007/978-981-10-3932-4_36

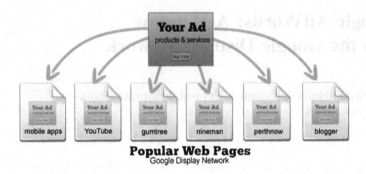

Fig. 1 Overview of Display Network [11]

display network. The different ways include reaching a new customer or fascinate users with layout or one other option is to choose the position of the visibility of your advertisement. The Display Network lets you place your content in front of spirituous customers at the correct place and at the appropriate time in various ways. One of the ways is to target users using the keywords and topics. Other ways include selecting distinct pages or Web sites and trying to search for the users who are already intensified in the stuff that you are proposing. Display Network has a variety of advertisement layouts which includes text advertisements, rich media advertisements, image advertisements, and video advertisements. Ad auction [2] in Display Network is quite like the regular auction except the feature of incremental clicks and an additional service fee for interest category (Fig. 1).

To know whether you are achieving success with your business targets, Display Network provides reports [3] to see on which web pages your advertisements are succeeding and this way you will be able to know which Web sites are giving you higher profits at lower costs. If the advertisement is not doing according to your favor, then the Google Display Network automatically decreases proposal for that Web site.

2 Google Display Network Campaign Setup

The effectiveness of a Google AdWords [4] campaign can be increased by understanding the various campaign settings. The campaign settings selected by the user will have an effect on all ads that belong to the same campaign. Various settings are available depending upon the type of campaign chosen by the user. The settings decide three basic attributes of any ad. Namely

- Budget and bid: Setting the budget and bid will decide the price you pay for your ad.
- Extra features that the user can add in his ad, apart from text and links.
  This can be done with the help of ad extensions.

- Deciding the perfect location for your ad using the geographical location, and language settings.

Three main components have to be kept in mind while deciding the content of any ad group, namely a set of specific keywords [5], ads, and the budget/bid. More than one ad group forms an ad campaign. Grouping your keywords with specific ads can be done by ad groups. This makes sure that customers find the ads that are most relevant to them, when they search using a specific keyword or a phrase.

2.1 Structure of an AdWords Account

AdWords is subdivided into three basic sections: the account, campaign, and the ad group.

- Unique email address, password, and billing details are assigned to an account.
- A fixed budget and specific settings are provided to the ad campaign. These details decide the location where your ad will appear.
- The keywords/phrases and sets of related ads that you wish to be displayed in your advertisement are contained in an ad group.

An organized account can help with effective ad targeting [6] and quick changes which will eventually result in achieving your ad goals. Tips to consider while structuring the account:

- Build your campaign in such a way that it reflects the core idea of your Web site.
- While advertising in different regions, it is a good practice to create different campaigns corresponding to the different regions.
- Manage your campaigns effectively with the help of AdWords editor.

2.2 AdWords Ad Gallery

The Ad gallery is a tool for creating advertisements which provides various formats for displaying your ads. Many ad formats are available such as image ads, video ads, and dynamic ads. Advantages are as follows:

- Differentiated products and services: By setting appropriate and clever product images, colors, and logo that reflects the brand you wish to advertise, you can tempt the customers to engage with the ad.
- Increase campaign effectiveness: By using display ads that provide visual effects, you can achieve higher click-through rates and witness an increasing conversion volume on the Google DN.

- Free ad templates: try and choose different color palettes, images, etc., for your ad at zero cost.
- Customize ad styles: You can create your own custom ad in any available size and color and create your own variations of the style. You will be charged only for the number of clicks received by your ad, or the impression your ad made on sites across Google's DN.
- Business ad templates: Create clear-cut and precise ads with limited but relevant text information.

2.3 Features for Bidding on DN

There are various ways to set the bids in your campaign when advertising your ad on the DN.

- Default Bids: If you have not placed a custom bid on your ad, AdWords will assign your ad group default bid value to your ad, in case your ad appears in a location which matches the area you were targeting.
- Custom Bids: By enabling custom bids for a single targeting method, such as keywords, AdWords will assign the custom bid value to your ad when it appears on Web sites that match the keywords of your ad.
- Adjusting Bids: By adjusting the bid values at the campaign and ad group stage, you can control the time and location for which your ad will appear.

The type of bids provided by the Google Display Network can help you in increasing the effectiveness of your campaign by adjusting and setting the perfect price for your ads. Two bid types are provided by Google Display Network: (1) You can set your ad group default bids, your own custom bids, or choose the bid adjustment feature, for the campaigns targeted specially for the Display Networks. (2) Automatic bid adjustment will be done for the campaigns that fall under the Search Network and Display Select categories.

2.4 Viewable Impressions Bidding with Viewable CPM

With the viewable cost-per-thousand impressions, you can choose to pay for only those ads that are marked as viewable. In the case of display ads, the ad is considered as "viewable," if more than 50% of the ad appears on the screen for more than one second; and for video ads, the ad is considered "viewable," if it stays on the screen for two or more seconds. If you are using CPM bidding method for the campaigns meant only for the Display Networks, it is advisable to choose the VCPM as your bidding strategy. For ads belonging to the campaigns that target mobile applications, you cannot use the VCPM bidding strategy. Ads for app installment or app engagement cannot use the VCPM. By using VCPM, you will only have to pay for those impressions that are considered to be "viewable." It also optimizes your bids in such a way that your ads appear in the slots that are most viewed. If you are looking to maximize the views on your ads, and not the clicks, VCPM is the go-to strategy.

2.5 Flexible Bidding Strategies

Flexible bidding strategies provide automatic bidding facilities for the exact time and location that you want. This can be done for multiple campaigns or within a single campaign. Many flexible bidding strategies are available:

- Clicks Maximization: This helps you by automatically setting bids to acquire the maximum clicks, but within the specified spend time.
- Target search page location: Adjusts the bids in a manner which can raise the chances of your ad appearing on Google search results first page, or on top of a page. This strategy increases the ad visibility.
- Enhanced Cost per click: Increases or decreases the manual bid automatically on the basis of a click conversion probability.
- Target outranking share: adjusts bids to achieve greater ad visibility and rank, relative to ads from other domains.
- Target Cost per acquisition: increases click conversions, while simultaneously achieving average CPA.

2.6 Conversion Optimizer

CO focuses on increasing/maximizing the conversions, instead of targeting CPC, or CPM. Its main focus is to get the maximum possible conversions, while keeping a close track on the cost-per-acquisition specifications. Conversion optimizer completely eliminates the need to adjust the bids manually to achieve the conversion targets. It takes into account information related to your previous campaigns and auto sets the best CPC bid for your ad, every time it is eligible to appear.

2.7 Elevate Your Ad Using Ad Extensions

Ad extensions give customers more reasons to click on your ad. Ad extensions are basically formats for your ads which you can use to add extra information about your business. This can be done either manually or automatically. Ad extensions are excellent tools for increasing the CTR and visibility of your ads. If adding the extension can elevate the performance of your campaign and if your "Ad Rank" is high, AdWords can associate extensions with your ad. There are two categories of ad extensions: manual and automated.

2.8   Campaigns for Mobile App Installs

Use campaigns to increase app downloads. The feature "Mobile app installs" on DNs and "Trueview" on YouTube allow you to make your own app install ads, specifically targeted for mobiles and tablets. AdWords use the logo of your app, and customer reviews to create the app install ad. This ad can redirect customers straight to the app store to download your app. Connect your AdWords account to Google play developer account to easily monitor and track installs. There are two types of app campaigns: mobile app installs and universal app campaign.

2.9   Dynamic Display Ads

You can promote your products and services from your feeds by creating dynamic display ads for them. Such ads can be created while setting up a dynamic re-marketing campaign from the ad gallery of your account. You can specify the type of layout you want. If you choose a custom layout, you will have to specify the attributes that you wish. There are many layout features to choose from such as multiple product carousels, image cropping, and star ratings. Apart from the layout, you can also customize the text that will appear in your dynamic ads. You can decide how the different attributes should be arranged in your dynamic ad.

2.10   Lightbox Ads

Lightbox ads are a perfect way to target new customers by using vivid ads that have interactive formats. You can decide your own billing model based on your needs and objectives. Two kinds of Lightbox ads are available:

- Ready Lightbox ads: This allows you to add your brand images, videos, maps, product feeds, etc., to the already available set of templates that are pre-designed by AdWords.
- Custom Lightbox ads: These ads belong to ads served by third parties.
  The ads are professionally and entirely developed by creative agencies.

Google assigns a family status attribute to all ads. Any ads that are Lightbox ads must be family safe as the audience targeted by such ads belong to all ages. Ready

Lightbox ads take only a few minutes to make and v = can be published well within a day. Ads that use studio layouts can take anywhere between 3–4 days to be developed and published. Lightbox ads are available in standard sizes, decided by the Interactive Advertising Bureau (IAB). Once customers engage with such ads, new media is added to the ads either directly within the ad, or in a separate expandable canvas. This helps reduce the accidental clicks and increase the customer satisfaction. Two different formats are available for Lightbox ads: Lightbox ads with many videos and Lightbox ads with a combination of images and videos gallery.

2.11 Ad Publishing Requirements

All ads must meet certain standards in order to be published. This ensures excellent user experience. An ad can be rejected for errors such as:

- Ambiguous promotion: when the ad is difficult to understand, inaccurate, or all together nonsense.
- Ambiguous relevance: when the ads are not relevant to the landing pages where they appear.
- Unreliable claims: when the claims made in the ads have not been verified by the party on whose page the ad appears.
- Grammatical/spelling errors: ads that contain grammar or spellings that are not commonly recognized or when the ads use improper punctuations, capitalization, spacing, etc.

If your ad has been disapproved or rejected, you should try to find the reason for the rejection, modify your ad, and then wait for it to be reviewed again.

3 Ads on Display Network

On the Google Display Network, you can achieve an extensive variety of clients, pick which destinations or pages your advertisements show up on, and connect with individuals with an assortment of engaging promotion groups. To promote on the Display Network, utilize any campaign "Display Network only" or "Search Network with Display Select." The nuts and bolts get your promotions on to the Display Network targeting strategies such as keywords, and proclivity gatherings of people get your advertisements on to the Display Network.

Focusing on settings such as "Target and bid" and "Bid only" additionally decides the range of your promotions. Consider the controls that do the following:

Target and bid: confines your advertisements to appearing for the focusing on strategy you have chosen, as catchphrases. Permits you to set bids for individual focusing on, as $2.00 for purchasing soccer balls.

Bid only: confines your advertisements to appearing for the focusing on strategy you have chosen, as catchphrases. Permits you to set bids for individual focusing on, as $2.00 for purchasing soccer balls. Does not confine your advertisements to appearing for the focusing on technique you have chosen. Permits you to set bids for individual focusing on, as $2.00 for soccer fans.

Placements are areas on the Google Display Network where your advertisements can show up. They can be a site or a particular page on a site, a portable application, video content, or even an individual ad unit. What makes a placement a "managed placement" is that you have focused on a site, versatile application, or ad unit particularly.

With negative keywords [7], you can keep your advertisement from appearing to individuals scanning for or going to, sites about things that you do not offer, demonstrate your ads to individuals who will probably tap on them, reduce costs by excluding keywords where you may be spending cash but not getting a return.

AdWords gives you a chance to do the following with negative keywords keeping in mind the end goal to enhance your campaigns:

- Add negative keywords.
- Edit, remove, or download negative keywords.
- Exclude keywords.
- Get negative keyword ideas.

In the event that you attempt to speak with other people who do not talk the same dialect, then you may think that it is intense to convey the desired information. So also with AdWords, you need your ads to show up for clients who can comprehend them. Language targeting permits you to pick the dialect of the sites that you would like your ads to show up on. We will demonstrate your advertisements to clients who use Google items (e.g., Search or Gmail) or visit sites on the Google Display Network (GDN) in that same language. Remember that AdWords does not interpret ads or keywords. Every Google area has a default language. For instance, Google.com defaults to English and Google.fr defaults to French. On the Google Display Network, AdWords may take a gander at the language of the pages that somebody reviews or has as of late seen to figure out which ads to show. This implies that we may identify the language from either pages that the individual had seen before, or the page that they are at present survey.

Display Planner is a free AdWords tool that you can use to arrange your Display Network ad campaigns. You will simply require a couple of fundamental points of interest to begin, similar to your clients intrigues or your landing page. Display Planner then produces targeting ideas alongside impression estimates and historical expenses to guide your plan.

4 Reaching Audience on Display Network

You can add different audiences to the ad groups and target groups on the basis of your advertisement goals and the stage of their purchase process. There are three main types of audiences: affinity audiences, custom affinity audiences, and in-market audiences.

You can decide what advertisements to show the audience based on the content of a Web site/page when they visit a Google partner Web site. You can use this content to show relevant ads. The data from third-party companies and the major topics from the web page visited can be used to associate the audiences' interest with their anonymous cookie ID.

Google's re-marketing feature, as the name suggests, can help you in reaching customers that have already visited your Web site and hence makes it easier to connect with an "Already Interested" audience. The similar audience feature can help you in discovering new customers that share similar interests with your existing audience. This opens up a new way to find potential and qualified customers. This feature, however, is available in the Google Display Network only.

By observing the browsing activities of up to 30 days on Display Network sites and by using the contextual engine, AdWords can understand the similarities and interests of the audience in your re-marketing category. By using this information, AdWords [8] can find new audiences with similar interests and characteristics to the already existing people on your re-marketing list. Benefits of using the similar audience feature are it simplifies audience targeting and attracting new and potential customers.

With the demographics targeting feature of AdWords, you can target customers in a specific demographic group such as gender: male, female, or other; age: below 18, 18–30, 30–45, 46–60, or other; parental status: parent, not a parent, and other. This kind of demographic targeting on Display Networks can help in reaching the right audience, customizing your ads, and in target refining.

For demographics on mobile applications, instead of using cookies, this feature uses an anonymous identifier to remember that person's preference and the apps visited, by keep tracking of browsing history and app activities. This identifier can be linked to the customer's cell phone and associated with a certain demographic category.

Re-marketing helps you to target those people who have previously visited your mobile app or Web site, so that you can show your ads to them. Re-marketing is limited to only previously visited apps and web pages. A new feature called "Dynamic re-marketing" adds another level of sophistication by targeting those products and services that people were attracted to, "within" the ads. There are many forms of re-marketing supported by Google, namely standard re-marketing, dynamic re-marketing, search ads re-marketing lists, video re-marketing, and mobile app re-marketing.

5 Google Display Network Versus Google Search Network

5.1 Google Display Network

It can create all types of advertising such as text, image, and video. In DN, you can place where your ads are displayed on the Web site that are relevant to what you are selling. It also manages and tracks your budget. It is similar to the Google Search Network [1] but with additional feature of the displaying ads in network sites and videos. Advertising format includes pictures, video, text ads with sounds. It targets the search with advance category option. It is used for displaying branding ads. In Google Display Network, you can choose place where your ads appear on the particular Web site and particular geographic area.

5.2 Google Search Network

Ads can be placed in Google search results and search partner sites linked with it. Advertising format is only based on text ads. It only targets the search keywords. Bidding is done with CPC (cost per click) [9] basis. It is used for call to action ads. In Google Search Network [10], the costing of campaign ads is 5–10% higher than display ads. In Google search network, you can decide how much amount to spend and after that you will not charged more than that.

6 Conclusion

Google Display Network has brought a new approach to the advertisers by displaying ads in such a way that the ads reach to the 90% of the people worldwide. This paper has discussed the in-depths of the Google Display Network platform of the Google AdWords. The paper also discusses setting up an advertiser's campaign which also includes the details of the various types of advertisements and their format. The various bidding strategies and the different display options for the ads are covered along with various options to reach the diverse customers.

References

1. https://support.google.com/adwords/answer/2404190?hl=en
2. Yoon S, Koehler J, Ghobarah A (2010) Prediction of advertiser churn for Google AdWords. JSM Proc
3. Marshall P (2004) Playing to win in Google AdWords: how to structure your campaigns for maximum results from the very beginning

4. Jacobson H (2011) AdWords for dummies. Wiley
5. Geddes B (2014) Advanced Google AdWords. Wiley
6. Mehta A et al (2007) AdWords and generalized online matching. J ACM (JACM) 54(5):22
7. Joshi A, Motwani R (2006) Keyword generation for search engine advertising
8. Tan A (2010) Google AdWords: trademark infringer or trade liberalizer. Michigan Telecommun Technol Law Rev 16(2)
9. Davis H (2006) Google advertising tools: cashing in with AdSense, AdWords, and the Google APIs. O'Reilly Media, Inc.
10. Jahan F, Fruitwala P, Vyas T (2016) A study of working of ad auctioning by Google AdWords. In: Proceedings of international conference on ICT for sustainable development. Springer, Singapore
11. Semrush: www.semrush.com/blog/choosing-websites-for-google-ads-at-the-google-display-network/

Clustering to Enhance Network Traffic Forecasting

Theyazn H.H. Aldhyani and Manish R. Joshi

Abstract Network traffic forecasting has become more and more vital and important in present days for monitoring the network traffic. The number of users that are connecting to network utilization is experiencing exponential growth. The accurate of modeling and forecasting network traffic is increasingly becoming significant in achieving guaranteed quality of service (QoS) in network. The enhanced QoS is maintained in the network by modeling and forecasting network. In this paper, we propose an integrated model that combines clustering with linear and nonlinear time series forecasting models, namely, Weighted Exponential Smoothing (WES), Holt-Trend Exponential Smoothing (HTES), autoregressive moving average (ARMA), Hybrid model (Wavelet with WES), and AutoRegressive Neural Network (NAR-NET) models to enhance forecasting of loading packets in network. The experimental results show that the integrated model can be an effective way to enhance forecasting accuracy attained with assist of derived centriods. The performance measures MSE, RMSE, and MAPE are used to evaluate the results of conventional time series models and proposed model.

Keywords Network traffic · Forecasting · Clustering · Time series models

1 Introduction

Nowadays, the broadcast speed of wired and wireless network has increased manifold. They have created the concepts such as cloud and always remain connected, which in turn have generated high volumes of data on network. This increasing transmission speed of different types of network, the Internet of things (IoT) and machine

T.H.H. Aldhyani (✉) · M.R. Joshi
School of Computer Sciences, North Maharashtra University, Jalgaon, India
e-mail: th0ha0@yahoo.com

M.R. Joshi
e-mail: joshmanish@gmail.com

© Springer Nature Singapore Pte Ltd. 2018
D.K. Mishra et al. (eds.), *Information and Communication Technology
for Sustainable Development*, Lecture Notes in Networks and Systems 9,
https://doi.org/10.1007/978-981-10-3932-4_37

to machine (M2M) have gained momentum. In future, this data volume will grow substantially and experts are forecasting it to grow to tenfold by 2019 [1].

Consequently, forecasting of network traffic plays a vital role in planning, management, and optimization of modern telecommunication networks. On the other hand, accurate and dependable forecasting permits planning the capacity of a network on time and sustaining the necessary level of quality of service (QoS). Network traffic forecasting is the key issue to ensure effective bandwidth allocation, speeding of network, congestion control, and engineering flow. The network traffic data has more fluctuations among entire data. According to the network traffic data, we need to establish the suitable mathematical model to forecast network traffic. The network traffic forecasting models can prevent network congestion, and the utilization rate of the network can effectively improved. There has report of a large work focused on developing forecasting models for computer data networks. The most popular and common time series models are linear time series models such as single moving average, Holt-Winters, exponential smoothing, and autoregressive moving average (ARMA) models and nonlinear time series models such as artificial neural network (ANN) and wavelet models.

In this paper, we discuss the use of linear and nonlinear time series models for forecasting of loading packets volume in network traffic. Moreover, in order to enhance forecasting accuracy, we applied k-means clustering approaches. An integrated model is proposed for network traffic forecasting that integrate the results of various time series models with derived centriods that obtained from clustering approaches. The model is tested using real network traffic data.

The rest of this paper is organized as follows: Section 2 discusses related work. The description of generic framework is discussed in Sect. 3. Section 4 presents integrated model, followed by results analysis in Sect. 5. Finally, Sect. 6 concludes this paper.

2 Related Work

The first stage to achieve a good forecasting algorithm in order to optimize network traffic with a objective to sustain network utilization is an accurate model for the network traffic. The major objective for Internet traffic modeling and forecasting is to determine appropriate model, which can best represent the network traffic behavior and incorporate the model for enhancing loading packets despite the fact that network traffic is complex in nature. The motivation of proposed work is from previous research where researchers have used several time series models based on the behaviors and characteristics of traffic data and also keeping in mind the loading packets in network traffic. In the following sections, we will discuss some of the research works done so far, which has motivated this work.

Feng and Shu [2] studied ARMA model to predict network. Di et al. [3] and Li et al. [4] used wavelet approach with linear and nonlinear time series model to forecast network traffic for network management and design. Rutka [5], Wang and Liu [6], and Chabaa et al. [7] proposed neural network approaches to predict network traffic for enhancing QoS in network.

Kuang et al. [8] proposed a hybrid model (two-dimensional correction and single exponential smoothing) for prediction mobile network.

In this research works, we are concerned to use clustering approaches for enhancing time series models for forecasting network traffic. Nevertheless, we come up with a new method that can help to enhance forecast time series models. Our new method is focused on the centriods of clustering for improving the forecasting of time series models.

3 Generic Framework

Figure 1 displays the architecture of our integrated model that combines forecasting of conventional models with the derived centriods obtained from clustering. The various conventional forecasting models, namely, Weighted Exponential Smoothing (WES), Holt-Trend Exponential Smoothing, autoregressive moving average (ARMA), Hybrid model (Wavelet with WES), and AutoRegressive Neural Network (NARNET) models are implemented to forecast future loading packets in network traffic. Limitation description of the methodology and procedures of these conventional forecasting models can be read from [2, 9].

Further, the derived centriods are used to improve forecasting. The novelty of integrated model is an integration of the forecasting results that are obtained from time series conventional models with the derived centriods that are obtained from clustering approaches. We applied k-means clustering approaches to enhance time series forecasting models for forecasting loading packets of network traffic. The explanation of theory and mathematical model of these clustering approaches is available in [9].

Fig. 1 Generic model for proposed network traffic forecasting model

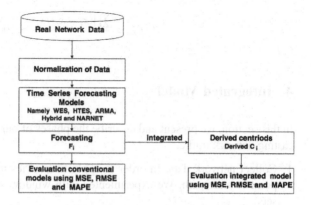

Finally, the integrated model is tested with online data attained from WIDE backbone network. It is observed that the integrated model has enhanced the accuracy of conventional models.

3.1 Data Set

The data set of our research is attained from real network traffic from the WIDE backbone network. WIDE backbone network repository is maintained by the MAWI working group. This data daily trace at the transit link of WIDE (150 Mbps) to the upstream Internet service provider (ISP). For our analysis, we used nine year data from 2008 to 2016 which is aggregated every one hour. We extracted the numbers of packets by using wireshark software [10].

3.2 Evaluation Metrics

To evaluate our forecasting models, we used three error indicators. Mean square error (MSE), root mean square error (RMSE), and mean absolute percentage error (MAPE) are applied as performance indices. These standard indicators methods are defined as follows:

$$MSE = \frac{1}{N} \sum_{k=1}^{n} (x_t - \overline{x_t})^2 \tag{1}$$

$$RMSE = \sqrt{\frac{1}{N} \sum_{k=1}^{n} (x_t - \overline{x_t})^2} \tag{2}$$

$$MAPE = \frac{\sum |(x_t - \overline{x_t})/x_t|}{n} * 100 \tag{3}$$

4 Integrated Model

In this section, we present and describe the phases of our integrated model for forecasting network traffic.

1. Collect network data. In order to apply time series model, we need a few data point to start with. We experimented with window size of last six month data sets.

2. Data of last six months from August 2015 to January 2016, monthly data is used for forecasting $F_{feb2016} = F(D_{Aug2015}, \ldots, D_{Jan2016})$, where D_m is a data obtained from months m and F_m is a forecasting for month m.

3. Normalize the data.

4. **Deriving cluster centriods**: Obtain cluster centriods ($Derived_{C_i}$).

5. **Forecasting**: Apply conventional time series models, namely, WES, HTES, ARMA, hybrid, and NARNET models for forecasting future loading packets of network traffic (F_i).

6. **Enhance forecasting**: An integration of forecasting of each conventional time series models with the derived centriods. The integration function is $EF_i = f(F_i, Derived_{C_i})$.

7. Evaluation of conventional models and enhanced model using MSE, RMSE, and MAPE metrics.

The Step 4 and Step 6 are important steps of the overall model. The steps are described in detail in separate subsections below:

4.1 Deriving Cluster Centriods

In case of prediction, we only cross check our prediction with already known data. Hence, for existing data, clustering can be applied and accordingly an integrated model is developed [9]. However, for forecasting, we have to develop a model that would predict network traffic for unseen period.

Unlike prediction, the centriods information for the period for which the traffic is to be forecasted is not available. A mechanism is devised to overcome this issue. According to [11], clustering of time series subsequences is meaningless. More concretely, clusters extracted from the time series are forced to obey a certain constraint that is pathologically unlikely to be satisfied by any data set. Hence, we can use the previous clusters knowledge to build new anticipated centriods.

When applied k-means clustering approaches to eight years data from 2008 to 2015 that are obtained from WIDE backbone network. Five centriods for each year are obtained. It is decided to use these centriods for improving future loading packets in network traffic. Consequently, it is the proposed mechanism that can implicitly help to improve forecasting loading packets in network traffic. The newly derived centriods are used to enhance conventional forecasting models. The procedural steps for deriving centriods are as follows:

1. The network data from 2008 to 2015 is clustered using k-means clustering. We obtained centriods of these clusters for all years.

2. Organized the centriods into five groups. Group j contains jth centriods of all years. ($1 \leq j \leq 5$)

3. Sort all centriods of all years in ascending order. This rearrangement of clusters centriods enables proper grouping of appropriate centriods.

Table 1 Derived centriods that are obtained from k-means clustering

Derived centriods obtained from k-means approach
18.39
18.52
18.22
18.29
18.301

4. New set of centriods is derived by applying single exponential smoothing technique to all centriods in each group.

Table 1 shows the derived centriods that are obtained by applying the single exponential smoothing model. We used these centriods to improve forecasting of future loading packets of network traffic. We obtained five derived centriods from observation centriods that belong to k-means clustering approaches.

4.2 Enhance Forecasting

In this step, we used derived cluster centriods along with result obtained from conventional time series forecasting models. The forecasting phase forecasts loading packets of February 2016 in real network for congestion control. The enhanced forecasting is a function of existing forecasting value F_i and appropriate derived centroid values $Derived_{C_i}$ as follows: $EF_i = f(F_i, Derived_{C_i})$ where F_i is a forecasting value that is obtained by various conventional models. $Derived_{C_i}$ is the derived centriods of the ith cluster ($1 \leq i \leq 5$). The function used is average of forecasting results that are obtained from conventional time series models and derived centriods values that are obtained from clustering approaches.

5 Results and Analysis

We test our integrated model on real network data for forecasting of loading packets in network traffic. Our integrated model is implemented in MATLAB. We forecast only packet's numbers from WIDE real network in order to provide quality of services (QoS) for network.

The conventional time series, namely, WES, HTES, ARMA, Hybrid, and NAR-NET models have been applied for forecasting loading packets in network traffic. We demonstrate the generalized performance of conventional time series model and integrated model when applied to forecast loading packets in network traffic. Real network traffic data set is collected from WIDE backbone network. It has been decided to gather network data in window size last 6 months to forecast February 2016 with

Table 2 Results of conventional models for forecasting of network traffic using last six month data

	WES	HTES	ARMA	Hybrid model	NARNET
MSE	0.0372	0.2240	0.0366	0.0345	0.0999
RMSE	0.1929	0.4733	0.1913	0.1858	0.3161
MAPE	0.9515	2.2980	0.9502	0.9080	1.4084

Table 3 Results of integrated model (k-means with conventional time series models) for forecasting of network traffic using last six month data

	WES	HTES	ARMA	Hybrid model	NARNET
MSE (Improvement %)	0.0343	0.0891	0.0270	0.0331	0.0386
	7.08	60.23	7.66	4.6	61.37
RMSE (Improvement %)	0.1851	0.2985	0.1644	0.1819	0.1964
	4.05	36.94	3.82	2.99	37.87
MAPE (Improvement %)	0.9026	1.4547	0.7479	0.8809	0.7231
	4.06	36.07	5.09	2.99	48.66

Fig. 2 Performance improvement with respect to MSE using last six month data as window size

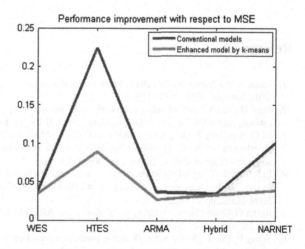

time interval one month. In order to enhance the conventional time series models for forecasting loading packets in network traffic, we proposed the integrated model that combines derived cluster centriods with conventional time series models to enhance forecasting of loading packets in network traffic.

In this experiment, we collected last six month data from August 2015 to January 2016 to forecast February 2016. Table 2 shows the results that are obtained from conventional time series models, namely, WES, Holt-Trend, ARMA, Hybrid, and NARNET. Whereas the Table 3 shows the results that are obtained from integrated model. We observed that the integrated model improves the performance of conventional forecasting network traffic models. MSE, RMSE, and MAPE metrics show the average of 28.18, 17.13, and 19.37

6 Conclusions

For network traffic forecasting, an innovative model is proposed which can best represent the network traffic data and its characteristics. The proposed model combines the conventional linear and nonlinear time series models with clustering to improve the forecasting of loading packets in network traffic. The standard live network data is attained from WIDE backbone that is used for experiments. The MSE, RMSE, and MAPE performance measures are applied to evaluate the proposed model.

We used k-means clustering techniques to cluster the existing network traffic data to capture the traffic characteristics in the form of cluster centriods. These derived cluster centriods are used further to enhance the forecasting obtained from conventional time series models. We forecast network traffic for February 2016 using window size last 6 months network data. The proposed forecasting model outperformed the conventional models. For k-means clustering augmentation average of 28.18% of improvement in MSE is recorded.

This model demonstrates the feasibility and effectiveness of clustering approaches to enhance the quality of services(QoS) of network. In future, we shall attempt to apply another soft clustering approaches in order to obtain better results.

References

1. Saleh AAM, Simmons JM (2011) Technology and architecture to enable the explosive growth of the internet. IEEE 49(1):126–132
2. Feng H, Shu Y (2005) Study on network traffic prediction techniques. In: Wireless communications, networking and mobile computing, 2005. IEEE, pp 1041–1044
3. Di C, Hai-Hang F, Qing-jia L, Chun-xiao C (2006) Multi-scale internet traffic prediction using wavelet neural network combined model. In: First international conference on communications and networking in China, ChinaCom 2006. IEEE, pp 1–5
4. Li J, Shen L, Tong Y (2009) Prediction of network flow based on wavelet analysis and ARIMA model. In: International conference on wireless networks and information systems, WNIS 2009. IEEE, pp 217–220
5. Rutka G (2008) Network traffic prediction using ARIMA and neural networks models. Int J Res Inf Sys 311(4):47–52
6. Wang P, Liu Y (2008) Network traffic prediction based on improved BP wavelet neural network. In: 4th international conference on wireless communications networking and mobile computing, WiCOM 2008. IEEE, pp 1–5
7. Chabaa S, Zeroual A, Antari J (2009) ANFIS method for forecasting internet traffic time series. In: Mediterranean microwave symposium (MMS). IEEE, pp 1–4
8. Kuang J, Zhai D, Wu X, Wang Y (2013) A network traffic prediction method using two-dimensional correlation and single exponential smoothing. IEEE, pp 403–406
9. Aldhyani THH, Joshi MR (2016) An integrated model for prediction of loading packets in network traffic. In: Second international conference on information and communication technology for competitive strategies. ACM
10. MAWI Working Group Traffic Archive, Packet traces from WIDE backbone. http://mawi.wide.ad.jp/mawi/
11. Keogh E, Lin J (2005) Clustering of time-series subsequences is meaningless: implications for previous and future research. In: Knowledge and information systems. Springer, pp 154–177

Load Balancing Mechanism Using Fuzzy Row Penalty Method in Cloud Computing Environment

Narander Kumar and Diksha Shukla

Abstract Nowadays cloud computing becomes an emerging standard for accessing and organizing large computing application over network. In cloud environment, two entities plays major role such as cloud users and cloud service providers. Cloud users submit request, and cloud providers provide services to end users. Load balancing is one of the major concern in cloud computing environment means how to distribute load efficiently among all the nodes. For solving such problem, we need some load balancing algorithm, so this paper proposed a solution, fuzzy row penalty method, for solving load balancing problem in fuzzy cloud computing environment. Here, fuzzy technique is used for solving uncertain response time in fuzzy cloud environment. Fuzzy row penalty method is used for solving both balanced fuzzy load balancing problem and unbalanced fuzzy load balancing problem in cloud environment. The proposed algorithm is implemented on CloudSim, and the obtained result is used for solving load balancing problem in terms of response time and space complexity in order to increase performance and scalability, minimize associated overheads, and avoid bottleneck problem.

Keywords Fuzzy load balancing problem · Row penalty method
Virtual machine · Cloud computing

1 Introduction

Cloud computing means to access computing resources over the Internet on pay per use basis, and in cloud computing environment, data is stored remotely and cloud service providers provide cloud resources. Nowadays, cloud computing is growing

N. Kumar (✉) · D. Shukla
Department of Computer Science, Babasaheb Bhimrao Ambedkar University
(A Central University), Lucknow 226025, India
e-mail: nk_iet@yahoo.co.in

D. Shukla
e-mail: dikshashukla73@gmail.com

© Springer Nature Singapore Pte Ltd. 2018
D.K. Mishra et al. (eds.), *Information and Communication Technology
for Sustainable Development*, Lecture Notes in Networks and Systems 9,
https://doi.org/10.1007/978-981-10-3932-4_38

rapidly so cloud users request for more and more services and want to achieve better result, and this will increase the load on the network. There are three types of loads, i.e., network load, CPU load, and memory capacity. Load balancing is the key issue in cloud computing environment. As if workload is not distributed properly among all the nodes then some nodes gets heavy load and some nodes gets low load, this will create communication overheads and leading to bottleneck problem. For solving such problem, we need some load balancing algorithms in order to increase performance, minimize response time and communication overheads, increase scalability, and minimize failover in cloud computing environment. Balancing algorithms are of two types such as static balancing algorithm and dynamic balancing algorithm. Dynamic balancing algorithm does not require prior knowledge of all the system applications and resources and also depends upon the current state of the system. Dynamic balancing algorithms are more popular for solving load balancing algorithm as compared to static load balancing algorithm, as dynamic balancing algorithms are more efficient and provide better result. There are many dynamic load balancing algorithms available in cloud computing environment; for example, fuzzy Hungarian algorithm is more popular dynamic algorithm for solving both fuzzy load balancing balanced problem and fuzzy load balancing unbalanced problem in fuzzy cloud computing environment. Here, fuzzy technique is used for solving uncertain parameters such as uncertain response time and uncertain cost but fuzzy Hungarian method is not working well in fuzzy cloud computing environment for solving unbalanced fuzzy load balancing problem as it gives high response time and gives high space complexity. So this paper proposed an algorithm, fuzzy row penalty method, for solving load balancing problem in fuzzy cloud computing environment, and row penalty method solves unbalanced fuzzy load balancing problem effectively and takes less execution time as compared to fuzzy Hungarian method. The proposed algorithm is implemented on CloudSim, and the obtained result is used for showing variation between fuzzy row penalty method and fuzzy Hungarian method in terms of execution time and space complexity and shows that row penalty method is more efficient for solving fuzzy load balancing problem in fuzzy cloud computing environment.

2 Related Work

In [1–3], the author gives brief discussion on existing round-robin and throttled scheduling algorithms for load balancing in cloud computing environment and then compares those using parameters such as response time and processing time. Different types of dynamic load balancing mechanism for dynamically distributing workload in cloud computing environment have been discussed in [4]. In [5], the author proposed fuzzy controller which is used to allocate virtual machine to each application in cloud environment in order to enhance quality of service. In [6], the author describes fuzzy logic-based novel load balancing algorithm for balancing load in cloud environment. In [7], this paper describes various techniques for efficiently managing virtual machines in cloud environment and then compares these techniques on various parameters. In [8], the author discusses load balancing strategy based on cloud partitioning concept in order to improve response time and throughput of public clouds. In [9], the author describes the method for partitioning cloud and then compares various load balancing algorithms in order to balance dynamic load in cloud environment. In [10], this paper proposed method to balance load on various nodes in order to reduce movement cost. Existing load balancing mechanism in cloud environment are compares on the basis of various parameters like scalability, energy consumption etc. in [11, 12]. In [13, 14], this paper gives an overview on load balancing and load balancing algorithm, and compares load balancing algorithm in order to achieve maximum resource utilization. In [15, 16], the author describes various load balancing algorithms for solving issues related to load balancing in cloud environment. Three distributed solutions biased random sampling, and active clustering mechanisms to balance the load in cloud environment are discussed in [17]. In [18], the author compares various load balancing algorithms in order to increase reliability and user satisfaction. A survey of dynamic load balancing mechanism in cloud environment and its comparison on the basis of various balancing parameters are given in [19]. In [20], the author describes queue algorithm with ACBLA algorithm which can be used to schedule tasks among all processing nodes in order to reduce communication overheads. In [21], this paper uses switch mechanism which is used to divide public cloud into cloud partitions and then uses improved round-robin algorithm and game theory to balance load in cloud environment.

Generally, there are so many linear programming algorithms available for solving load balancing problem in cloud environment. Fuzzy Hungarian method is used to solve fuzzy load balancing problem, but this algorithm is not working well in fuzzy environment; due to high response time and space complexity. So to resolve all the limitations associated with existing fuzzy Hungarian algorithm, this paper proposed dynamic fuzzy row penalty method for allocating workload efficiently across all the nodes in order to solve uncertainty problem in fuzzy cloud computing environment and gives less response time and space complexity as compared to fuzzy Hungarian algorithm in order to increase performance of cloud computing environment.

This research paper is organized as follows: Sect. 1 presents introduction. Section 2 represents related work. Proposed mechanism is given in Sect. 3. Section 4 represents result and discussion. Section 5 represents the conclusion.

3 Proposed Mechanism

This paper proposed fuzzy row penalty mechanism for solving load balancing problem in fuzzy cloud computing environment. Fuzzy row penalty method is used to solve uncertain response time and memory capacity in fuzzy environment. This mechanism solves fuzzy balanced and unbalanced load balancing problem effectively, takes less iteration so as to improve response time and space complexity, and gives minimum cost in order to enhance performance and user satisfaction. In this paper, load balancing problem can be solved in terms of response time and space complexity. The proposed algorithm is implemented on CloudSim, and the obtained result shows that fuzzy row penalty method gives least response time and space complexity as compared to fuzzy Hungarian method in order to increase performance and minimize bottleneck in cloud environment (Fig. 1).

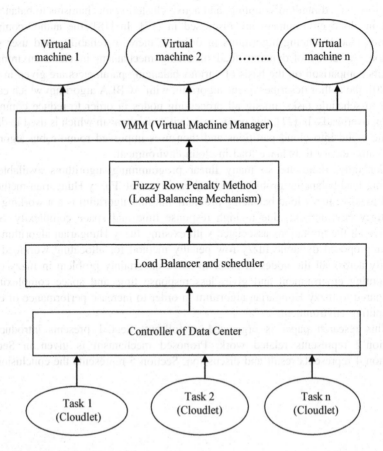

Fig. 1 Fuzzy load balancing model

3.1 Fuzzy Row Penalty Mechanism

Step 1: Create matrix with fuzzy values
Matrix[i][j] = number of cloudlets/workload response rate.

Step 2: Calculate the ranking value of each fuzzy response time Xij:
If response time is in fuzzy triangular number (b1, b2, b3), then ranking value is calculated as Mag ($\sim$ Xij) = (b1 + 10b2 + b3)/12.
If response time is in fuzzy trapezoidal number (b1, b2, b3, b4), then ranking value is calculated as Mag ($\sim$ Xij) = (b1 + 5b2 + 5b3 + b4)/12.

Step 3: In matrix, check each row and circle the response time whose corresponding ranking value is less than other response time in that row.

Step 4: Optimal workload

(a) Check all the circled fuzzy response time in the matrix and then examine the fuzzy response time which is uniquely circled in column as well as in row. Assign this fuzzy response time and then remove the corresponding column and row.

(b) In the above steps, if no response time is found, then examine single row I which contains single fuzzy response time but corresponding column j contained more than one circled response time. Then, calculate fuzzy row penalty. Row penalty is calculated by subtracting ranking value of minimum and next to minimum fuzzy response time in same row. Select maximum penalty row and then assign circled fuzzy response time to that row. Then, remove corresponding column and row from matrix table.

Step 5: Repeat third and fourth steps until all columns and rows are crossed out. Now, calculate optimal fuzzy response time.

$$\text{Optimize} \sim T = \sum_{i=1}^{N} \sum_{j=1}^{N} \sim Xij \ Yij$$

4 Result and Discussion

Figures 2, 3, 4 and 5 are self explanatory and show the comparisons between Fuzzy Hungarian and Row penalty method. From the results we can say that Fuzzy Row penalty Method is better. Therefore, proposed mechanism is good.

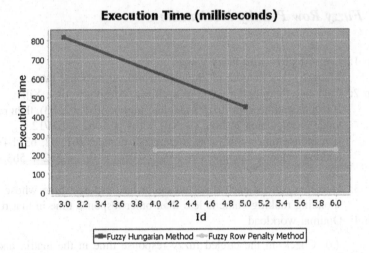

Fig. 2 Comparison between Fuzzy Hungarian and row penalty method in terms of response time
for solving balanced fuzzy load balancing problem

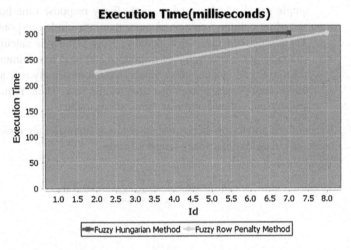

Fig. 3 Comparison between Fuzzy Hungarian and row penalty method in terms of response time
for solving unbalanced load balancing problem

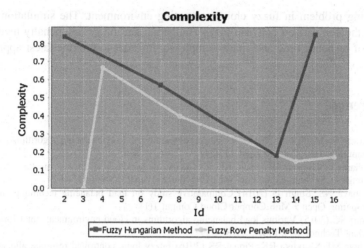

Fig. 4 Comparison between Fuzzy Hungarian and row penalty method in terms of space complexity for solving balanced load balancing problem

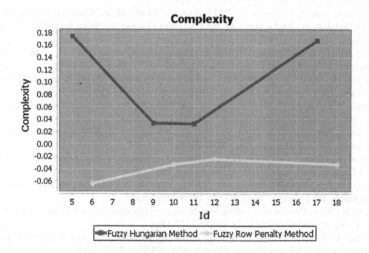

Fig. 5 Comparison between Fuzzy Hungarian and row penalty method in terms of space complexity for solving unbalanced load balancing problem

5 Conclusion

In cloud computing environment, the major issue is how to distribute workload across all processing nodes so that all nodes gets equal job at a given instant of time. So this paper proposed fuzzy load balancing algorithm for solving load balancing problem in fuzzy cloud computing environment. The simulation result shows the comparison between fuzzy Hungarian and fuzzy row penalty method in terms of response time and space complexity in order to provide latest approach.

References

1. Panwar R, Mallick B (2015) A comparative study of load balancing algorithms in cloud computing. Int J Comput Appl 117(24)
2. Mohamed Shameem P, Shaji RS (2013) A methodological survey on load balancing techniques in cloud computing. Int J Eng Technol 5(5). ISSN: 0975-4024
3. Raghava NS, Singh D (2014) Comparative study on load balancing techniques in cloud computing Open J Mob Comput Cloud Comput 1(1)
4. Dimri SC (2015) Various load balancing algorithms in cloud environment. Int J Emerg Res Manag Technol 4(7). ISSN: 2278-9359
5. Akhtar MJ, Sanodiya RK, Pippal RS (2016) Fuzzy logic controlled resource allocation for efficient load balancing in cloud computing environment. Int J Eng. Sci Res Technol 5(1). ISSN: 2277-9655
6. Sethi S, Sahu A, Jena SK (2012) Efficient load balancing in cloud computing using fuzzy logic. IOSR J Eng 2(7):65–71. ISSN: 2250-3021
7. Khan DH, Kapgate D, Prasad PS (2013) A review on virtual machine management techniques and scheduling in cloud computing. Int J Adv Res Comput Sci Softw Eng 3(12). ISSN: 2277 128X
8. More MSD, Mohanpurkar A (2015) Load balancing strategy based on cloud partitioning concept. Multidiscip J Res Eng Technol 2(2):424–431. ISSN: 2348-6953
9. Nadaph A, Maral V (2014) Cloud computing-partitioning algorithm and load balancing algorithm Int J Comput Sci Eng Inf Technol 4(5). doi:10.5121/ijcseit.2014.4504
10. Dobale RG, Sonar RP (2015) Review of load balancing for distributed system in cloud. Int J Adv Res Comput Sci Softw Eng 5(2). ISSN: 2277 128x
11. Begum S, Prashanth CSR (2013) Review of load balancing in cloud computing. Int J Comput Sci Issues 10(1):2. ISSN(online): 1694-0814
12. Kansal NJ, Chana I (2012) Cloud load balancing techniques: a step towards green computing. Int J Comput Sci Issues 9(1):1. ISSN(online): 1694-0814
13. Ramya R, Krishsanth M, Arockiam L (2014) A state-of-art load balancing algorithms in cloud computing. Int J Comput Appl 95(19). ISSN: 0975-8887
14. Gupta R (2014) Review on existing load balancing techniques of cloud computing. Int J Adv Res Comput Sci Softw Eng 4(2). ISSN: 2277 128x
15. Kumar V, Prakash S (2014) A load balancing based cloud computing techniques and challenges. Int J Sci Res Manag 2(5):815–824. ISSN(e): 2321-3418
16. Nipane NS, Dhande NM (2014) ABC—load balancing techniques—in cloud computing. Int J Innov Res Adv Eng 1(2). ISSN:2278-2311
17. Randles M, Lamb D, Taleb-Bendiab A (2010) A comparative study into distributed load balancing algorithms for cloud computing. In: 2010 IEEE 24th international conference on advanced information networking and applications workshops (WAINA), pp 551–556. ISBN: 978-1-4244-6701-3. doi:10.1109/WAINA.2010.85. 20–23 April 2010
18. Aruna M, Bhanu D, Punithagowri R (2013) A survey on load balancing algorithms in cloud environment. Int J Comput Appl 82(16). doi:10.5120/14251-2472
19. Kumar S, Rana DS (2015) Various dynamic load balancing algorithms in cloud environment: a survey. Int J Comput Appl Found Comput Sci (FCS) NY, USA, 129(6). doi:10.5120/ijca2015906927

20. Prasanna Kumar K, Arun Kumar S, Jagadeeshan (2013) Effective load balancing for dynamic resource allocation in cloud computing. Int J Innov Res Comput Commun Eng 2(3). ISSN (online): 2320-9801
21. Divya Bharathi VV, Nivedha S, Priyanka T, Daya Kanimozhi Rani M (2014) A survey on load balancing in public cloud. Int J Adv Res Comput Sci Softw Eng 4(3). ISSN: 2277 128x

Performance Evaluation of Data Mining Techniques

Mani, Bharti Suri and Manoj Kumar

Abstract Data mining has gained immense popularity in various fields of medical, education and industry as well. Data mining is a process of predicting the result and extraction of useful information from huge dataset. In this paper, we have surveyed various data mining techniques. Further, performance of various data mining techniques, namely decision tree, random forest, naive Bayes, AdaBoost, multilayer perception neural network, radial basis function, sequential minimal optimization and decision stump, have been evaluated using UCI communities and crime dataset for classifying crime in US states. On the basis of results obtained, we found that the decision tree outperforms with 96.4% accuracy and minimal false-positive rate.

Keywords Data mining · Classification techniques · Decision tree

1 Introduction

Data mining is an analytical process which handles huge data. It is defined as a process of extraction of meaningful information from a huge, unclean, ambiguous and inconsistent data set. Data mining has shown its applicability in numerous fields like education, health care, tourism, marketing and fraud detection. Data analysis is being done in these areas so as to make managerial decision for success in marketing, to analyze previous data of patients to predict the disease, to detect the

Mani (✉) · Bharti Suri
USICT, Guru Gobind Singh Indraprastha University, Sector-16C, Dwarka,
New Delhi, India
e-mail: manideol90@gmail.com

Bharti Suri
e-mail: bhartisuri@gmail.com

Manoj Kumar
Ambedkar Institute of Advanced Communication Technologies and Research,
Geeta Colony, New Delhi, India
e-mail: manojgaur@yahoo.com

© Springer Nature Singapore Pte Ltd. 2018   375
D.K. Mishra et al. (eds.), *Information and Communication Technology
for Sustainable Development*, Lecture Notes in Networks and Systems 9,
https://doi.org/10.1007/978-981-10-3932-4_39

fraud in banks [1], to analyze or improve the student performance in education [2]. The data mining process has become a supporting system [3] for understanding requirements of stakeholders and analyzes their behavior. The process of data mining follows a conceptual architecture [4] which illustrates framework where the database or data repositories undergo cleaning process followed by requesting a relevant data from data server which are then used to extract the interesting result pattern on the basis of knowledge domain and at the end presents the result to user through graphical user interface. Before applying any data mining algorithm, dataset undergoes filtering process [5] like removing missing values in data attributes and non-relevant data, thereby empowering the accuracy of classification techniques. This motivates us to review various data mining techniques and evaluate their performance on communities and crime unnormalized dataset [6]. The rest of the paper is organized as follows. Sect. 1 contains the related work, Sect. 2 discusses framework of a process, and Sect. 3 discusses the performance evaluation of data mining techniques followed by conclusion in Sect. 5.

2 Related Work

Various data mining methods have been implemented and prove to be a beneficial predictor in real world [2, 7–9]. Every business process has a common framework of data mining process. In [10], CRM (customer relationship management) data mining framework is proposed to predict the behavior of the customer in order to intensify the decision-making process, thereby retaining the valued customer. The data so acquired from CRM framework are used to evaluate the effectiveness of two classification method, i.e., naive Bayes and neural network. Merceron and Yacef [2] have proposed a education data mining system to uncover the pertinent facts contained in a database acquired by Web-based educational system. The relevant knowledge for such system is extracted using data mining algorithm such as clustering, classification and association. This system helps teacher to direct and retain their class and comprehend their student's learning ability. Mai et al. [8] have shown that data mining has gained its importance in healthcare sector. The objective of mining data of healthcare sector includes early disease diagnose and prevention of disease and analyzes the better health policy making. As we know that health awareness is growing among the people and they have become more caution about their health status. The major group which suffers from health issues is old age group people, and to keep track of their health status MyPHI [7] is one of the successful systems developed to predict the personal health index of elderly people. Kumar and Toshniwal [9] have proposed framework that includes clustering technique to categorize the road accidents and perform regression analysis. Every system has a main concern of extraction of knowledge from a database, and process to discover such useful information is termed as knowledge discovery [11].

3 Data Mining Framework

Data mining process follows a systematic approach which is shown in Fig. 1. The basic approach of mining process is as follows:

- Initially, the business objectives are set, which formulate the reason to mine the useful knowledge for constructing profitable strategies and decisions.
- After setting out the objective, business model is designed.
- Objective supporting data are collected as per requirements followed by cleansing and reduction.
- For the ease of user, analysis can be represented in graphical format to visualize the data behavior or to see the critical point in analysis that can benefit or may lead to loss in future circumstances.
- In the end, the data mining techniques are evaluated for the performance measures.

3.1 Dataset Used

We have used communities and crime unnormalized dataset [6] that contains the 147 attributes and 2216 instances. We have selected only 16 attributes including class label which classify the country or community with rate of crime as high or low. The data consist of detailed information of the crime in some US states. The attributes used in dataset are as follows:

- String attributes: communityname, state.
- Numeric attributes: countrycode, communitycode, population, murders, murdperpop, rapes, rapeperpop, robberies, robbperpop, assaults, assaultsperpop, burglaries, burglperpop.
- Categorical attribute: crime (class label).

3.2 Data Preprocessing

Data available is usually raw, noisy and inconsistent in nature. There are many factors which affect the quality of data under mining process. Concerned quality factors [12] of data are accuracy, completeness, consistency and interpretability. So,

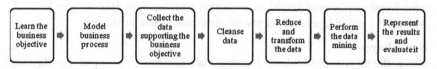

Fig. 1 Basic data mining framework

we can say that the data in real world consist of various discrepancies and these should be removed before performing the data mining techniques. To make the data suitable for the mining process, every relational dataset undergoes cleaning process followed by data reduction and data transformation. Data cleansing [12] is process where missing values are removed by replacing it with mean or median value, and noisy data are removed by binning methods. After cleaning the data, the dataset is subjected to reduction process [13] where the attributes are reduced or selected using filters [14] in WEKA. Data preprocessing has great influence [13] on accuracy of data mining techniques, and hence, it is recommended to preprocess the dataset in order to speed up the process or increase the accuracy rate. While classifying the crime dataset [6] by considering all attributes and instances of it, the resulting accuracy of classifiers came to be inaccurate. So for the sake of accuracy we have used only 16 relevant attributes and reduced the data instances number from 2215 to 533 using unsupervised instance filter in WEKA.

3.3   Data Mining Techniques

The major categories of the data mining methods are association rule mining, clustering and classification. In this paper, we have used supervised classification method which is as follows:

Decision Tree
Decision tree [12] is a classifier structured as tree with the root and leaf nodes. It is a rule-based method and easily comprehended. Decision tree is highly advantageous and based upon greedy top-down approach [15]. Internal node and leaf nodes are represented as attribute and class label, respectively. Basic algorithm of decision tree [12] is used for attribute selection, and there are various measures [16, 17] adopted for its selection. Decision tree uses information gain as measure for attribute selection, and attribute with highest information gain value is selected as the splitting attribute node. Information gain is an average amount of information to classify data instance as class label.

Random Forest
It is ensemble learning method and also known as nearest neighbor predictor for classification and regression. Random forest [18, 19] conceptually based upon decision tree and produces numerous classifying tree. Ensemble is depicted as a set of multiple classifiers that contributes toward the overall classification tree and created by two methods bagging [20] and boosting [21]. Mathematically, the random forest can be represented as collection tree classifier $\{f(x, \theta_k), k = 1, \ldots, k\}$ where $\{\theta_k\}$ are independent identically distributed random vector.

Naive Bayes
Naive Bayes is a probabilistic learning classifier and based upon the Bayesian network [12]. It predicts the class membership probabilities, such as the probability

that a given tuple belongs to a particular class. It assumes that all variables or features x_i are mutually independent for given C class label.

Sequential Minimal Optimization

Sequential minimal optimization (SMO) [22] is a method that overcomes the drawback of support vector machine of solving large quadratic programming (QP) optimization problems. SMO is an iterative process that parts the complex QP problems in smaller QP problems, and these small problems are analyzed, thus avoiding time-consuming optimization. Another advantage of SMO is that it does not require extra memory.

Multilayer Perceptron

Multilayer perceptron (MLP) [23] is a supervised and nonparametric method of classification. MLP structure [24] consists of three layers: input, hidden and output layer. The input layer units are multiplied by a weight with a threshold added and then is passed through an activation function that may be linear or nonlinear.

Radial Basis Function Network

Radial basis function (RBF) [25] is artificial neural network and characterized by the several features. RBF network comprises of three layers [26]: the input layer, the hidden layer and output layer. Input layer consists of n component of input vector that is fed into proceeding input units of hidden layer. Each unit in hidden layer computes the radial Gaussian activation function. Every stimulus of input component is evaluated by activation function associated with the hidden units and weight associated with links between hidden layer and the output layer.

AdaBoost

AdaBoost [27] is an machine learning algorithm developed by Yoav Freund and Robert Scaphire. The motive of this algorithm is to enhance the accuracy of the prediction rule by combining the multiple weak and inaccurate rules. It is a boosting algorithm which takes input as training set, $s = \{(x_1, y_1) \dots (x_m, y_m)\}$ where x_i is instance drawn from some space X and represented in some manner and $y_i \in Y$ is the class label associated with x_i.

Decision Stump

Decision stumps [28] are one-layered decision tree and are useful for classifying small volume dataset. Decision stumps are easy to build than compared to decision tree. In one-layered decision tree [29], only one feature or attribute is selected for class label, and for each attribute, it computes the score measuring the capability of separating the classes.

4 Performance Evaluation

The effectiveness of the classifiers has been estimated using the following measures [12]:

True-positive rate (TPR): It is defined as number of positive data tuples that are classified correctly, which is given as:

$$TPR = \frac{TP}{TP + FP}$$

False-positive rate (FPR): It is defined as number of negative data tuples that are incorrectly classified as positive data tuples, which is given as:

$$FPR = \frac{FP}{TN + FP}$$

True negative (TN): It is defined as number of negative data tuples that are correctly classified.

False negative (FN): It is defined as number positive data tuples incorrectly classified as negative data tuples.

Accuracy: It is defined as the percentage of correctly classifying test dataset.

$$accuracy = \frac{TN + TP}{P + N}$$

where $P = TP + FN$ and $N = FP + TN$.

Precision: It is defined as degree of exactness (i.e., the percentage of tuples classified positive are actually positive).

$$precision = \frac{TP}{TP + FP}$$

Recall: It is defined as degree of completeness

$$recall = \frac{TP}{TP + FN}$$

F-measure: It is defined as measure of test's accuracy and takes both recall and precision to compute its value.

$$F - measure = \frac{2\{(precision * recall)\}}{(precision + recall)}$$

Receiver operating characteristics (ROC) curve: It is a plot between the true-positive rate and false-positive rate, and hence it, illustrates the performance of a classifier.

Here, comparison of the classifiers is being done on the basis of above-mentioned measures and depicted their performance in bar graph (see Fig. 2). We have tabulated (see Table 1) the results of each classifier and thereby analyzed their capability. The results have given us a clear view that decision tree is

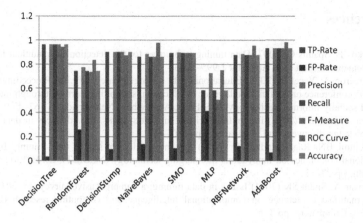

Fig. 2 Bar graph for performance measures of classifiers

Table 1 Performance measure of classifiers

Techniques	TP rate	FP rate	Precision	Recall	F-measure	ROC curve	Accuracy
Decision tree	0.964	0.036	0.964	0.964	0.964	0.945	0.964
Random forest	0.743	0.258	0.773	0.743	0.736	0.837	0.743
Decision stump	0.902	0.098	0.904	0.902	0.902	0.874	0.902
Naive Bayes	0.864	0.137	0.888	0.864	0.862	0.879	0.864
SMO	0.896	0.104	0.898	0.896	0.896	0.894	0.896
MLP	0.583	0.414	0.728	0.583	0.506	0.75	0.583
RBF network	0.887	0.123	0.886	0.877	0.877	0.956	0.887
AdaBoost	0.932	0.068	0.932	0.932	0.932	0.985	0.932

highly accurate (see Fig. 2) with higher degree of exactness (precision) and completeness (recall), whereas MLP has minimal accuracy with high false-positive rate, and hence, it is an inefficient classifier. Another most efficient method is AdaBoost with highest ROC performance curve.

5   Conclusion

We have done a comparison of data mining classification techniques and evaluated results of performance measures of each classification techniques. As our dataset has predefined class label, we have used supervised learning or classification methods for evaluation. Our main purpose of applying techniques or method is to classify the crime rate with high accuracy and see which technique does it with best results (see Table 1). The result we have concluded from the table is that decision tree outperforms with 96.4% accuracy and has minimal false-positive rate.

References

1. Kirkos E, Spathis C (2007) Data mining techniques for the detection of fraudulent financial statements. Expert Syst Appl: Int J 995–1003
2. Merceron A, Yacef K (2005) Educational data mining: a case study. In: Proceedings of the 2005 conference on artificial intelligence in education: supporting learning through intelligent and socially informed technology. IOS Press, Amsterdam, Netherland, pp 467–474
3. Bâra A, Lungu I (2012) Improving decision support systems. In: Advances in data mining knowledge discovery and applications, pp 397–417
4. Lakshmi BN, Raghunandhan G (2011) A conceptual overview of data mining. In: 2011 national conference on innovations in emerging technology (NCOIET). IEEE, Erode, Tamil Nadu, pp 27–32
5. Purwar A, Singh SK (2014) Issues in data mining: a comprehensive survey. In: 2014 IEEE international conference on computational intelligence and computing research (ICCIC). IEEE, Coimbatore, pp 1–6
6. http://archive.ics.uci.edu/ml/
7. Chen L, Li X, Yang Y (2016) Personal health indexing based on medical examinations. Decis Support Syst 54–65
8. Shouman M, Turner T (2012) Using data mining techniques in heart disease diagnosis and treatment. In: 2012 Japan-Egypt conference electronics, communications and computers (JEC-ECC). IEEE, Alexandria, pp 173–177
9. Kumar S, Toshniwal D (2015) A data mining framework to analyze road accident data. J Big Data
10. Bahari TF, Elayidom MS (2015) An efficient CRM-data mining framework for the prediction of customer behaviour. In: Proceedings of the international conference on information and communication technologies, ICICT 2014. Elsevier, Kochi, pp 725–731
11. Anand SS, Grobelnik M (2007) Knowledge discovery standards. Artif Intell Rev 21–56
12. Han J, Kamber M (2012) Data mining concept and techniques. Elsevier, USA
13. Crone SF, Lessmann S (2006) The impact of preprocessing on data mining: an evaluation of classifier sensitivity in direct marketing. Eur J Oper Res 781–800
14. Ramaswami M, Bhaskaran R (2009) A study on feature selection techniques in educational data mining. J Comput 7–11
15. Barros RC, Basgalupp MP (2012) A survey of evolutionary algorithms for decision-tree induction. IEEE Trans Syst Man Cybern Part C: Appl Rev
16. Mantaras RL (1991) A distance-based attribute selection measure. Mach Learn 81–92
17. Quinlan JR (1986) Induction of decision trees. Mach Learn 81–106
18. Breiman L (2001) Random forests. Mach Learn 5–32
19. Kulkarni VY, Sinha PK (2013) Random forest classifiers: a survey and future research direction. Int J Adv Comput
20. Breiman L (1996) Bagging predictor. Mach Learn 123–140
21. Schapire RE (2002) The boosting approach to machine learning: an overview. In: Nonlinear estimation and classification, pp 149–171
22. Platt JC (1999) Fast training of support vector machines using sequential minimal optimization. MIT Press, Cambridge, MA, USA
23. Verma B (2002) Fast training of multilayer perceptrons. IEEE Trans Neural Netw 1314–1320
24. Delashmit WH, Manry MT (2005) Recent developments in multilayer perceptron neural networks. In: Proceedings of the 7th annual memphis area engineering and science conference
25. Orr MJ (1996) Introduction to radial basis function network
26. Oyang Y-J, Hwang S-C, Ou Y-Y, Chen CY, Chen ZW (2005) Data classification with radial basis function networks based on a novel kernel density estimation algorithm. IEEE Trans Neural Netw 225–236

27. Schapire RE (2013) Explaining AdaBoost. In: Empirical inference, pp 37–52
28. Choy M (2010) Building decision trees from decision stumps
29. Iba W, Langley P (1992) Induction of one-level decision tree. In: Proceedings of the ninth international workshop on machine learning, ML '92, USA, pp 233–240
30. Akinola OS, Afolabi AC (2012) Evaluating classification effectiveness on sequential minimal optimization (SMO) algorithm chemical parameterization of granitoids. IJRRAS

Automatic White Blood Cell Segmentation for Detecting Leukemia

Pooja Deshmukh, C.R. Jadhav and N. Usha Rani

Abstract White blood cells are used to detect different diseases infected to human body. The classification and segmentation of white blood cells for detection of leukemia are one of the important and complex steps. It allows detection of acute lymphoblastic leukemia (ALL). Partially automated system do not give accurate results also manual diagnosis process results are depend on operators ability. These problems can be resolved using fully automated system. This system uses computerized segmentation and classification techniques for detection of leukemia accurately and within less time period. Segmentation scheme segment WBC's into nucleus and cytoplasm, classification is used to classify WBC's into various as per different characteristics also, features of nucleus and cytoplasm extracted.

Keywords Image processing · White blood cell · Acute lymphoblastic leukemia · Segmentation · Classification · RGB (Red Green Blue)

1 Introduction

The White blood cells are present in whole human body that is in blood as well as lymphatic system. These WBC's protect human body from various diseases. White blood cells also known as leukocytes help in diagnosis of various diseases infected to human body. White blood cells are important cells used for detection of leukemia. Leukemia is classified into two types [1] as (1) Acute leukemia and (2) Chronic leukemia. Acute leukemia grows quickly whereas, chronic leukemia

P. Deshmukh (✉) · C.R. Jadhav · N. Usha Rani
Department of Computer Engineering, D.Y. Patil Institute of Engineering & Technology, Pimpri, Pune, India
e-mail: poojadeshmukh454@gmail.com

C.R. Jadhav
e-mail: chaya123jadhav@gmail.com

© Springer Nature Singapore Pte Ltd. 2018
D.K. Mishra et al. (eds.), *Information and Communication Technology for Sustainable Development*, Lecture Notes in Networks and Systems 9, https://doi.org/10.1007/978-981-10-3932-4_40

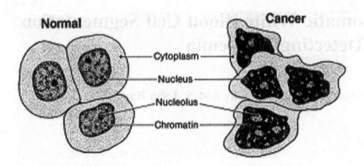

Fig. 1 Normal cell versus cancer cell [13]

grows slowly. This leukemia is curable if detected at early stage. WBC's play important role leukemia detection. These white blood cells consist of nuclei and cytoplasm. Features of nuclei and cytoplasm also differentiate between normal/healthy cell and abnormal or unhealthy cell. Characteristics of nuclei and cytoplasm of normal cell are different than infected cell. Also, WBC count is indicator of disease. Existing techniques include number of systems including hardware and software. Some of those software systems are partially automated systems which are time consuming. Currently most of the hospitals using hardware machines for detection which provide accurate results but time required for results is in hours. Also, automated system of detection uses classification methods such as SVM, k-NN classifier for classifying normal and abnormal cells. Whereas, proposed system includes clustering technique to form clusters of normal and abnormal clusters (Fig. 1).

2 Basic Concept

Above figure shows normal cells and cancer cells difference. It shows that normal cell nucleus is of regular size and shape, whereas, cancer cell nucleus is of irregular shape and size. Cancer cell nucleus size is more than normal cell nucleus size.

There are five different types of leukocytes which are Neutrophil, Eosinophil, Basophil, Lymphocyte, Monocyte [1]. These types are distinguishable by their functional characteristics.

Section 2 is about basic concept of cell and WBC. Section 3 describes related work of leukemia detection and related information. After background details Sect. 4 is of proposed scheme which includes methodologies used. Section 5 describes expected results. Section 6 is devoted to conclusion.

3 Related Work

Donida et al. [2] proposed freely available dataset which includes blood sample images. This dataset was specially designed to compare and evaluate classification and segmentation algorithms. The dataset designed which is publically available is known as ALL-IDB.

Madhloom [3] Each WBC have nucleus so only nucleated were considered and other unnecessary data was removed. To separate nucleus from remaining cell body arithmetic operations on cell and global threshold and filter was used. Otsu's method was used for threshold calculation. Histogram equalization was used to highlight the objects in blood image.

Scotti [4] presented segmentation and measurement technique in order to diagnose disease such as leukemia. L * a * b color space used to convert input image into grayscale. Fuzzy k-means forms clusters of blood cells. Histogram threshold technique is segmentation scheme used to segment nucleus and cytoplasm.

Piuri and Scotti [5] focused classification and identification of white blood cells from microscopic images. Firstly, white blood cells were differentiated from other types of cells. Then, morphological analysis performed to get indexes for classification of cells. Also, explained classifier selection process from set of classifier to classify cells into various types of WBC's. There are 5 different types of leucocytes as Basophil, Eosinophil, Lymphocyte, Monocyte and Neutrophil.

Halim et al. [6] proposed a method which automatically count the number of blasts in blood sample. This method helped in segmentation which was most difficult step in detection process. HSV color space based segmentation was used to separate WBC's from background. Morphological operations were used to extract shape information based on geometry properties. Dilation and erosion are two basic morphological operations for extraction process. In this method, only erosion was used.

Mohapatra et al. [7] explained independent method to recognize lymphoblasts. Independent methods were required to recognize types of blasts as myeloblast and lymphoblast. In this first step was sub imaging given blood image. Then, SCM clustering algorithm was used to segment identified leucocytes. Segmentation was done as nucleus and cytoplasm. Group of classifiers were used for classification based on statistical features of nucleus and cytoplasm.

Cheewatanon [8] proposed new segmentation algorithm. There were two tasks in an algorithm region growing algorithm and mean shift (MS) filter algorithm. MS filter was used to reduce noise. Region growing algorithm based on first assumption. First assumption was that the WBC images represented as a set of regions. The observed colors of set of region change gradually, but they change unexpectedly across the boundary between the regions. These algorithms were tested using RGB color space and CIE L * a * b color space. In this method cytoplasm was over segmented, this over segmented region was reduced using CIE L * a * b color space.

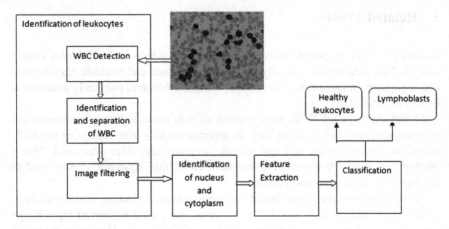

Fig. 2 System architecture

Hsu et al. [9] described a classification technique: support vector machine (SVM). SVM produces a model which calculates the target values which is based on training set. The purpose of this classifier was to evaluate target value for test data as per test data attributes. This technique did not give satisfactory result for number of features. If there are many features then set of features are selected for classification. If given data is in attributed form then it is converted into numerical form.

Rezatofighi and Soltanian-Zadeh [10]: Different image processing algorithms were proposed to distinguish different types of white blood cells. A method was used for segmentation of cytoplasm and nucleus of cells. Then segmented parts were used for number of feature extraction. Sequential Forward Selection algorithm was used to select distinct features from segmented region. Features were extracted based on LBP. Classification algorithms ANN and SVM are compared which shows SVM was more accurate classification algorithm.

Nazlibilek et al. [11]: Automatic counting of white blood cells and their sizes were evaluated for easy detection. Further features were extracted using PCA algorithm and classifier was used to classify different types of white blood cells. Five types of WBC's were classified as lymphocyte, monocyte, basophil and neutrophil. To convert grayscale image into binary image Otsu's method was used. Otsu's method uses threshold calculation for conversion of image into binary image (Fig. 2).

4 Proposed Scheme

The proposed system includes WBC detection process, Filtering, nucleus and cytoplasm selection, feature extraction of nucleus and final step is classification.

4.1 WBC Detection Process

WBC detection takes input of blood image. Then, further processes input blood image to detect WBC's from an image.

4.2 Color Conversion

In this process colored blood image is converted into CMYK color model. RGB color space is used for conversion process. An RGB color space considers intensity of each pixel and converts into CMYK for further processing. Because white cells are more clearly visible in yellow color. Figure 3 shows original image and image after conversion.

4.3 Histogram Equalization

Enhancement process enhances image by increasing intensity of objects of blood image. This method highlights the objects and borders of image. It equalizes the intensities of cells present in blood image. This method takes converted y plane image as input and produce result after enhancement. Figure 3 shows enhancement of image.

4.4 Filtering

There are various filters available which remove noise from an image. One of those filters is median filter. Median filter reduces unnecessary data from given image. It considers pixel values under structuring element and sort and arranges pixel values linearly. Then, median of those pixels is calculated and center pixel value is replaced by computed median value.

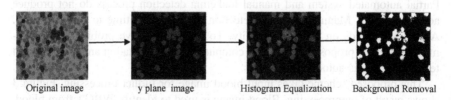

Original image   y plane image   Histogram Equalization   Background Removal

Fig. 3 Original image, converted image, enhanced image, background removal

4.5   Nucleus and Cytoplasm Selection

Nucleus and cytoplasm are identified and segmented using segmentation algorithm. Each WBC has nucleus and cytoplasm so segmentation algorithm helps to identify those WBC parts easily. Green component of RGB color space is used to get binary image and binary image obtained from a* component of CIELab color space are combined [1] to identify nucleus and cytoplasm.

4.6   Feature Extraction

Feature extraction process is done on segmented nucleus and cytoplasm. Features of nucleus and cytoplasm are considered to differentiate between normal and abnormal cells. There are three different features are considered as shape, color and texture. Shape descriptors include area, perimeter, convex are etc. [1]. These measures are used to compute solidity and compactness. Also, color features include mean, standard deviation, smoothness [1]. There are number of texture features.

4.7   Classification

Automatic classification technique is important to classify cells into normal cells and abnormal cells. SVM (Support Vector Machine) classifier [12] uses large number of variables for separation of classes of WBC's for detection of leukemia. SVM is suitable for binary classification [12]. K-means clustering is used to cluster cells in healthy and unhealthy cells. Clustering is done until all cells go in correct cluster and no cell should remain unclustered.

5   Expected Result

Partial automated system and manual leukemia detection process do not produce accurate results. Manual leukemia detection process takes time to detect disease. Accuracy depends on operator's ability. To overcome above problem fully automated system is proposed which uses computerized segmentation and classification techniques that is automated schemes.

This proposed scheme takes input blood image for further processing. Figure 3 shows result of preprocessing. Blood image is used to identify WBC's from blood. After identification process image will be cleaned using feasible filter that is median filter will remove unnecessary data from a blood image. Figure 4 shows result of

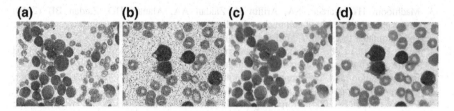

Fig. 4 **a, b** Original noisy blood image; **c, d** images with median filter

median filter. Identification of nucleus from WBC's will separate nucleus and cytoplasm from each WBC. These separated nucleus and cytoplasm will be used for feature extraction process. Feature extraction process will consider color, shape and texture features for classification of healthy and unhealthy cells. This proposed scheme will give almost 80–90% accuracy using median filter and k-means clustering algorithms. Implementation of this system is done using Matlab.

6 Conclusion

Blood images are segmented and detection of leukemia is done by feature extraction process. Further features of cells are considered for accurate results of detection. These features are shape features such as area, circularity, perimeter etc. Classification of extracted features for leukemia detection is done using k-means clustering algorithm and comparative study of algorithm. Automatic segmentation and classification scheme produce accurate results by detecting leukemia using blood image and classifying cells as normal cells and abnormal or cancer cells. Feature extraction process extracts shape, color and texture features for identification. Moreover the proposed technique should provide accurate results to unnecessary and grouped cells. Also, it is important to study and analyze new features that help to improve detection process. Blood images required for detection process are taken from ALL-IDB1 and ALL-IDB2 datasets which are freely available.

References

1. Putzua L, Caocci G, Di Ruberto C (2014) Leucocyte classification for leukaemia detection using image processing techniques. Artif Intell Med 62:179–191
2. Labati RD, Piuri V, Scotti F (2011) ALL-IDB: the acute lymphoblastic leukemia image data base for image processing. In: Benot M, Schelkens P (eds) Proceedings of the 18th IEEE ICIP international conference on image processing, 11–14 Sept. IEEE Publisher, Brussels, Belgium, pp 2045–2048

3. Madhloom HT, Kareem SA, Ariffin H, Zaidan AA, Alanazi HO, Zaidan BB (2010) Anautomated white blood cell nucleus localization and segmentation using image arithmetic and automated threshold. J Appl Sci 10(11):959–966
4. Scotti F (2006) Robust segmentation and measurements techniques of white cells in blood microscope images. In: Daponte P, Linnenbrink T, editors. In: Proceedings of the IEEE instrumentation and measurement technology conference, 24–27 Apr. IEEE Publisher, Sorrento, Italy, pp 43–48
5. Piuri V, Scotti F (2004) Morphological classification of blood leucocytes by microscope images. In: Proceedings of the IEEE international conference on computational intelligence for measurement systems and applications, 14–16 July. IEEE Publisher, Boston, MA, USA, pp 103–108
6. Halim NHA, Mashor MY, Hassan R (2011) Automatic blasts counting for acute leukemia based on blood samples. Int J Res Rev Comput Sci 2(4)
7. Mohapatra S, Patra D, Satpathy S (2014) An ensemble classifier system for early diagnosis of acute lymphoblastic leukemia in blood microscopic images. J Neural Comput Appl 24:1887–1904
8. Cheewatanon J, Leauhatong T, Airpaiboon S, Sangwarasilp M (2011) A new white blood cell segmentation using mean shift filter and region growing algorithm. Int J Appl Biomed Eng 4:30–35
9. Hsu C-W, Chang C-C, Lin C-J (2003) A practical guide to support vector classification. http://www.csie.ntu.edu.tw/cjlin/papers/guide/guide.pdf
10. Rezatofighia SH, Soltanian-Zadeh H (2011) Automatic recognition of five types of white blood cells in peripheral blood. Comput Med Imag Graph 35:333–343
11. Nazlibilek S, Karacor D, Ercan T, Sazli MH, Kalender O, Ege Y (2014) Automatic segmentation, counting, size determination and classification of white blood cells. Measurement 55:58–65
12. Putzu L, Di Ruberto C (2013) White blood cells identification and classification from leukemic blood image. In: Rojas I, Ortuo Guzman FM (eds) Proceedings of the IWBBIO international work-conference on bioinformatics and biomedical engineering, 18–20 Mar. Copicentro Editorial, Granada, Spain, pp 99–106
13. https://visualsonline.cancer.gov/details.cfm?imageid=2512

Indian Script Encoding Technique (ISET): A Hindi Text Steganography Approach

Sunita Chaudhary, Meenu Dave, Amit Sanghi and Hansraj Sidh

Abstract A prehistoric technique of concealing clandestine information is steganography which takes use of a shelter media as text, audio, video, etc. The most important thing related to same is the capacity of hiding information under the cover. The typical individual in the same concern is linguistic steganography or text steganography due to the fact that text is lacking in terms of redundant data as in audio or video steganography, this property comes in abundance and serves as the basis of stego process. Here we put forward a linguistic text steganography technique. This technique implants secret information into the shelter media for searching novel probabilities for occupying a language Hindi which is apart from English. The proposed approach namely Indian Script Encoding Technique (ISET) will work by implementing the linguistic attributes of Hindi language. In this approach, there is no difference between vowels and consonants. In this new approach, we worked with random and Hindi texts where Hindi text is used as a secret message and random text as a cover medium. Secret message will be hidden in cover medium. Our aim to propose this new ISET method is to increase robustness of the data and also increase in capacity of hidden data.

Keywords Linguistic · Text · Steganography · Random · Hindi
Robustness · ISET

S. Chaudhary (✉) · M. Dave
Department of Computer Science and Information Technology,
Jagannath University, Jaipur, India
e-mail: er.sunita03@gmail.com

M. Dave
e-mail: meenu.s.dave@gmail.com

A. Sanghi · H. Sidh
Department of Computer Science and Engineering,
Marudhar Engineering College, Bikaner, India
e-mail: dr.amitsanghi@gmail.com

H. Sidh
e-mail: hansrajsidh@gmail.com

© Springer Nature Singapore Pte Ltd. 2018
D.K. Mishra et al. (eds.), *Information and Communication Technology for Sustainable Development*, Lecture Notes in Networks and Systems 9,
https://doi.org/10.1007/978-981-10-3932-4_41

1 Introduction

As the Internet and wireless networks are rapidly spreading day by day in which multimedia streams are used for information exchange. For this, issue of security is a major concerned area for the researchers. From the past several years, maintaining the security of data and maintaining confidentiality over the Internet have now become the necessity. To maintain the security of data and confidentiality, there are many techniques among which steganography is one of the most effective technique, which is gained through obscurity. The steganography refers to the method of sending data into another data in such a way that no one can suspect the original hidden data, as in between the communication intruder can see the cover media only, not the original data. We can use various types of data, viz image, audio, video, and text as a cover media to hide the original data. Further, text can be of any language such as Hindi, English, Urdu, Marathi, Persian, and Chinese[1]. There are various areas of applications such as watermarking, SMSing, CAPTCHA, e-business, defense, bank, and portal security. Thus, we can very promptly able to say that there is no area where security is not needed; hence, steganography touches every area of information networks.

2 Text/Linguistic Steganography

Text steganography utilizes the properties of text features for communicating the secure messages over the Internet. It can be implemented mainly through the random and statistical generation methods, format-based method, and linguistic method; this is given in the Fig. 1.

In random and statistical generation methods, statistical and random attributes of the particular language are basically deployed for hiding the original information and to generate cover content that means we can generate random character sequence, random sentence sequence, and random word sequence to use as cover text to avoid attacks like known plain text attack [17]. Because of randomness, this method is less suspicious, and in our research also, we have used this method along

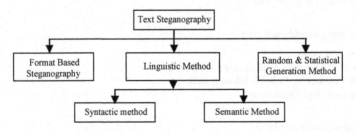

Fig. 1 Types of text steganography

with the linguistic properties of India's national language Hindi to conceal the original information into cover media.

In format-based method, format attributes of the particular language are used to generate the cover text. These format attributes are included white spaces, deliberate misspelling, changes in color, and size or type of fonts [18].

In linguistic-based steganography, the linguistic attribute of the particular language is used to generate cover text or to hide the original data into the cover text. Linguistic attributes are included commas, punctuation marks, and synonyms of the words [2].

Language-based steganographic scheme gives more highly developed methods to hide the original secret messages in cover media [3]. As this scheme handles specific property of the particular language which may not be available in other language or not suitable to other languages means language-based methods not work on the identical properties of all the languages.

Linguistic scheme can be implemented by two ways, one is syntactic method and another one is semantic method. If we include some commas, punctuation marks in this way that no one can suspect it, then it is called syntactic method [4, 5].

If we put synonyms of a particular word in such a way that the meaning of the sentences will not change, then it comes in semantic method.

3 Related Works in Indian Language Steganography

Hiding information in plain text is called text steganography. Previously, various researchers have done work in this field. In next section, we give brief idea of previous work.

In paper [6], the focus is on Urdu and Arabic languages. Urdu and Arabic languages have a most important feature of having redundant dot symbols. These dot symbols are used for hiding the information. To hide the data, one can change or shift the position of dot by some degree in horizontal or vertical position in references to the customary position of point in the text.

The paper [7] reflects the attributes of Hindi language. Every language has its own features. Hindi language has its own vowel, consonants, and compound letters which are combination of vowel and consonants. In this technique, author made two groups: one is for simple pure letters means only vowel and consonants, and another is compound letters means group of vowel and consonants.

In paper [8], the author has used three different techniques, namely THK, Core-Noncore character and using matraye, which has given new idea to hide the Telugu data.

In paper [1], researchers have used diacritics of Arabic script as cover text to put out of sight the Chinese original secret data. This is a combination of Arabic and Chinese languages. So, one can suspect it if he or she is aware of both the languages deeply.

The paper [9] provides an approach which is basically designed for sending secret message in Bengali language. Here, the researchers convert the secret message into the binary stream using grammar structure of Bengali script and encode it in meaningful sentence.

In paper [10], approaches designed by the researchers are for sending the secret information of Guajarati language. In this method, researches have used their own revised SSCE code and passkey to encode the data, so one can give input in any digital form.

In paper [11], the vowels have assigned 0 bit and consonant characters assigned 1 bit for diacritic and compound characters of Hindi language.

In paper [12], the researchers have implemented feature coding technique to hide and embed the secret message. In feature coding technique, one can shift the letters by some degree or can add some extra white space into the sentences and these white spaces can be used to hide the data.

In paper [13], the authors had given a new approach in which they have assigned 4-bit binary value to every Hindi letter. And by this binary value, the authors have generated a numerical code to hide the data. But the drawback of this method is that it is very time-consuming.

In paper [14], there is an approach designed by the researchers, where they have used the technique for sending the secret information in Oriya language. Thus, this technique uses the concept of inputting the data in any digital form, as they have used quantum truth table to map the secret message and generate the cover text.

In paper [15], special characters like inverted commas of Punjabi language are used to the secret data, and cover text is generated by the quantum truth table. The researchers embed the secret information into cover information according to their ASCII representative alphabet.

In paper [16], an approach is designed for hiding the secret message in Telugu language. Two approaches are implemented by the researchers. In first approach, the secret message is hided in compound letters of the Telugu language; in second approach, punctuation marks are accustomed to conceal the covert message. Original message is hided according to the Unicode value of punctuation marks.

After reviewing and analyzing above mentioned techniques it has been observed that still there is a need to do research in the area of linguistic steganography, as security issues are still a major concern.

4 Proposed Work

The problems encountered with different existing methodologies of text steganography is that if we change the format of text or retype the text, then it results in loss of secret information [17]. In our research, we have proposed a new method which conceals information written in Hindi language by using the linguistic method of steganography. We named this technique as Indian Script

Encoding Technique (ISET). In this technique, we have used the random character sequence as a cover medium and feature coding method of text steganography.

We use two processes to implement the proposed method. First, we encode all the characters of the secret message with new encoding technique which is based on cataloging of Hindi fonts. Second is hiding the secret information into the cover content. Proposed method reduces the memory consumption and size of cover text used for stenography in compare to other methods. As in this method, one letter can hide total eight bits. Because of this hiding capacity, the development of proposed approach to hide the message as well as to retrieve the message proved to be efficient. In proposed method, we classify the Hindi language characters and diacritics into eight groups. The grouping is based on the tongue position in the pronunciation of Hindi script grammar, namely Kanth, Talu, Murdha, Dant, Hont, Naak, Sayunktakshar, and Ayogyavah. In this approach, we used the random content as a cover text and secret message, which will be hidden in cover medium, is in Hindi language. Table 1 shows 8-bit encoding format of Hindi letters and diacritics. In this method, two schemes are used: one is encoding and another is decoding.

Encoding Algorithm:

In this approach, we introduce new encoding technique to conceal secret information in cover content. In our proposed technique, every character of secret Hindi message is encoded in the form of 8-bit binary number after that the equivalent ASCII character is replaced the original character. In this, the left most 0th and 1st bit will always remain '0' for Hindi characters. That means it shows that the input is in Hindi language not in others.

While making group, we consider only Hindi letters and diacritics of Hindi script. The left most 2nd, 3rd, and 4th bit of 8-bit number represents the group number these are from 1 to 8. In this approach, all letters and diacritics are divided into groups and every letters and diacritics has its position in corresponding group. This position will represent the last three bits of 8-bit number. By using this approach, we can encode all eight bits of one letter of secret message into one letter of cover text at a time.

After encoding the secret character, we get 8-bit binary data of the recreate character then the corresponding ASCII equivalent of this 8-bit binary data will be fetched. And this fetched character is embedded with the cover text according to the key.

Table 1 8-bit encoding format of character

Zero bit	First bit	Second bit	Third bit	Fourth bit	Fifth bit	Sixth bit	Seventh bit
Always remain '0'	Group number (1–8)				Character position in particular group		

Algorithm for Message Encoding:

1. Read character one by one.
2. for each character:
   2.1 Find the character in the group table and convert into 8-bit binary value.
   2.1 If character is compound statement (have diacritics then), then we make two separate binary streams, one for character and another for diacritics or next character.
   2.3 Convert each 8-bit binary sequence into its ASCII equivalent.
   2.4 This new letter is now embedded into the cover text according to the key.

Decoding Algorithm:
Decoding is the reverse of encoding. Following decoding algorithm has been used to decode the secret message.

1. Get the letter from the embedding position (according to the key).
2. Get the equivalent ASCII code.
3. Translate the ASCII code into its equivalent 8-bit binary format.
4. Map 8-bit Binary value and find the character from the group table.

5 Experiment Result

The experimental results are on the basis of linguistic features of Hindi language; we develop proposed method ISET through Java language and Netbeans profiler.

Through our new method, we are able to achieve the high security for the transmitted text, since we have initially encoded the text with the help of our new generated algorithm and then embedded the encoded characters into the cover content according to the key, and that key is having a separate method of formation. Another strong point is that our cover text is randomly created so no one can easily suspect it, as generally normal people take this embedded cover file as a corrupted file and ignore or delete it. The unauthorized person who wants to access the data has to first find out the key and then he has to crack the algorithm. Thus, it will be not an easy task. Here we show each algorithm interface individually. Figure 2 is a GUI, shows example of generation of random character sequence of cover text with secret message. It contains encode, reset, and generate buttons through which one can encode, reset, and generate cover text. In secret message window, secret information can be written. Similarly, a window for cover text is given. This window can be used to randomly type the cover text or cover text can be randomly generated by generate button. Now, secret information and cover text are encoded by the encode button. Reset button is used to reset the main window to its original.

Figure 3 shows message after encoding and embedding of original message into cover content.

After decoding the message, we get the original message which is shown in Fig. 4.

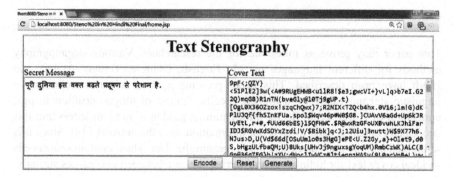

Fig. 2 Generation of cover text and secret message

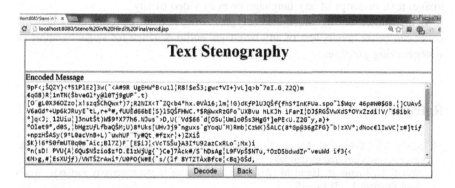

Fig. 3 Encoding of the secret message

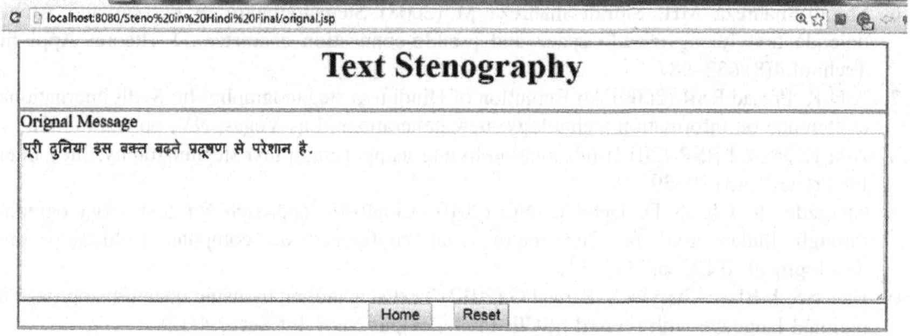

Fig. 4 Decoding of secret message

6 Conclusion

This paper may prove as milestone for the researchers. Various steganography methods for different languages namely English, Chinese, Telugu, Arabic, and Hindi have been described. This paper presents a novel advance technique to linguistic steganography by using the specific feature of tongue position in pronunciation of Hindi script. Secret information is hidden in random letters and this concealed content is used to send out information over the network [18]. Since it is scrambled text, no one can expect it as a meaningful data. Thus, random sequence is an ideal cover text for hiding and transferring the data between two parties strongly. Our new method helps to reduce the bit required to hide each character along with their diacritics and compounds. In this way with minimum efforts, large amount of message can be hidden and ensure better security of the secret contents.

Further, improvement in this method enables us to use cover media such as image, text, or script of any language or even video of any.

In the future, we can embed the encrypted string into any cover media, where the cover can be an image, text of any language, audio or video for steganographic or watermarking purposes.

References

1. Shakir AC, Xuemai G, Min J (2010) Chinese language steganography using the Arabic diacritics as a covered media. Int J Comput Appl 11(1):43–46
2. Nosrati M, Karimi R, Hariri M (2011) An introduction to steganography methods. World Appl Program J 1(3):191–195
3. Wyseur B, Wouters K, Preneel B (2008) Lexical natural language steganographic system with human interaction
4. Kaleem MK (2012) An overview of various forms of linguistic steganography and their applications in protecting data. J Glob Res Comput Sci 3:33–38
5. Bennett K (2004) Linguistic steganography: survey, analysis, and robustness concerns for hiding information in text. CERIAS technical report
6. Shirali-Shahreza MH, Shirali-Shahreza M (2008) Steganography in Persian and Arabic unicode texts using pseudo space and pseudo-connection characters. J Theoret Appl Inf Technol 4(8):682–687
7. Alla K, Prasad RSR (2009) An Evolution of Hindi text steganography. In: Sixth international conference on information technology: new generations, Las Vegas, NV, pp 1577–1578
8. Alla K, Prasad RSR (2012) Information hiding using Telugu text steganography. Int J Intell Inf Process 3(3):30–39
9. Changder S, Ghosh D, Debnath NC (2010) Linguistic approach for text steganography through Indian text. In: 2nd international conference on computer technology and development. IEEE, pp 318–322
10. Banerjee I, Bhattacharyya S, Sanyal G (2012) Text steganography using quantum approach in regional language with revised SSCE. Int J Comput Netw Inf Secur 4(12)
11. Changder S, Debnath NC, Ghosh D (2010) Hindi text steganography: an approach to information hiding. J Comput Methods Sci Eng 10(1)

12. Pathak M (2010) A new approach for text steganography using Hindi numerical code. Int J Comput Appl 1(8)
13. Changder S, Debnath NC, Ghosh D (2009) A new approach to Hindi text steganography by shifting Matra. In: International conference on advances in recent technologies in communication and computing. IEEE, pp 199–202
14. Banerjee I, Bhattacharyya S, Sanyal G (2012) A procedure of text steganography using Indian regional language. Int J Comput Netw Inf Secur 8:65–73
15. Banerjee I, Bhattacharyya S, Sanyal G (2007) Text steganography using one Indian language. CSI J Comput
16. Prasad RSR, Alla K (2011) A new approach to Telugu text steganography. In: IEEE symposium on wireless technology and applications, pp 60–65
17. Chaudhary S, Mathur P, Kumar T, Sharma R (2013) A capital alphabet shape encoding (CASE) based text steganography. In: Conference on advances in communication and control systems, Atlantis Press, pp 120–124
18. Chaudhary S, Dave M, Sanghi A (2016) An elucidation on steganography and cryptography. In: Second international conference on information and communication technology for competitive strategies, Mar 2016. ACM, Udaipur, India

IC Technology to Support Children with Autism Spectrum Disorder

Nara Kalyani and Katta Shubhankar Reddy

Abstract The research in computing technology enables many areas to have a tremendous advantage. Education is one field where the traditional classrooms are replaced by E-classrooms and is proved to be effective in showing student's progress. The study carried out by International Clinical Epidemiology Network Trust (INCLEN) shows that there are ten millions of children in India who suffer from autism. Autism is a kind of development disorder of an individual with a deficit in learning ability, social interaction and language skills. In the recent past, there are many research scholars working across the world to provide supporting aids which are personalized based on the assessment of the child. This paper provides the details of the app developed that is personalized a solution to kids of age group 3–4 years to overcome the learning disability through a comfortable handheld device.

Keywords Autism · Vocabulary building · Improving pronunciation
Speech synthesis device · Sentence training

1 Introduction

Autism is a social and skill development disorder which is a kind of inability in kids toward social, mechanical and language learning skills. It is not similar to an organic disease that can be diagnosed and cured with medication. This requires a team of psychology experts who would assess the kid regarding the actual disorder and suggest a personalized solution in terms of training. This problem if properly identified and addressed in right time would help in molding the kid in the right direction with required skills. This will help individual to lead an independent life

N. Kalyani (✉)
G. Narayanamma Institute of Technology and Science, Hyderabad, Telangana, India
e-mail: nara.kalyani@gnits.ac.in

K. Shubhankar Reddy
Chaitanya Bharathi Institute of Technology, Hyderabad, Telangana, India
e-mail: shubreddy01@gmail.com

© Springer Nature Singapore Pte Ltd. 2018   403
D.K. Mishra et al. (eds.), *Information and Communication Technology
for Sustainable Development*, Lecture Notes in Networks and Systems 9,
https://doi.org/10.1007/978-981-10-3932-4_42

and also be a part of society. At the same time, if it is not addressed properly, the individual would develop psychological disorders and the impact is unpredictable.

The role of teacher and parent plays a major part in diagnosing the children with autism. This requires a keen observation of the response (physical/communication) given by the child in suitable activities carried out on at the very early age of childhood. There are many universities with remarkable research contributions since 50 years. To name few institutes, National Institute of Health, Southwest Autism Research and Resource Centre (SARRC), Autism Speaks and Autism Society of America are in lead addressing this issue. In India, the awareness of such disorder is coming into light and even the doctors are giving possible guidelines to parents to enable them to use the suitable techniques at the early stage of childhood.

ASD are a group of behaviorally defined disorder with abnormalities or impaired development in mainly two directions. First is relating to the continuous deficit in social communication or social interaction and the second being in restricted interests, activities or in repetitive patterns of behavior. In the recent past, it is observed that advancements in computer science engineering are said to have a better platform for scientists in providing supporting models using Artificial Intelligence Contemporary Techniques. Multimodal social–emotional interactions would help the child to develop the ability to identify, correct the true means of communication [1]. It improves the ability of children in terms of understanding the intentions, emotions of other people and gives the spontaneous and appropriate emotions as response in terms of usage of verbal and facial expressions [2, 3].

The authors in [4–8] carried out experiments using robots for associating labels to different facial expressions and categorizing the emotions into "happiness", "sadness", "anger" etc. They extended their work by providing a sensory motor architecture for recognition of facial expressions. These robots can interact with humans and produce required facial expression or emotions and also imitate them even online. This facilitated in developing the online games between the robot and the human. Similar architecture can be used to interact with the child and based on the style of interaction and response, the child can be tracked and analyzed to understand the level of understanding.

There is a very interesting multidisciplinary research carried out using different technologies, and it is successful in providing a supporting system for specially challenged people. The authors in [9] explored and carried out an experiment to use robots in providing therapy solutions. They devised an ICT device for treating ASD. This device provided a new paradigm of learning through imitation that explored the learning of different participants. It also provides systems that analyze and conceptualize through cognitive computation.

Duquette et al. [10] made an experimental setup and conducted experiments on two pairs of children with autism: one pair interacting with the robot and another pair interacting with the experimenter. The objective was to explore whether mobile robot toy could provide equivalent social interaction in situations where the robot was predictable. The results were interesting and proved that the body movements and actions of children were more when they were interacting with humans rather than with robot. The two children interacting with the robot had better-shared attention in terms of eye contact and physical proximity. They were also able to mimic facial expressions than the children that were interacting with a human partner.

Fujimoto et al. [11] designed and used techniques that could mimic and evaluate the human motions in real time. They designed a therapeutic humanoid robot to perform practical experiments for testing the interaction of ASD children with robots and to evaluate the improvement of children's imitation skills.

Feil-Seifer and Mataric [12] designed a mobile robot and experimented on a group of eight children with ASD and observed tremendous variability in the valence of an effective response. This study reveals that the robot automatically distinguished between positive and negative reactions of children with ASD. Studies proposed by Dautenhahn and Werry [13] have shown that some children with ASD preferred interacting with robots compared to non-robotic toys or human partners. However, it is also found that there are few individual differences in whether children with ASD preferred robots to non-robotic toys. Two of the four participants exhibited more eye gazes toward the robot and more physical contact with the robot than with a toy.

Flores et al. in [14] showed that it is possible to improve both communications, social skills of children with ASD by providing Augmentative and Alternative Communication (AAC). The latest forms of AAC are available as cell phones, MP3 Players, and personal computer tablets. They investigated the utility of iPad as a communication device was more effective when compared its use to a simple system that incorporates communication using picture cards.

Recently Serret [15] developed a serious game, "Jestimule", to improve social cognition in ASD. This game was designed with consideration for heterogeneity of ASD. Young children and children with developmental delays can easily play this game, and it also helps these kids to recognize facial emotions, emotion gestures and emotional situations.

2   Android Solutions for Individuals with ASD

In this busy world, parents and teachers hardly find extra time to spend with children with special needs. This is a serious issue when the child suffers from Autism Spectrum Disorder. For children with ASD, multimodal social–emotional interaction plays an important role in the development of skills. Based on the statistical analysis, it is observed that computers are user-friendly for children with autism due to their predictability and consistency, compared with the unpredictable nature of human responses.

There are many assistive devices to help an autistic person in society and help them, and their families achieve a better quality of life. The existing technology is suitable for early intervention but lacks the facility to track the performance of the individual. In September 2014, i-Autism and Foundation Orange have launched APPYAUTISM, a new website devoted to apps for people with ASD. There are

various apps that are useful in developing communication, learning skills, understand the social behavior and many more features. With this motivation, this
research is carried out where the kid is given a scope in developing the required
skills, the parent is informed about the progress of the kid and the activities are
regularly monitored by the expert psychologists. Since from time to time there is
regular monitoring on the kid, the training process can be modified from time to
time based on requirement.

This work proposes an architecture that would perform three tasks simultaneously and is shown in Fig. 1. The app is developed with three main modules one for
training, second for testing and third for tracking the performance of the kid. The
app is focused on the kids of young age groups. Usually, kids of 3–4 years would
be very comfortable in framing good sentences. In the case of kids with autism,
there could be problems in regard to proper usage of vocabulary, proper framing of
sentences and expressing the proper emotions while communicating with others.

For example even in spite of repeated training, ASD kids find it difficult in
mapping the words with objects. They would have confusion regarding proper
recognition of characters and also spelling the words that are not phonetically same
in all cases. The best example is: "know" and "no" pronunciation is same, but the
written form is different. Such words always leave a trace of confusion in the minds
of ASD kids. The second example is: when a question is framed to ASD kid, the
response is also in the form of a question. They find it difficult in processing and
framing the correct response. Even if they understood, they may not put the words
in the correct order using proper prepositions wherever required. These issues are
addressed in this project based on simple language models. The application has two
levels of learning. The first level is regarding the vocabulary pronunciation and
image identification. The second level is regarding framing sentence using images.

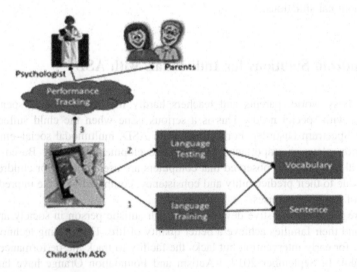

Fig. 1 System architecture

3 Vocabulary Training

Training for words is done as an incremental process. When the child gets exposed to the surroundings and the scenarios, he gets a chance to increase his knowledge of using the words. It is essential to pay special attention regarding the words, what is its meaning and how it is used. There are few words that differ in primary meaning when used as a noun and secondary meaning when they are used as a verb. Some words can also be replaced with synonyms retaining the meaning. Use of antonyms changes the meaning of the statement. The process of building word knowledge is done in following ways.

- Adapt the child to recognize the word using a picture at initial step.
- The child is trained for proper pronunciation of the word using SSDs.
- The child will be able to understand the meaning of the word in primary form when it has multiple meanings.
- A new word is introduced that has similar meaning, form, adds purpose or attribute to the primary word.
- The child will be able to use the word appropriately after training.

For example:
Process of building the vocabulary relating to bird
The first step is to identify the given picture is a bird. This can be done by attributes the bird would have. All birds would have feathers, wings, and beak. This information can be used for differentiating a bird with the others. At second level what the birds can do. Some birds can fly and some can swim. Once it is clear then train on different types of birds and the words that can be used for describing them.

1. First, display images of birds to train for the word bird.
2. Second, display the images of wings, feathers, and beaks to train these words.
3. Next, display the images relating to flying birds like a crow, sparrow, owl etc.
4. Then, train with the words relating to birds that can swim like a duck, swan etc.

This process enables the child to learn without any ambiguity what the birds can do and the difference between various birds.

The process of improving pronunciation Incorporating Speech synthesis device (SSD) facilitates to control the voice output. There are many voice output devices that are commercially available that can be used for verbal communications which are close to natural speech. Such devices may not be helpful as there is special attention to be given for the autistic kids as they cannot repeat the utterance at once if specified. Most of the kids require a partial word to utter at a time and later should have the facility to combine the subunits to get the acquaintance for the complete utterance. Otherwise, the usual problem that would arise is mispronunciation learning which is difficult to correct.

The SSDs are designed which is customized to the child based on the level of assistance needed to each individual. The system is adaptive and also has the facility to synthesize the part of the desired unit when touched with finger or head stick. In various researches, it is observed that if the child is given a communicative device, it will support to overcoming speech disorders. This may not be an efficient substitute to humans but would be a possible solution for improving language skills, as there is a scarcity of trained people and is not available at affordable price.

SSD can serve as a first step in exploring communication with children who have ASD. The process of tracking the activities would help in identifying the level where the child is and the enhancement that can be made to the device from time to time to take care in improving the vocabulary building.

This module is designed which would make the device very useful for the kid in the initial stages of childhood. The android app has a module which takes care of training the style of pronouncing a word. Selecting an image the word is identified and its corresponding phonetic representation is taken from the dictionary. The word is divided into subunits such that, selection of each unit, the pronunciation is given in a slow manner so that the child can get acquainted with spelling and the corresponding utterance. If the child wants to repeat, he can select the unit; otherwise, it displays set of videos which contain related word's use. This helps the child to register the word, image, and the corresponding pronunciation. If he wants to continue he can, otherwise terminate the process. During the entire session, the activities of the child are tracked in terms of time for each activity, the way the selection is made and the hand pressure applied by him while selections are made. The following Fig. 2 explains the flow of the process.

Fig. 2 Process of training vocabulary

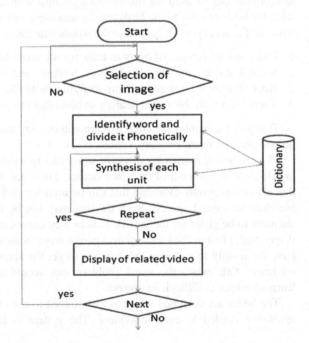

The recorded information is sent to the parent and the trainer/psychologist. This information helps the attitude of the child to be tracked and take appropriate measures from time to time.

4 Sentence Training

Kids find it difficult in framing sentences: they would find it difficult to answer simple questions that are framed. For such kids, this device would be very useful. Training the kids to frame sentences can be done level by level by using structured context-free grammars. Figure 3 explains the process of training the sentences.

Research observations show that some children with ASD find it difficult to frame sentence or have the tendency to imitate the same when asked a question rather than answering. These kids try to frame sentences by picking words from the question itself. Sometimes we need to frame set of questions in sequence to enable him to answer using single words. Some children may find it difficult to combine words. There may be a need to break down the words into small articulation units like syllables so that the child attempts to articulate the unit and combine more than one syllable to give the complete utterance of the word or multiple words. Slowly

Fig. 3 Process of training sentences

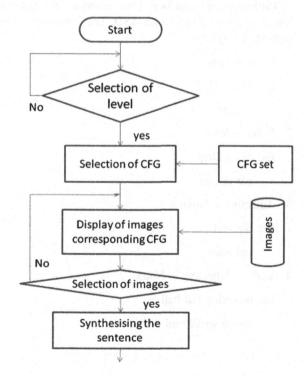

increasing the length of the sentence would enable the child to produce the utterance for the complex sentence.

Initially, the child starts framing the sentence with one or two words using incorrect grammar and gradually learns to use proper word order in the sentence. In the case of a child with ASD, once the pattern is trained, it is difficult to change the pattern. For such children, it is preferred to teach grammatical forms from the beginning. The main aim of the second module is to identify the words and the complexity within the word. Based on this information, decision may be taken whether to produce the sounds as a sequence of syllable units or words combined together.

For example, when the sentences to be trained are

- I saw ball
- I saw blade
- I saw hippopotamus

In these sentence, there is no complexity in the first and second statement there is a need to give a pause at the blade as there are two consonants together which may be little difficult. In the third case, the word is complex for the child as there are five syllables and two syllables share same articulation. This kind of word is difficult for the kid to utter. Hence, there is a need to separate each syllable and train it as individual units and then combine two or more units at the convenience of the child.

Partington and Sundberg [16] recommended few word orders that can be followed to train the child with ASD. The following are few such sequence that was used in this system.

1. Noun + Noun

   Ex: Mom, Dad

       cat, rat

2. Noun + Verb

   Ex: ball rolling

       car driving

3. Adjective + Noun

   Ex: big ball,

       red rose

4. Verb + Adjective + Noun

   Ex: bouncing red ball

       flying small bird

5. Adjective + Noun + verb

   Ex: white bear running

      Bad dog barking

The teaching techniques adapted for training the kid toward sentence formation are listed below.

1. *Mapping of perception to response*: the child is encouraged to respond to the query in one word. The response given by the child shows the clarity regarding the transfer of perception to response. It is essential to make sure that the response is appropriate.
2. *Building appropriate word combinations*: this procedure uses the expertise the child has acquired to build simple sentences. These sentences should be framed such that the verbs are used with appropriate adjectives/adverbs.
3. *Expanding utterances*: this technique is used when the kid is able to consistently form simple phrases to respond to the request. It helps to reinforce the correct utterances by adding one more word to the phrase he is using based on his simple activities.
4. *Contingent comments*: once he has learned to make simple sentences the other similar sentences can be taught easily. It is always preferred to encourage the child to respond with variety forms. For instance, when training on adjectives it is preferred that child gets a chance to get exposed to multiple objectives with the same adjective or same object with multiple adjectives.

   Example: red rose, pink rose, beautiful rose, etc.
   Beautiful bird, beautiful flower, beautiful butterfly... etc.

These are the techniques used in the model at three levels in which the context-free grammars are used to track the level of training and step-by-step procedure to expand the sentence. The first level is suitable when the kid is aware of minimum 40–50 words and is capable of pronouncing the words in a comfortable zone. The framing of the sentence is trained in a standard format with using two words at a time. In the first level, the sentence is framed by a simple rule which contains the form "Noun + Noun, Noun + Verb, Adjective + Noun" Sentences are of the form a *big ball, I ate, red rose, give me*, etc. In Level 2 training is done using three words in correct order that contains the form "Verb + Adjective + Noun, Adjective + Noun + Verb". For example, *I blow a balloon, bouncing red ball, I like sweet, I want fruit.* etc. In Level 3, a complete sentence is framed using the prepositions wherever necessary like *I blow a red balloon, I saw the blue bag,* etc. The advantage of this system is even if the kid is not aware of the words, images are provided in the appropriate order so that on selecting these images, correct sentence is framed for the kid to practice.

5 Testing

The app is provided with options where testing can also be done at vocabulary and sentence levels. The child is allowed with a separate interface where he can make a choice to test his/her vocabulary and sentence framing skill. Sequences of images are displayed, and these are to be mapped with the words that are listed. It facilitates the pronunciation of the words when clicked on the words. For testing at sentence level, the kid is given the option to select suitable words in order of forming a valid sentence. If the selection order is valid, then an appreciation sound is generated. The activities are tracked from time to time in terms of the order of selection and the pressure applied while performing the activities. This information is pooled and stored in log records and from time to time the complete details are sent to the parent and psychologist for required analysis. The system was tested with special children, selected from special school that was associated with NIMH, Hyderabad, following the special curriculum. Totally 16 students of 5–8 years age group were selected who required repeated training. Of these students when allowed to use this app, 9 students tried using it with ease and showed interest. The recorded evidence provides a means to understand the learning capability of the kid and helps in updating the unit which is personalized to the kid.

6 Conclusion

The increase in Autism rate affects the nation's human resources. The development of ICT-based approaches improves therapy facilities and supports the education of children with ASD. Integration of interfaces, sensor technology and algorithms would provide the kid with ASD a platform that can simulate the role played by a trainer. This work is an initial contribution in the direction of providing a solution for children with ASD and also for tracking the activities. These small contributions are valuable and find a place in future. It is essential to develop ICT architectures and devices that are clinically accepted. More studies need to be performed to generate a more reliable, attentive, emotional device that can tailor the special education methods that are personalized or can be adapted to the children with ASD.

References

1. Carpendale J, Lewis C (2004) Constructing an understanding of the mind: the development of children's social understanding within social interaction. Behav Brain Sci 27:79–151
2. Decety J, Jackson P (2004) The functional architecture of human empathy. Behav Cogn Neurosci Rev 3(2):71–100

3. Narzisi A, Muratori F, Calderoni S, Fabbro F, Urgesi C (2013) Neuropsychological profile in high functioning autism spectrum disorders. J Autism Dev Dis 43(8):1895–1909
4. Boucenna S, Gaussier P, Andry P, Hafemeister L (2010) Imitation as a communication tool for online facial expression learning and recognition. In: IEEE/RSJ international conference on Intelligent robots and systems (IROS), 2010. IEEE, pp 5323–5328
5. Boucenna S, Gaussier P, Hafemeister L (2011) Development of joint attention and social referencing. In: 2011 IEEE international conference on development and learning (ICDL), vol 2. IEEE, pp 1–6
6. Boucenna S, Delaherche E, Chetouani M, Gaussier P (2012) A new approach for learning postures: an imitation game between a human and a robot. In: IEEE/RSJ international conference on robots and systems, (IROS 2012). Workshop: cognitive neuroscience, robotics, pp 1–4
7. Boucenna S, Anzalone S, Tilmont E, Cohen D, Chetouani M (2014a) Learning of social signatures through imitation game between a robot and a human partner. IEEE Trans Auton Ment Dev (in revision)
8. Boucenna S, Gaussier P, Andry P, Hafemeister L (2014b) A robot learns the facial expressions recognition and face/non-face discrimination through an imitation game. Int J Soc Robot (to appear)
9. Ricks D, Colton M (2010) Trends and considerations in robot-assisted autism therapy. In: 2010 IEEE International Conference on robotics and automation (ICRA). IEEE, pp 4354–4359
10. Duquette A, Michaud F, Mercier H (2008) Exploring the use of a mobile robot as an imitation agent with children with low functioning autism. Auton Robots 24(2):147–157
11. Fujimoto I, Matsumoto T, De Silva P, Kobayashi M, Higashi M (2011) Mimicking and evaluating a human motion to improve the imitation skill of children with autism through a robot. Int J Soc Robot. 3(4):349–357
12. Feil-Seifer D, Mataric M (2011) Automated detection and classification of positive vs. negative robot interactions with children with autism using distance-based features. In: 2011 6th ACM/IEEE international conference on human-robot interaction (HRI). IEEE, pp 323–330
13. Dautenhahn K, Werry I (2004) Towards interactive robots in autism therapy: background, motivation and challenges. Pragmat Cogn 12(1):1–35
14. Flores M, Musgrove K, Renner S, Hinton V, Strozier S, Franklin S, Hil D (2012) A comparison of communication using the apple iPad and a picture-based system. Augment Altern Commun 28(2):74–84
15. Serret S (2012) Jestimule, a serious game for autism spectrum disorders. Neuropsychiatr Enfance Adolesc 60(5):59
16. Sundberg ML, Partungton JW (1998) Teaching language to children with autism or other developmental disabilities. Behavior Analysts Inc., Danville, CA

Using Hybrid Cryptography and Improved EAACK Develop Secure Intrusion Detection System for MANETs

Sharad Awatade and Pankaj Chandre

Abstract MANET is widely used in different areas, because it supports lots of feature. Day by day importance of wireless network increases. The issue in mantes are security, broadcasting, Packet dropping there for giving security to mantes is important. There are different types of attack such as misbehaving attack, wormhole attack, gray whole attack and flooding attack. In this paper, we study how the attack is detected or prevented. And finally using specific parameter, we measure performance. We study the EAACK algorithm, and it uses hybrid cryptography technique for reducing network overhead.

Keywords Digital signature · Intrusion detection systems (IDS)
Denial of service (DOS) · MANET · EAACK

1 Introduction

MANETs support different features such as dynamic network topology, do not support centralized control, it has small memory size and also battery storage. It is used in military battlefield, fire, earthquake and PAN. In broadcasting, there is problem occurred such as collision and redundancy in the network. Remove the problem in MANETs like clustering, broadcasting, power and bandwidth management is necessary. Classification of MANET Protocol is given below:

- Reactive protocol
- Proactive protocol
- Hybrid protocol

S. Awatade (✉) · P. Chandre
Department of Computer Engineering, Flora Institute of Technology, Pune, India
e-mail: Sharad.awatade@gmail.com

P. Chandre
e-mail: Pankajchandre30@gmail.com

© Springer Nature Singapore Pte Ltd. 2018   415
D.K. Mishra et al. (eds.), *Information and Communication Technology
for Sustainable Development*, Lecture Notes in Networks and Systems 9,
https://doi.org/10.1007/978-981-10-3932-4_43

Examples of reactive protocol are AODV and DSR. Example of proactive protocol is DSDV. Example of hybrid protocol is ZRP [1]. Most commonly used protocol is AODV in which it maintains new root when it demands. It uses three message formats, i.e., RREQ, RREP and RERR. Static sink and mobile sink are the two type of sinks present in the MANET. In static sink, information is collected from only the source node; mobile sink information is collected from whole network. Giving security is a real test or requirement for securing the system [2]. There are number of interruption identification frameworks have been produced, however there are number of downsides for expelling these disadvantages. In this paper, we anticipated new framework, i.e., improved EAACK. MANET is used in emergency requirements, because it allows easy deployment, minimal configuration and low cost. It has restricted battery power and resources.

2   Background

Providing security is very challenging task in MANETs. There are numerous IDS have been produced for giving security. In this area, we fundamentally portray three exhibited method in which some have weakness explained below.

2.1   Watchdog

Marti anticipated method watchdog for detecting misbehaving node which is unsafe for network. It operates in two phase, first is watchdog and second is pathrater. It uses its next hops transmission for detecting the misbehaving attack which is present in the network. This algorithm has certain week point such as collusion, power limitation and dropping packet [3].

2.2   TWOACK

Some of the drawbacks which are present in previous IDS to avoid these limitations and to increase the performance of network TWOACK schema are proposed. It uses three consecutive nodes to transfer packet from source to destination [4, 5].

2.3   AACK

It consists the combination of TWOACK and TACK. It transfers packet from the first node to last node. Destination node gives feedback to first node. It gives the better performance than the watchdog and TWOACK. But drawbacks of the AACK are it is not suitable for when there is number node in the network is large [6, 7].

3 Literature Review

Fourth watchdog and pathrater is designed by S. Marti and T.J. Giuli, et al. It improves performance of the network. It is first IDS for MANET but it suffers from certain drawbacks [8].

N. Kang, E. Shakshuki and T. Sheltami develop new IDS for MANETS called Enhanced Adaptive Acknowledgment (EAACK) [9]. Main goal for establishing of EAACK is to overcome drawbacks of traditional IDS, i.e., watchdog. MANETS having number of advantages such as convenience, deployment, faster and cost consuming as compared to any wired network. But security is a very important task because attacker can easily attack on MANETS.

Leady E. Shakshuki Proposed EACCK using DSA Routing protocol it has advantageous such as root cache store multiple root source node first check root cache then path between source and destination is established. It uses RSA and DSA algorithm. But there are disadvantages like link breakage. After digital signature is used for preventing from forging acknowledgment [5].

Now, we use hybrid cryptography and improved EAACK algorithm for prevention of attack or to eliminate requirement of redistributed keys. We use key exchange mechanism such as RSA & DSA. And finally, testing the performance of mobile ad hoc network in real environments by skipping in simulation [10].

4 System Description

4.1 ACK

It is circular acknowledgment schema packet is send originator node to target node. Then target node sends an acknowledgment packet to source within fixed time, otherwise again the same packet is sent once more.

4.2 S-ACK

It uses three consecutive nodes for detecting misbehaving node which attacks the system. Transmitting packet from first to last, then it gives acknowledgment back to first node. For sending S-ACK packet to source node, third node is used. It removes limitation of TWOACK schema. It urgently creates misbehavior report, and it is the primary step of misbehavior report authentication. Main purpose of this algorithm is detecting the misbehaving nodes in the network channel.

4.3 MRA

First, check destination side dropped message is reached or not. Then, source node sends misbehavior report to MRA node. Then, MRA node sends same packet which was being sent but at this time it uses a different routes for sending packet. For sending packet, it searches path using its own local knowledge base table, it contains information about route path selection. Then, it checks the result if the same packet reaches target node for early time, then misbehavior report generated is correct. On the other hand, if same packet is already designated, then false report is generated and which node generates this report is marked as folksy Node or misbehaving node or malicious node and removing these node for securing the network. All the above three are the part of EAACK algorithm [7].

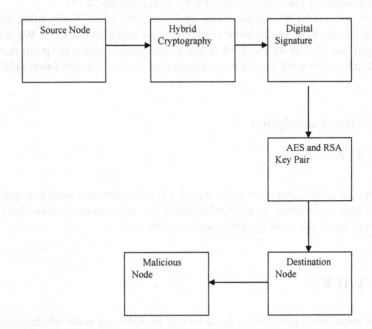

Fig. 1 Proposed system architecture

5 System Architecture

Source node is used to send packets to the number of destination nodes. Source node sends packet to its neighboring node, then it gives quick response. When packet is sent from next node, then back acknowledgment is directly sent to source node; in this way, we can create a structure of MANETs. Digital signature technique is used for authentication purpose. Secure data transmission is done by using RSA and AES key pair. Combination of cryptography and EAACK gives advantages [9]. After to check what type of attack arises on the network and according to this classify this attack (Fig. 1).

6 Detection of Attack

6.1 Gray Hole Attack Detection

Detecting Gray Hole attack is easy node showing correct direction for transiting data but after it dropping the data. Misbehaving is occurred in gray hole attack [4].

6.2 DOS Attack

When attacker avoids to access particular information to authorized user, then this attack is detected.

Detection step:

Step 1: First sender sends the packet.
Step 2: Node between sender and receiver works as attacker node
Step 3: Packets which are sent by the sender capture the attacker node.
Step 4: If it is not forwarding the information at receiving node, then DOS attack is generated.

7 Attack Prevention

To build up the security in the mobile ad hoc network, concept of elliptic curve cryptography (ECC) is used for providing more security. ECC algorithm basically utilized on behalf of encryption and decryption of the data, text, and packets [11]. While using the concept of encryption, it creates secured connection, when data is to be sent through the network and the observer cannot see the relevant information. The attack prevention system takes the precaution of providing prevention from the malicious node in the MANET. This system probably utilizes either RSA or DES for the concept of providing security data for the function of key exchange and encryption or decryption [20].

8 Performance Evaluation

For measuring the outcome of our system, we use the different parameters or technique which are given below:

(1) **Packet Delivery Ratio**: It gives information about how many packets send and how many packets received.
(2) **Delay**: Time required for traveling packet from first node to last node of the network.
(3) **Routing Overhead**: It gives information about overflow rate.

1. Result Scenario I
In scenario I, we perform our system from 1 node to 3 nodes in which PDR, retransmission count, underflow rate, and overflow rate is to be calculated which is shown in the Fig. 2.

2. Result Scenario II
In scenario II, we perform our system from 1 node to 7 nodes in which PDR, retransmission count, underflow rate, and overflow rate is to be calculated which is shown in the Fig. 3.

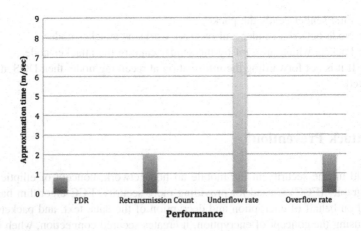

Fig. 2 Result Performance I

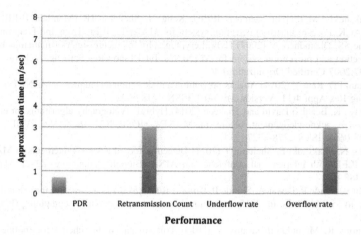

Fig. 3 Result Performance II

9 Conclusion

For securing any system there is need to develop strong mechanism of protecting network there for in this paper we develop secure intrusion detection system for MANET named EAACK. Also, we study formation structure of MANET and study of previous intrusion detection system.

Study some attack occur in the network, which can be harmful to the system. In this paper, we study broadcasting in MANETS, different attacks on MANET; also; we study the EAACK algorithm using hybrid cryptography technique. Using this technique, we can cover all the drawbacks of traditional IDS. In future scope need to develop faster system used for increase the speed of network and give more throughput than existing technique.

Acknowledgements I am mainly thankful to my guide Prof. Pankaj Chandre, who has provided guidance, expertise, and encouragement. Thanks to all those who helped me in completion of this work knowingly or unknowingly.

References

1. Lakshmi SM, Bhavana S, Sujata T (2014) Enhancement of security levels using intrusion detection system in MANETs. IOSR–JCE. e-ISSN 2278-0661, ISSN 2278-8727
2. Anantvalee T, Wu J (2006) A survey on intrusion detection in mobile ad hoc networks. In: Wireless/mobile security. Springer, New York
3. Ruchita M, Seema L (2014) Review paper on flooding attack in MANET. Int J Eng Res Appl 4(1, Version 2):39–46. ISSN 2248-9622
4. Hally K, Gupta A, Pal D (2014) Mitigation of HTTP-GET flood attack. Int J Res Appl Sci Eng 2(XI). ISSN 2321-9653

5. Shakshuki LM (Senior Member, IEEE), Kang N, Sheltami TR (Member, IEEE) (2013) EAACK a secure intrusion-detection system for MANETs. IEEE Trans Ind Electron 60(3)
6. Jathe SS, Dhamdhere V (2007) Hybrid cryptography for secure superior malicious behavior detection and prevention system for MANET's. Int J Innov Res Sci Eng Technol (An ISO 3297:2007 Certified Organization) 4(7)
7. Nikam PD, Raut V (2014) Attacks prevention and detection techniques in MANET: a survey. J Eng Res Appl 4(11, Version 2):15–19. ISSN 2248-9622
8. Ramya K, Beaulah David and Shaheen (2014) Hybrid cryptography algorithms for enhanced adaptive acknowledgement secure in MANET. IOSR–JCE 16(1, Ver. VIII). e-ISSN 2278-0661, ISSN 2278-8727
9. Kang N, Shakshuki E, Sheltami T (2011) Detecting forged acknowledgements in MANETs. In: IEEE 25th international conference on AINA, Biopolis, Singapore, 22–25 Mar 2011, pp 488–494
10. Joshi P, Nande P, Pawar A, Shinde P, Umbrae R (2015) EAACK—secure intrusion detection system for MANETs. In: 2015 international conference on pervasive computing (ICPC), Jan 2015
11. Ferdous R, Muthukkumarasamy V (2010) Trust management scheme for mobile ad-hoc networks. In: 2010 10th IEEE international conference on computer and information technology (CIT 2010)
12. Suruthi B, Kumar NVR (2014) An enhanced intrusion detection system for MANETs using hybrid key cryptography. IJCSIT 5
13. Pattnaik O, Pani S (2012) Application of IDS in WSN: a survey. Int J Res Comput Commun Technol IJRCCT 1(7). ISSN 2278-5814
14. Thanikaivel B, Pranisa B (2012) Fast and secure data transmission in MANET. In: 2012 international conference on computer communication and informatics (ICCCI-2012), 10–12 Jan 2012, Coimbatore, India
15. Andhiya D, Sangeetha K, Latha RS (2012) Adaptive acknowledgement technique with key exchange mechanism for MANET
16. Liu K, Deng J, Varshney PK, Balakrishnan K (2007) An acknowledgment-based approach for the detection of routing misbehavior in MANETs. IEEE Trans Mobile Comput 6(5):536–550
17. Buttyan L, Hubaux JP (2007) Security and cooperation in wireless networks. Cambridge University Press, Cambridge, U.K.
18. Pahlevanzadeh B, Samsudin A (2007) Distributed hierarchical IDS for MANET over AODV. In: IEEE international conference on telecommunications an communications, May 2007, pp 99–104
19. Jayakumar G, Gopinath G (2007) Ad hoc mobile wireless networks routing protocol—a review. J Comput Sci 3(8):574–582

Diversification in Tag Recommendation System Using Binomial Framework

Jayeeta Chakraborty and Vijay Verma

Abstract Diversity has been recently identified to be one of the major contributors for improving performance of a recommendation system in terms of user satisfaction. In social tagging-based systems, tag recommendation is used to suggest tags to a user for a resource. Diversity in tag recommendations has been overlooked in traditional tag recommendation techniques so far. In this paper, we propose a novel tag recommendation system that recommends a diverse yet relevant set of tags to the user. Our system utilizes a simple greedy-based algorithm that optimizes an objective function, defined using recently proposed binomial framework for diversification that considers high coverage and penalizes redundancy. To incorporate user preference to the recommendations, tags are suggested based on user's affinity with tags along with diversity. We experimented with the MovieLens 10M dataset. Effectiveness of the system has been evaluated off-line with respect to both relevance and diversity.

Keywords Tag recommendation · Diversity · Coverage · Redundancy
Recommendation system · Tags

1 Introduction

Users need to be guided properly so that they can find the products based on their interests. Recommendation system (RS) suggests a user to easily find their preferred items among other items in a broad system where large amount of information is available on the Web [1]. Researches in recommendations have traditionally focused only to improve accuracy of a RS [2–4]. But it has been observed that only

J. Chakraborty (✉) · V. Verma
Computer Engineering Department, National Institute of Technology,
Kurukshetra, India
e-mail: jayeeta.iem2012@gmail.com

V. Verma
e-mail: vermavijay1986@gmail.com

© Springer Nature Singapore Pte Ltd. 2018   423
D.K. Mishra et al. (eds.), *Information and Communication Technology
for Sustainable Development*, Lecture Notes in Networks and Systems 9,
https://doi.org/10.1007/978-981-10-3932-4_44

accuracy of the predictions may not always satisfy user needs [5]. There exist metrics beyond accuracy, e.g., diversity, novelty, and serendipity that are considered major dimensions to evaluate performance of a recommendation system to actually satisfy a user.

Diversity represents the dissimilarity between the items recommended [1]. It comes at the expense of accuracy. With increasing diversity, accuracy of the recommendations decreases. So a trade-off between accuracy and diversity has to be maintained to keep the balance. In this paper, we focus on increasing diversity of specifically a tag recommendation system. Tag recommendation to users is an integral part of social tagging-based systems. In social tagging-based systems (STS), also known as the participative Web, users can contribute by annotating and categorizing resources of interests with freely chosen keywords called tags [1]. These tags can be treated as meta-data for the resources. So, proper tag recommendation is vital as it can play potential roles in information retrieval and data mining domains. As users are free to associate any tags to a resource, it creates the problem of information overload. It has become essential to improvise tag recommendations to overcome this situation. Unlike resource recommendation or user recommendation, small size of tag recommendation list is a major factor. As users are not interested to select from a long list of tags, it is important that the list contains as much different varieties as possible and covers different aspects with no redundant recommendation. Our approach to diversify the recommendations provided by the tag recommendation system is based on a binomial framework that uses a greedy algorithm to optimize an objective function that considers both coverage and redundancy. This framework penalizes redundant recommendations and maximizes the coverage by recommended tags.

There are very few diversification techniques available for tag recommendations. We can categorize diversification techniques into two categories: implicit diversification and explicit diversification. In implicit diversification, diversification is done along with the relevance ranking. Reference [6] presented a framework for tag recommendations using genetic programming, a nonlinear method. It applies a global search mechanism on individual that represents a solution to the target problem. A fitness function is optimized over multiple generations. The individuals with highest fitness function value are chosen to evolve. For the RS, each individual represents a ranking function and the fitness is measured using NDCG (Normalized Discounted Cumulative Gain). This fitness function formerly considered only relevance. But in [6], authors extended the function to include diversity of the tags using implicit diversification. Explicit diversification is a very common technique. It is also known as re-ranking technique. Reference [7] used explicit diversification to diversify recommended tags to the user. In the first step, genetic programming and random forest are used as baseline recommender systems to generate relevance score of tags. In the second step, a greedy approach-based algorithm, xTRed, is used for diversification. This xTRed algorithm is adapted from xQuAD algorithm [8] used in search result diversification considering user profile aspect analogous to query intent. In next section, we will thoroughly discuss explicit diversification.

2 Explicit Diversification

As discussed in previous section, diversification techniques can be classified into two categories based on the process of diversification: implicit diversification and explicit diversification. Explicit diversification technique is more popular approach than the former one. This is a two-step process. First, a list of items is found using any relevance-driven baseline recommendation system. Second, a diversifying method is applied that selects and re-ranks items to produce a list with higher diversity. The re-ranking step is done using a simple greedy approach algorithm that selects the tag with maximum objective function value among the candidate tags, similar to [7]. The algorithm takes a baseline ranking R as input and re-ranks into a new ranking S by iteratively selecting item $i*$ belonging to the set R-S that maximizes an objective function F. The output of this algorithm is the final tag list S for recommendation. The steps are shown below:

Algorithm 1
1. Initialize$S = \phi$
2. while $|R| > 0$ do
3. $i* = argmax^{i \in R - S} F(i, S)$
4. $R = R/i*$
5. $S = S \cup i*$
6. end while
7. return S

3 Binomial Framework

To define the objective function in our tag RS, we used binomial framework. A binomial framework is used to model random selection from a list of relevant recommendations to incorporate diversity. The objective function F is defined as the combination of relevance and binomial diversity,

$$F(i, S) = \lambda * rel(i) + (1 - \lambda) * dvr(i, S) \tag{1}$$

where $rel(i)$ is the relevance score obtained from baseline recommender, λ is the balancing factor between relevance and diversity, and $dvr(i, S)$ is the difference between the binomial diversity of the list S including item i and the binomial diversity of the list S excluding item i. Diversification of i to the list S has been defined as follows:

$$dvr(i, S) = binDiv(i \cup S) - binDiv(S) \qquad (2)$$

The binomial diversity incorporates maximum coverage and minimum redundant recommendations. These are termed as coverage score and non-redundancy score. Binomial diversity of a list S is defined as follows:

$$binDiv(S) = coverage(S) * nonRed(S) \qquad (3)$$

The coverage score and redundancy score are defined using binomial distribution. The binomial distribution is the discrete probability distribution of the number k of successes in a sequence of N-independent Bernoulli trials with the same probability of success p. A random variable X following binomial distribution has the following probability mass function:

$$P(X=k) = \binom{N}{k} * p^k * (1-p)^{N-k} \qquad (4)$$

This framework has been recently introduced by Saul Vargas et al. for item recommendation system [9]. We have adapted this framework for tag recommendation system. We considered the selection of tags covering different movies as a Bernoulli trial. If a tag t covers a set of movies $M(t)$, the Bernoulli trial is considered whether a randomly sampled movie m belongs to $M(t)$. The number of successes is the number of set of tags belonging to each movie in $M(t)$.

$$S_m^t = |t \in T : m \in M(t)| \qquad (5)$$

The probability of success of this trial has two aspects: global aspect and user aspect. So the probability is defined as follows:

$$p_m = \alpha * p_m' + (1 - \alpha) * p_m'' \qquad (6)$$

p_m' represents the user-specific probability, and p_m'' represents the global probability. α is the personalization factor to balance personal aspect and global aspect diversity.

$$p_m' = S_m |I_u|, p_m'' = \sum_u S_m^{I_u} \sum_u |I_u| \qquad (7)$$

Here, I_u represents user's interaction with the system. To ensure maximum coverage, the coverage score is defined as the product for the movies not represented in the recommendation list of their probabilities of not being selected according to random variable X_m.

$$coverage(S) = \prod_{m \notin M(R)} P(X_m = 0)^{1/|M|} \qquad (8)$$

The redundancy is represented as the product of the penalty scores on each movie appearing in the list more than at least k times. Both scores are normalized using $|M|$th root.

$$nonRed(S) = \prod_{m \in M(R)} P(X_m \geq k_m | X_m > 0) \qquad (9)$$

The main advantage of using this framework is that it is able to penalize redundant recommendations and maximize the coverage by recommended tags. As tag recommendation has the limitation of small size, it is important that each recommendation has coverage of different aspects as much as possible and no redundant data are entertained.

4 Experiments

4.1 Dataset and Evaluation Methodology

The dataset we have chosen for our experiment is MovieLens 10M dataset, which is publicly available by GroupLens. It consists of 10 million ratings and 95K tags for 10,681 movies by 71,567 users. The dataset includes information about the title, and genres of the movies and tags associated with it. But here, we are only concerned about the tags which are user-generated meta-data about movies.

To evaluate the performance of our model tag recommendation system, we performed experiments with different test datasets and varying parameters for each case. We split the total MovieLens 10M dataset into segments of size 10. For each experiment, all segments except one segment are used for training and the selected segment is used for testing. Result of each segment test data has been collected, and average performance of the system has been calculated. As the framework suggests, a baseline algorithm is needed for the first phase of the system. We have used the TagRec project available online on GitHub [10]. We selected the following algorithms as our baseline algorithms for comparison: BLL (base-level learning) [2], 3L (3 Layers) [3], and LDA (latent Dirichlet allocation) [4]. There are mainly two parameters based on which we ran a series of experiments to find out the best values needed for the required results. First one is the trade-off parameter between accuracy and diversity, lambda, and the second one is the personalization factor alpha, used to balance between global probability and user-specific probability of a movie. We have varied these parameters to find out the optimal value for the desired result. For the results, different evaluation metrics have been used to measure both accuracy and diversity. The following metrics have been used: recall, precision (basic metrics for usage prediction), NDCG@10 (to measure ranking accuracy),

ERR-IA@10, and binomial diversity (for measuring diversity). Recall is the ratio of number of recommended useful items, i.e., true-positive predictions to the total number of useful items for a user. Precision is the ratio of number of recommended and useful items to the total number of recommended items. Useful items are those that a user selects or visits. NDCG@k and ERR-IA@k both are ranking measure metrics where user gain is discounted with increasing item rank. For NDCG@k, the discount is done logarithmically, i.e., $1/log(1+i)$, and for ERR-IA@k, the discount is $(1/i)$ for ith position of the ranking. The value of k is 10 to consider top 10 recommendations.

4.2 Result

As explicit diversification works in two stages, we used a separate diversification component, also referred as re-ranking component, for the framework. The results presented here show the difference in performance metrics for the cases where the system excludes diversity component and where the system includes the diversity component. We derived the results varying values of different parameters. It shows their effect on the performance of the system. The trade-off parameter we mentioned before, λ, is varied from 0.6 to 0.9. Only values, i.e., 0.7 and 0.9, for trade-off are shown in tables for easily visible comparison of results. The results for different values of λ are shown in two separate columns. For each trade-off value, we varied the value of α from 0 to 1 and tried to derive the contribution of personalization factor on the output for each baseline algorithms. Three separate tables, Table 1, Table 2, and Table 3, present results for $\alpha = 0.0$, 0.5, and 1.0, respectively. Each row represents different metrics as mentioned in each table. The values presented in the table show how much better or worse the recommender performed collectively with our re-ranking component than it would have performed individually for each metrics.

As it can be seen from the table, the values of precision and recall hardly change with change in trade-off parameter. This is due to the fact that recall and precision metrics are independent of ranking of items and only concerned about true-positive items or in this case tags. But increase in both recall and precision values can be observed with high personalization of the recommendations. On the other hand, as

Table 1 Results for $\alpha = 0.0$

Difference in	Trade_off = 0.7			Trade_off = 0.9		
	BLL	3L	LDA	BLL	3L	LDA
Precision	−0.19	−0.07	0.055	−0.183	0.21	0.055
Recall	−1.38	−1.36	0.5	−2.08	0	−0.6
NDCG@10	−4.03	−3.514	−2.22	−2.04	−0.09	−2.098
ERR-IA@10	2.8	2.73	1.9	2.82	2.29	1.86
BinDiv	3.67	3.625	3.33	3.62	3.37	3.33

Table 2 Results for $\alpha = 0.5$

Difference in	Trade_off = 0.7			Trade_off = 0.9		
	BLL	3L	LDA	BLL	3L	LDA
Precision	−0.17	−0.05	0.255	−0.346	0.09	0.155
Recall	−1.18	−1.16	0.7	−1.38	0.24	1.5
NDCG@10	−0.584	−2.134	−0.61	−0.74	1.8	1.16
ERR-IA@10	0.62	0.6	0.3	0.42	0.4	0.23
BinDiv	1.049	1.047	0.918	1.006	0.893	0.838

Table 3 Results for $\alpha = 1.0$

Difference in	Trade_off = 0.7			Trade_off = 0.9		
	BLL	3L	LDA	BLL	3L	LDA
Precision	−0.13	−0.01	0.755	−0.23	0.13	0.185
Recall	−0.78	−0.76	0.7	−0.38	0.74	3.5
NDCG@10	2.81	1.23	2.55	3.37	2.15	2.654
ERR-IA@10	0.005	0.005	0.002	−0.0005	−0.0013	−0.05
BinDiv	0.023	0.022	0.0075	0	−0.0001	−0.048

NDCG@10 is a ranking measure metric, its value changes with changing parameter, along with the diversity metric. For each algorithm, higher trade-off value leads to higher NDCG@10 value which means higher accuracy in prediction. It is observed that accuracy increases with increase in personalization factor. It indicates that personalization can play a major factor for more accurate prediction. Increase in trade-off causes increase in diversity which is predictable because trade-off value is used to stimulate accuracy and it's complement value is used to stimulate diversity. Also, we can make sure less personalization means more diversity in the result. From the overall result, it can be easily stated that diversity costs accuracy. But a recommendation having both accuracy and diversity is preferable. So, there needs to be a balance between diversity and relevance for a more accurate recommendation. In Table 3, it can be observed that for trade-off value 0.7 and personalization factor 1.0, both NDCG@10 values and binDiv values are positive. Though for trade-off value 0.9, accuracy value is higher, but binDiv values are negative. So easily the former one with less accuracy and higher diversity will be a better choice.

5 Conclusion

Our proposed tag recommendation system is able to recommend a diverse set of relevant tags for a resource to a user. It also incorporates user's affinity based on personalization factor for maximum coverage and penalization for redundancy. We can deduce from the result analysis that it improves accuracy and diversity according to the parameter values. But, off-line evaluation is not enough to evaluate a recommendation system. The effectiveness of a recommendation system, specially to know how diverse the recommendations are, is mainly user interaction dependent. That is the reason it is important to set up an environment where users can interact with the system. In future, we intend to design the system so that users can interact with it and evaluate the system based on online users' review and feedback.

References

1. Ricci F, Rokach L, Shapira B (2015) Recommender systems handbook. Springer, US, pp 280–309
2. Kowald D, Seitlinger P, Trattner C, Ley T (2014) Long time no see: the probability of reusing tags as a function of frequency and recency. In: Proceedings of the companion publication of the 23rd international conference on WWW companion. International WWW Conferences Steering Committee, pp 463–468
3. Seitlinger P, Kowald D, Trattner C, Ley T (2013) Recommending tags with a model of human categorization. In: Proceedings of the 22nd ACM international conference on information and knowledge management. ACM, pp 2381–2386
4. Krestel R, Fankhause P (2010) Language models and topic models for personalizing tag recommendation. In: 2010 IEEE/WIC/ACM international conference on web intelligence and intelligent agent technology (WI-IAT), vol 1. IEEE, pp 82–89
5. Ziegler CN, McNee SM, Konstan JA, Lausen G (2005) Improving recommendation lists through topic diversification. In: Proceedings of the 14th international conference on world wide web, May 2005. ACM, pp 22–32
6. Belém FM, Martins EF, Almeida JM, Gonçalves MA (2012) Exploiting relevance, novelty and diversity in tag recommendation. In: Proceedings of the 18th Brazilian symposium on multimedia and the web, Oct 2012. ACM, pp 297–300
7. Belém F, Batista C, Santos R, Almeida J, Gonçalves M (2016) Beyond relevance: explicitly promoting novelty and diversity in tag recommendation. ACM Trans Intell Syst Technol (TIST) 7(3):26
8. Santos R, Macdonald C, Ounis I (2010) Exploiting query reformulations for web search result diversification. In: Proceedings of the 19th international conference on world wide web. ACM, pp 881–890
9. Vargas S, Baltrunas L, Karatzoglou A, Castells P (2014) Coverage, redundancy and size-awareness in genre diversity for recommender systems. In: Proceedings of the 8th ACM conference on recommender systems, Oct 2014. ACM, pp 209–216
10. Kowald D, Lacic E, Trattner C (2014) TagRec: towards a standardized tag recommender benchmarking framework. In: Proceedings of the 25th ACM conference on hypertext and social media. ACM, pp 305–307

Optimized Task Scheduling Algorithm for Cloud Computing

Monika and Abhimanyu Jindal

Abstract Cloud computing is an extraordinary advancement which has changed the work culture of organizations. In cloud computing, a substantial number of assignments can be executed simultaneously using the resources of cloud yielding better results and in lesser time. This simultaneous and rapid execution of assignments is performed with the assistance of scheduling which is the demonstration of allocating tasks on resources in such a manner that same performance is achieved in least time with lesser utilization of resources. Nowadays various task scheduling algorithms are accessible. In this paper, a task scheduling algorithm is proposed that points reducing the turnaround time and waiting time of the tasks in cloud infrastructure. The algorithm is simulated using Cloudsim simulator, and experiments are carried out to demonstrate the results.

Keywords Cloud computing · Task scheduling · Waiting time
Turnaround time · CloudSim

1 Introduction

Cloud computing intends to give the way to creating reliable, practical, and very versatile applications as services. It empowers awesome changes to occur in today's IT with the worldwide fame of hybrid computing environments. Cloud computing infers controlling, getting to and organizing applications online and profiting

Monika (✉) · A. Jindal
Department of IT, University Institute of Engineering and Technology,
Panjab University, Chandigarh, India
e-mail: monikahsp@gmail.com

A. Jindal
e-mail: erabhi.puchd@gmail.com

© Springer Nature Singapore Pte Ltd. 2018
D.K. Mishra et al. (eds.), *Information and Communication Technology
for Sustainable Development*, Lecture Notes in Networks and Systems 9,
https://doi.org/10.1007/978-981-10-3932-4_45

applications, infrastructure, and information storage over the Internet. The users are not required to present an item on their PC, and they can have the organizations of programming over the framework through internet. It is less exorbitant in light of the way that customers do not need to purchase the whole system instead they make payment for resources as indicated by their usage. This additionally spares the labor of the organization and expansions the benefits [1]. Taking into account the level of reflection that it gives, cloud computing can be seen from various methodologies: Infrastructure as a service (IaaS), platform as a service (PaaS), and software as a service (SaaS). Because of the recorded performance of cloud infrastructures, the quantity of clients expanded radically amid the previous couple of years which have brought about increased traffic on the clouds. This expanded activity influences the performance of the cloud systems and can be understood with the assistance of different strategies and among them, one is scheduling. Task scheduling is an NP-hard problem [2]. It is the method of administering tasks onto accessible resources of the cloud in time taking after the restrictions depicted by the customer and cloud supplier. The fundamental intention of task scheduling algorithm is to apportion resources to tasks, minimize finishing time, and expand asset usage [3]. Research on task scheduling in cloud computing has transformed into a fascinating issue. The task scheduling ought to arrange the tasks in a way where equalization between enhancing the performance and quality of service and in the meantime keeping up the proficiency and reasonableness among the tasks [4]. Expanse is another parameter which should be remembered while chipping away at cloud and ought to be least [5]. Cloud frameworks have constrained assets, and they are all that much exorbitant. Scheduling algorithm enhances QoS parameters and aims to reduce the costs of client by effectively planning assignments on the resources.

Load balancing is another fundamental challenge in cloud computing. It is a strategy which is required to flow the dynamic workload over various VMs (virtual machines) to ensure that no single virtual machine is overburden. Load balancing systems help in perfect use of resources. The main target is to maximize the resource utilization which will encourage diminish power use and carbon spread rate that is basic need of cloud computing [6, 7]. Load balancing also helps in reducing deadlocks and reducing the number of job rejections [8].

With the increasing trend of cloud computing, cloud service providers prefer to store and share data. Thus, there is a need to secure the data over the cloud. In [9], authors explain two cryptographic approaches to secure data; one is based on software and another on hardware. Hardware-based approach provides robustness and software-based approach provides flexibility. As number of cloud users are increasing more and more, it is necessary to provide bug-free services to user. While identifying the bugs, it is mandatory to check all the factors; in [10], authors study various cloud testing tools.

Users prefer to store their data on cloud and may experience loss of data. Thus, there is a privacy issue as user data is visible to the cloud service provider. In [11], authors provide security measures in which client data is encrypted and data flow is under the control of client. It builds faith in cloud customers regarding cloud services and encourages them to adopt cloud services.

The rest of this paper is organized as follows: The Sect. 2 describes some similar work related to the task scheduling, Sect. 3 describes the proposed work, Sect. 4 describes the experiment and results, and the paper is concluded in Sect. 5.

2 Related Work

A cloud scheduler acts as a cloud-empower distributed resource manager. The principal thought process of task scheduling in cloud computing is to plan tasks onto the accessible resources so that the resources are used and client solicitations are fulfilled. Many task scheduling algorithms have been introduced.

Zhang et al. [12] proposed an intelligent workload factoring scheme for hybrid cloud computing model. The scheme parts the framework workload into two sections, the base burden and the trespassing burden. The base burden can be taken care of by a personally owned datacenter, while the trespassing burden is taken care by an open cloud administration. Grounds et al. [13] presented an algorithm of cost-minimizing scheduling of workloads on a cloud. The clients submit their workflows to the scheduler, and a deadline is given to them which defines the time by which all the tasks of the workflow must be completed. The sum of the expenses for all workflows represents the cost on operation and maintenance. In [14], an algorithm is presented which works on the basis of task priority and completion time. The priority of the cloudlet is dictated by the properties such as customer benefits, length of the task and its workload. The cloudlet having least completion time is assigned first so that all the cloudlets may be executed in time. A similar algorithm PISA (Priority Impact Scheduling Algorithm) is defined in [15] which execute tasks according to the priority given by the client itself, and the task with highest priority is executed first. The algorithm shows that the highest priority tasks are executed successfully with a higher ratio than FIFO by 26.7%. In [16], authors have proposed an algorithm for scheduling the cloudlets according to the priority of both the tasks and the VMs. The priority of the tasks is calculated according to their QoS, and the priority of the VMs is determined on the basis of their MIPS value. A similar priority-based algorithm is presented in [17]. In [18], a two-level scheduling algorithm is used. The algorithm takes care of both the client requirements and load over the cloud resources. This algorithm efficiently meets the client requirements and enhances the resource utilization. Saranu and Jaganathan [19] proposed a scheduling algorithm which is based on load balancing in cloud, and the algorithm improves the performance of the system significantly. A similar load balancing algorithm is presented in [20] which reduced the makespan as well as completion time of the tasks. In [21], a scheduler named SRT-Xen is presented which focuses on scheduling soft real-time applications and also improves the management of VMs queuing in order to provide fair scheduling of both the real-time and non-real-time tasks.

Ding et al. [22] proposed a resource scheduling algorithm empowered with a relevant feedback network. The mechanism schedules the resources according to the previous feedback of the customers. The algorithm works in three stages: resource matching, resource selection, and feedback integration. The resources relevant to client needs are gathered and then matched with the tasks. Then, the resource scheduling feedback history of user is considered for resource allocation. This algorithm results in high customer satisfaction and efficient use of the resources of the cloud. In [23], authors proposed a resource management policy which keeps on monitoring the resources of the cloud continuously and migrates them on the tasks. It also switches off the resources, when they are not in use. This mechanism saves the energy and resources of the system.

Shin et al. [24] proposed a scheduling algorithm which aims at enhancing both the deadline achievement and resource utilization. The algorithm utilizes two algorithms that are earliest deadline first (EDF) and largest weight first (LWF). In the event that the cloud has enough resources to execute tasks, then the VM will be allotted to the occupation, occupation will be removed from the holding up line, and machine status will be changed to occupied from idle. Be that as it may, if the required resources are not available, it will use LWF as backfilling algorithm. Sindhu and Mukherjee [25] proposed two algorithms, namely, LCFP (Longest Cloudlet Fastest Processing Element) and SCFP (Shortest Cloudlet Fastest Processing Element). In LCFP, longest cloudlet is given priority and is assigned to the PE (processing element) with high power which reduces the makespan. In SCFP, shortest cloudlet is scheduled on the PE with high power which reduced the flow time of the system. In [26], authors proposed a backfilling algorithm to provision cloud resources so as to satisfy user-defined QOS parameters. To schedule tasks, three different queues are made according to their requests and then resources are provisioned. In [27], authors proposed a workflow scheduling algorithm in which algorithm tasks of the workflow are scheduled sequentially and in distributed way so that estimation of number of required resources can be made so that workflow can be completed within the defined deadline. In [28], an algorithm is proposed to reduce power consumption and load on the datacenters. Cloud computing services get executed on datacenters; when the load on datacenters is on its peak, datacenters consume more power. The authors conclude that power consumption can be reduced by using least number of hosts in datacenter. In [29], the authors state that when there are workflows in which task gets executed on various virtual machines in different time instances, then there is highest average waiting time and elapsed time in case of FCFS. The authors proposed an optimized way to generate the sequence so that average waiting time and elapsed time is reduced significantly as compared to FCFS. In [30], authors proposed a scheduling approach by considering the idle time of task on resources, and results show that this algorithm performs better as compared to min-min algorithm.

3 Proposed Work

3.1 Introduction

The clients of cloud systems want their work to be completed in least time with high QoS and, on the other hand, the cloud administration suppliers attempt to utilize lesser resources for a particular task. This task of achieving high performance in lesser time by using least resources is done with the help of a scheduling algorithm. The algorithm presented is based on the arriving pattern of the tasks in the system. It sends the tasks to the system in such a manner that the waiting time of the tasks is reduced resulting in the reduction of the turnaround time.

3.2 Problem Statement

In multi-objective task scheduling algorithm [16], all the tasks are dispatched collectively to system and tasks keep on waiting in ready queue which keeps on increasing the waiting time of the tasks and results in increase in the turnaround time. In the current scenario, all the tasks are not dispatched collectively to the system instead all the tasks are sent one after other but with a time gap between them.

3.3 Proposed Algorithm

All the tasks that need to be executed are first assigned to some priority value and are sorted according to it. The tasks are then sent one by one to the system with some delay after dispatch of every task. The delay value is a fixed time interval which is added to the dispatch time of each task. The first task is sent on starting time of the system, then the delay value is added to the starting time and the next task in ready queue is sent on that time. Again the time at which previous task was sent is added to the delay value, and the task in ready queue is dispatched at the time computed after adding the delay value and so on. The name given to proposed approach is "optimized task scheduling algorithm" and is described in Algorithm 1.

Algorithm 1: Optimized task scheduling algorithm.
1. Submit both VMs list of successfully created VMs in datacenter and task list to broker.
2. Create a received list of tasks.
3. Create a received list of VMs.
4. Define priority for each task.
5. Sort the tasks
    i ←0
    Create empty new_list
    new_list ←list of tasks
    initially put taski in the new_list
    for all i ←1 to size of task's list **do**
        for all j ←1 to size of new_list **do**
            if taskj has higher priority than taski **then**
                put taskj in new_list on higher position than taski
            else put the taskj on lower position in new_list
            end if
        end for
    end for
6. d ←delay
7. **for all** i ←0 to size of new_list **do**
        if i ≥0 **then**
            Send taski to VM at time t
            t = t + d ;
            i++ ;
        end if
    end for

---

4   Experimentation Results

The results of the algorithm are obtained by simulating it using Cloudsim 3.0.3 simulator which was run over Eclipse Juno. The operating system used was Windows 7 with core i3 processor. In this simulation scenario, different test cases given in Table 1 were used. A delay value was set initially and was same for each test case. The results obtained from the experiments are shown in Figs. 1 and 2.

Table 1 Test cases [16]

	Number of VMs	Number of tasks
Case 1	3	20
Case 2	3	50
Case 3	3	100
Case 4	5	50
Case 5	10	100

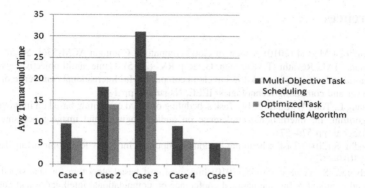

Fig. 1 Comparison of turnaround time

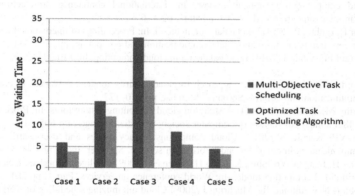

Fig. 2 Comparison of waiting time

The comparison of turnaround time is shown in Fig. 1, and comparison of waiting time is shown in Fig. 2. From both the figures, it is observed that the proposed algorithm performs better than the other algorithm.

5 Conclusion and Future Work

For productive utilization of resources of cloud taking all the QoS requirements into consideration, scheduling is required since cloud infrastructure has limited resources and they are costly too. The proposed optimized task scheduling algorithm is based on delaying the arrival of tasks to the system, and it results in reduced turnaround and waiting time. In future, the algorithm may be modified by introducing some more QoS parameters.

References

1. Armbrust M et al (2010) A view of cloud computing. Commun ACM 53(4):50–58
2. Sujana JAJ, Revathi T, Karthiga G, Raj RV (2015) Game multi-objective scheduling algorithm for scientific workflows in cloud computing. In: International conference on circuit, power and computing technologies. IEEE, Nagarcoil, pp 1–6
3. Zhang L, Tong W, Lu S (2014) Task scheduling of cloud computing based on improved CHC algorithm. In: International conference on audio, language and image processing. IEEE, Shanghai, pp 574–577
4. Sindhu S (2016) Task scheduling in cloud computing. Int J Adv Res Comput Eng Technol 4 (6):3019–3023
5. Selvarani S, Sadhasivam GS (2010) Improved cost based algorithm for task scheduling in cloud computing. In: International conference on computational intelligence and computing research. IEEE, Coimbatore, pp 1–5
6. Sreenivas V, Prathap M, Kernal M (2014) Load balancing techniques: major challenge in cloud computing-a systematic review. In: International conference on electronics and communication systems. IEEE, Coimbatore, pp 1–6
7. Kaur R, Luthra, P (2014) Load balancing in cloud. In: Proceedings of international conference on recent trends in information, telecommunication and computing, pp. 374–381
8. Kherani FF, Vania J (2014) Load balancing in cloud computing. Int J Eng Dev Res 2(1):907–912
9. Rohini T (2011) Comparative approach to cloud security models. Advances in computing, communication and control. Springer, Mumbai, pp 170–177
10. Murthy MSN, Suma V (2014) A study on cloud computing testing tools. In: Proceedings of the 48th annual convention of computer society of India. Springer, pp 605–612
11. Lijo VP, Kalady S (2011) Cloud computing privacy issues and user-centric solution. Communications in computer and information science. Springer, India, pp 448–456
12. Zhang H, Jiang G, Yoshihira K, Chen H, Saxena A (2009) Intelligent workload factoring or a hybrid cloud computing model. In: World conference on services, Part 1, pp 701–708
13. Grounds NG, Antonio JK, Muehring J (2009) Cloud minimizing scheduling of workflows on a cloud of memory managed multicore machines. Lecture notes in computer science. Springer, China, pp 435–450
14. Wu X, Deng M, Zhang R, Zeng B, Zhou S (2013) A task scheduling algorithm based on QoS-driven in cloud computing. Procedia Comput Sci 1162–1169
15. Wu H, Tang Z, Li R (2012) A priority constrained scheduling strategy of multiple workflows for cloud computing. In: International conference on advanced communication technology. IEEE, PyeongChang, pp 1086–1089
16. Lakra AV, Yadav DK (2015) Multi-objective task scheduling algorithm for cloud computing throughput optimization. In: International conference on computer, communication and convergence. ScienceDirect, India, pp 107–113
17. Lin C, Lu S (2011) Scheduling scientific workflows elastically for cloud computing. In: International conference on cloud computing. IEEE, Washington, DC, pp 746–747
18. Fang Y, Wang F, Ge J (2010) A task scheduling algorithm based on load balancing in cloud computing. Lecture notes in computer science. Springer, China, pp 271–277
19. Saranu KA, Jaganathan S (2015) Intensified scheduling algorithm for virtual machine tasks in cloud computing. Advances in intelligent systems and computing. Springer, India, pp 283–290
20. Banerjee S, Adhikari M, Kar S, Biswas U (2015) Development and analysis of a new cloudlet allocation strategy for QoS improvement in cloud. Arab J Sci Eng 1409–1425
21. Cheng K, Bai Y, Wang R, Ma Y (2015) Optimizing soft real-time scheduling performance for virtual machines with XRT-Xen. In: International symposium on cluster, cloud and grid computing. IEEE, Shenzhen, pp 169–178

22. Ding D, Fan X, Luo S (2015) User-oriented cloud resource scheduling with feedback integration. J Supercomput 1–22
23. Beloglazov A, Buyya R (2010) Energy efficient allocation of virtual machines in cloud data centers. In: IEEE/ACM international conference on cluster, cloud and grid computing. IEEE, Australia, pp 577–578
24. Shin S, Kim Y, Lee S (2015) Deadline-guaranteed scheduling algorithm with improved resource utilization for cloud computing. In: 12th annual IEEE consumer communications and networking conference. IEEE, Las Vegas, NV, pp 814–819
25. Sindhu S, Mukherjee S (2011) Efficient task scheduling algorithms for cloud computing environment. Communications in computer and information science. Springer, India, pp 79–83
26. Loganathan S, Mukherjee S (2013) Differentiated policy based job scheduling with queue model and advanced reservation technique in a private cloud environment. In: International conference on grid and pervasive computing. Springer, Korea, pp 32–39
27. Jung IY, Jeong CS (2014) Selective task scheduling for time-targeted workflow execution on cloud. In: 6th international symposium on cyberspace safety and security, 11th international conference on embedded software and system. IEEE, Paris, pp 1055–1059
28. Mehdi NA, Mamat A, Amer A, Abdul-Mehdi ZT (2011) Minimum completion time for power-aware scheduling in cloud computing. In: Developments in E-systems engineering. IEEE, Dubai, pp 484–489
29. Indukuri RKR, Varma PS, Moses GJ (2012) Performance measure of multi stage scheduling algorithm in cloud computing. In: International conference on cloud computing technologies, applications and management. IEEE, Dubai, pp 8–11
30. Anbazhagi, Tamilselvan L, Shakkeera (2014) QoS based dynamic task scheduling in IaaS cloud. In: International conference on recent trends in information technology, India, pp 1–8

"BOMEST" a Vital Approach to Extract the Propitious Information from the Big Data

V.K. Jain, Deepali Virmani, Preeti Arora and Ankit Arora

Abstract Cost-effective, innovational methods to extract information and to provide analytic solutions for intensify perspicacity is one of the biggest challenge of big data. So an effective system to extract the accurate information is required for big data. Thus, in this paper, an efficient system to extract the propitious information from the raw big data (BOMEST) is proposed. Proposed BOMEST works by taking the raw data, preprocesses it, and extracts the accurate information based on polarity assigned using POS tagging. BOMEST is applied on the twitter dataset. BOMEST algorithm enhances and improves the accuracy of results by 78% as compared to existing lexicon approach.

Keywords Big Data · Twitter · Rest API · Sentiments · Bag-of-words
POS · NLP and polarity

1 Introduction

Big data's huge volume, velocity, and variety information is known as 3-V's [1] that are generated from Sensor Networks, Human-generated Applications, ERP systems, and Web and Social Networks in the form of tweets, videos, images text, etc. Big data demands cost-effective, innovational, and creative forms of informa-

V.K. Jain (✉)
COER School of Management, Roorkee, India
e-mail: drvkjain72@gmail.com

D. Virmani · P. Arora
Bhagwan Parshuram Institute of Technology, New Delhi, India
e-mail: deepalivirmani@gmail.com

P. Arora
e-mail: erpreetiarora07@gmail.com

A. Arora
Mindfire Solutions, Noida, India
e-mail: ankitarora2688@gmail.com

© Springer Nature Singapore Pte Ltd. 2018
D.K. Mishra et al. (eds.), *Information and Communication Technology for Sustainable Development*, Lecture Notes in Networks and Systems 9,
https://doi.org/10.1007/978-981-10-3932-4_46

tion processing and analytics solutions for intensify perspicacity, decision making, customer's fulfillment, and happiness [2, 3]. With the hasty expansion of the Social Web media, all of us express our sentiments and thoughts for assorted activities such as blogs, tweets, products, and political issues [4]. All the data generated by these sources needs an effective algorithm to extract the accurate analysis of sentiments or opinions for unstructured data. Sentiment analysis is ascribed as opinion analysis to elaborate the emotions or feelings about some entities in place of simply squeeze the facts about these subjects [3]. It is a prolongation of data mining in the NLP (Natural Language Processing) domain. In this paper, social networking site such as Twitter and its REST API is used to access the tweets using the secure tokens obtained via OAuth as the source of big data which includes 100 and 500 million users, their blogs, tweets, and messages [5]. The vital handout of this paper is the layout and evolution of a peculiar BOMEST Big Data Analytics Framework and Algorithm to provide better solution to lexicon-based approach to evaluate the sentiments of each tweet and assign it the polarity for real time analysis.

The paper is organized as follows. In Sect. 2, the literature and study conducted in this domain is reviewed. In Sects. 3 and 4, our proposed BOMEST Big Data Analytics Framework and Algorithm for tweets extracted from Twitter are elaborated with the explanation of each layer in detail. Finally, the results of our experiments are validated in Sect. 5 and Sect. 6, present the conclusion.

2   Literature Review

In this paper [6], authors highlight the importance of preprocessing and various steps involve in preprocessing using Twitter data for the prediction of the US Presidential election. Firstly, all the uppercase words change to lower case, repeated words get eliminated, and all the punctuations marks get removed. By implementing the SVM model, they achieved about 69% accuracy.

In this paper [7], authors present the technique to collect data from Twitter using hashtags, emoticons, abbreviations, etc. Also shows that the POS tagging is not sufficient for sentimental analysis. Various datasets such as hashtagged datasets, iSieve dataset, and emoticon datasets with different models such as bigram, n-gram were be discussed.

3   Proposed Framework

Our proposed BOMEST Big Data Analytic Framework for unstructured stream tweets related to teacher for sentimental analysis is representing in Fig. 1. The framework consists of four layers; detailed explanation of each layer is as follows:

Data Storage Layer: In Step **1,** the tweets from search engine, Web and social media through Twitter Rest API using OAuth token keys is gathered. Then, these

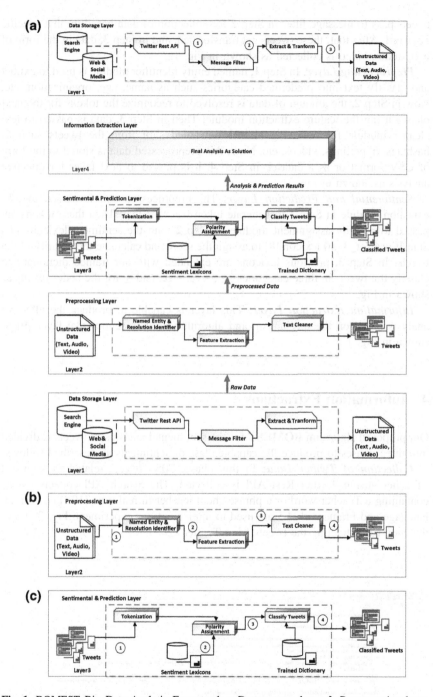

Fig. 1 BOMEST Big Data Analytic Framework **a** Data storage layer. **b** Preprocessing layer. **c** Sentimental and prediction layer

tweets pass to message filter in Step **2** for collection of data based on that specific keyword. After that, these tweets get transformed, and in Step **3,** data in the form of text, audio, video is collected as shown in Fig. 1a.

Preprocessing Layer: In Step **1,** named entity identifier module is used to extract and classify text into predefined categories such as name, age, organization, etc. Now in Step **2,** the amount of data is resolved to recognize the tokens for the next phases using the feature extraction module. Then in Step **3,** data is passed to text cleaner module for removing all the unwanted stuff from the tweets such as hashtags, hyperlinks, videos, etc. Now, this preprocessed data is stored in the form of CSV format or as a dataset; in Step **4,** it is passed to next layer for effective analysis as shown in Fig. 1b.

Sentimental and Prediction Layer: The preprocessed text is passed to tokenization module in Step **1** to generate token from the text. Then these tokens are passed to polarity assignment module in Step **2** with the sentiment lexicons i.e., SentiWordNet_3.0.0 lexicon [8] to assign the score and calculate the polarity of the tweets. In Step **3,** previous lexicons are extended with our trained dictionary to classify the tweets in three categories i.e., +ve, −ve, and ± for the better results as shown in Fig. 1c.

Information Extraction Layer: This layer is capable to produce the effective analysis based on the mechanism and algorithm used in sentimental layer [9] as shown in Fig. 1.

4 Information Extractions

Our proposed algorithm **BOMEST** for the sentimental analysis of tweets is divided into three phases to produce the capable class of sentiments as described follows.

Collection of Twitter Data: In this phase, 7054 tweets related to keyword "Teacher" using Twitter Rest API is collected. The Search API collects tweets containing a Teacher word or a phrase about teacher in real time [10, 11]. Dot-net language and Framework 4.5 are used to implement our experiment. Fig. 2 shows the code for extracting the tweets from Twitter.

```
public IEnumerable<string> GetTweets()
{
    var accessToken = GetAccessToken();

    var requestUserTimeline = new HttpRequestMessage(HttpMethod.Get,
    "https://api.twitter.com/1.1/search/tweets.json?q=teacher&count=100&lang=en");
    requestUserTimeline.Headers.Add("Authorization", "Bearer " + accessToken);
    var httpClient = new HttpClient();
    HttpResponseMessage responseUserTimeLine = httpClient.SendAsync(requestUserTimeline).GetAwaiter().GetResult();
    var serializer = new JavaScriptSerializer();
    dynamic json = serializer.Deserialize<object>(responseUserTimeLine.Content.ReadAsStringAsync().GetAwaiter().GetResult());
    var enumerableTweets = (json as Dictionary<string, object>).FirstOrDefault(c => c.Key == "statuses").Value as IEnumerable<dynamic>;

    return enumerableTweets.Select(t => (string)(t["text"].ToString()));
}
```

Fig. 2 Collection of twitter data

Cleaning of Data: This phase involves removal of errors and inconsistency from the tweets to improve the quality of the dataset prior to the process of analysis [12]. As shown in Fig. 3a, some tweets only contain hyperlinks or links, or mentions, etc., that did not provide any information and hence, should be removed before processing.

So with our algorithm, all the hyperlinks, URLs links video, photos that are attached with tweets are eliminated and then removed all the punctuations marks (:, @, "", comma, etc.) and stop words (for, is, the, to, of, etc.) from the tweets as shown in Fig. 3b.

After removing all the unnecessary stuff, blank lines, and URLs from our tweets, a clean dataset is prepared that passed for further analysis.

Proposed BOMEST Algorithm:

So, our designed new algorithm, the BOMEST, which takes input as clean dataset and gives better efficiency of POS tagging [3], assigning the score of polarity and enhancement in the analysis accuracy of the Twitter tweets [8] by analyzing each tweet and extracts the information in the real time by following the steps mentioned below and provide the improved result as shown in Fig. 4.

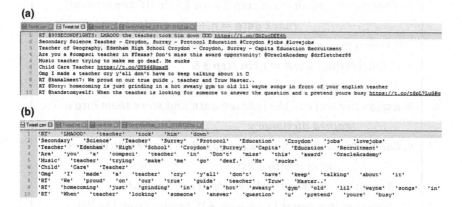

Fig. 3 **a** Twitter data before cleaning. **b** Twitter data after cleaning

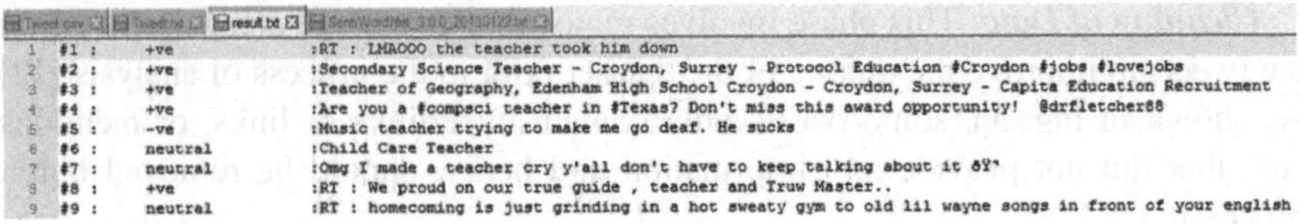

Fig. 4 Results of our experiment

Symbols Used: ±neutral, † Positive – Negative Polarity & ¶ Polarity

---

1. *Input line from the cleaned tweet data*

2. *Exact 1st noun i.e.* η = *noun*

3. η = η + *for all subsequent nouns*

4. *if(word preceding adj* = η) *then*

 4(a) *if (word preceding adj* = *adv)then*

4(b) *if (word preceding adv* = −*ve or* † *ve adv)then*

 ¶ = −*adv else* ¶ =† *adv*

 Else directly check for the ¶.

5. *Concatenate noun from step 3 with* ¶, *adv (if any)and adj*

 Index data = η + *adv* + *adj*

6.*if (word preceding* η = *adv)then repeat step 4(b)and 5*

7. *if (succeeding word* = η) *in step 3 &*

 repeat step 4 and 5 in succeeding direction

8. *Assign the score to the indexed data and store them into a*

 trained dictionary

9. *Calculate the final* ¶ *of tweet based on trained dictionary &*

 stores it into a list

10. *Repeat all steps until all tweets are processed*

11. *Now result* = *count* († *ve,* −*ve,* ±)¶ *tweets from the list*

 Print the final result of sentimental analysis as

 † *ve,* −*ve and* ±

---

5 Results

Here, two algorithms are implemented in our sentimental analysis experiments. In the first experiment, Bag-of-Words and SentiWordNet dictionary for tagging of POS to the tweets are used. The dictionary assigns a positive and negative score to

Table 1 Score of new synset based on (+, −) polarity

	Very bad	Bad	Avg	Good	Very good
+ve Inc	∞∞	0.3	0.6	0.75	0.8
−ve Dec	0.7	0.55	0.25	0.15	∞∞

each synset and the objective score is always the complement of the sum of the positive and negative scores [8].

$$Objective\ score = 1 - (+ve\ score + -ve\ score)$$

In second experiment, the BOMEST algorithm which is based on bigram model is used, with Bag-of-Words and tags all the POS in an efficient way as explained in Sect. 4 and uses trained dictionary that stores all the combination of words of nouns + adj, adv + adj, etc. The score to new synset is calculated by multiplying +ve polarity increases value by 0.45 whereas −ve polarity decreases value by 0.35 for the indexed data, and sum of all these lies in the range of 0.0–1.0 as shown in Table 1.

Comparison of both the algorithms is shown in Fig. 5. As proved by the result that sentiment accuracy depends on the size of dictionary. If dense dictionary with effective scoring schemes is used, it may result in enhancement of accuracy. Therefore, our trained dictionary provides accuracy greater than 78% using a BOMEST algorithm.

Fig. 5 Comparison of BOW + POS, BOMEST

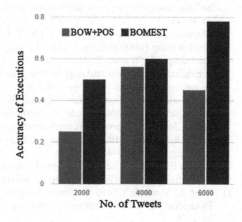

6 Conclusion

Big data demands cost-effective, innovational, and creative forms of information processing and analytic solutions to extract the propitious information from the raw data. In this paper, BOMEST approach was proposed to extract the accurate information based on sentiment analysis. BOMEST was applied to twitter dataset. First of all, raw data was entered then it was preprocessed (cleaning). Polarity was assigned to the processed data by POS tagging. Tokenized data was accessed by assigning +ve, −ve, and ± values. Based on these values, final information is extracted. An implementation result proves enhancement and increased accuracy up to 78% of proposed BOMEST over the existing lexicon approach with accuracy of 45%.

References

1. Boden C, Karnstedt M, Fernandez M, Markl V (2013) Large-scale social-media analytics on stratosphere. In: Proceedings of the 22nd international conference on world wide web companion, pp 257–260
2. Fang X, Zhan J (2015) Sentiment analysis using product review data. J Big Data (Springer). http://journalofbigdata.springeropen.com/articles/10.1186/s40537-015-0015-2
3. Virmani D, Taneja S, Bhatia P (2015) Maestro algorithm for sentiment evaluation. ACM, pp 244–249. doi:http://dx.doi.org/10.1145/2791405.2791479. ISBN 978-1-4503-3361-0/15/08.
4. Liu B (2012) Sentiment analysis and opinion mining. Morgan and Claypool Publishers, pp 18–19, 27–28, 44–45, 47, 90–101
5. Barbosa L, Feng J (2010) Robust sentiment detection on twitter from biased and noisy data. In: Proceedings of the 23rd international conference on computational linguistics: posters, pp 36–44
6. Chandrasekar S, Charon E, Ginet A (2012) CS229 project predicting the US presidential election using twitter data. In: CS229 machine learning course at Stanford University
7. Kouloumpis E, Wilson T, Moore J (2011) Association for the advancement of artificial intelligence (http://www.aaai.org). In: Proceedings of the fifth international AAAI conference on weblogs and social media, pp 538–541
8. SentiWordNet: http://sentiwordnet.isti.cnr.it, WordNet website: http://wordnet.princeton.edu/, SentiWordNet is distributed under the Attribution-ShareAlike 3.0 Unported (CC BY-SA 3.0) license. http://creativecommons.org/licenses/by-sa/3.0/
9. Buche A, Chandak MB, Zadgaonkar A (2013) Opinion mining and analysis: a survey. Int J Nat Lang Comput 39–48
10. Nielsen FÅ (2011) A new: evaluation of a word list for sentiment analysis in Microblogs. In: Proceedings of the ESWC workshop on making sense of Microposts, pp 41–45
11. Bifet A, Frank E (2010) Sentiment knowledge discovery in twitter streaming data. In: Proceedings of 13th international conference on discovery science. Streaming twitter data, pp 1–15
12. Jhaveri D, Chaudhari A, Kurup L (2011) Twitter sentiment analysis on e-commerce websites in India. Int J Comput Appl (0975–8887) 127(18):14–18

Information Privacy Using Stego-Data Element with Visual Cryptography

Manjeet Kantak and Sneha Birendra Tiwari Sharma

Abstract There exist many techniques for information hiding in images such that any alterations done in the image are perceptually invisible. Steganography and visual cryptography are most widely used methods for information hiding. Steganography uses a mask image and hides the information within the mask image whereas visual cryptography provides information hiding by creating shares of the image. The proffered algorithm performs steganography using region incrementing method combined with Floyd halftone approach to perform visual cryptography. The proffered method takes in the image to be classified and the other image to act as a mask for it. The mask image is converted into the CMY color model and each CMY component is then halftoned and reverse halftoned. A random select method is designed which processes every pixel of every region of the classified image in an incrementing method and embeds the processed pixels in each of the halftoned and reverse halftoned components. The shares are then generated by combining the halftoned and reverse halftoned components. The classified image is obtained only after the entire halftoned share is combined.

Keywords Steganography · Floyd halftone · Region incrementing
Visual cryptography · Mask image · CMY color model · Random select
Halftone · Reverse halftone

M. Kantak (✉) · S.B.T. Sharma
Computer Science and Engineering Department, Goa College of Engineering,
Farmagudi, Ponda, Goa, India
e-mail: manjeet959@gmail.com

S.B.T. Sharma
e-mail: snehabirendra@gmail.com

© Springer Nature Singapore Pte Ltd. 2018   449
D.K. Mishra et al. (eds.), *Information and Communication Technology
for Sustainable Development*, Lecture Notes in Networks and Systems 9,
https://doi.org/10.1007/978-981-10-3932-4_47

1 Introduction

The recent trends which display the very fact that Internet usage have increased in our day-to-day life for the very purpose to exchange information. Communicating parties are more concerned toward the privacy and security of the messages that is being exchanged over the Internet. To provide privacy to the messages many methods have been developed among which steganography and visual cryptography are the two most widely used techniques for providing privacy to the information. They hide the information by disrupting the visual appearance of the image that carries the information that is being shared among the communicating parties.

Steganography also termed as prisoner's problem [1, 2] defines how one can hide the valuable information or message behind an image formally termed as "mask image." The classified message could be a text or another image that is being shared among the communicating parties. The message to be classified is embedded into the mask message. Cryptography is another method to provide privacy to the information that is being shared over the Internet. Many cryptography algorithms such as the AES, DES, Blowfish, and genetic algorithms have been applied to provide information privacy. Visual cryptography is also been widely explored in the last decade by many research scholars. In visual cryptography [3, 4] the image that is to be classified is divided into multiple shares. The shares are biform image showing grayscale view. If the intruder gets access to one of the share, it will not be able to get access to the classified information because the classified message is engraved into both shares. The message is visually known when both the shares are overlapped.

In this paper, we have proffered a method that combines steganography with visual cryptography to provide privacy to the information. Steganography is performed using the region incrementing method, and visual cryptography is performed using the Floyd halftone method. In the proffered algorithm, the CMY components of the mask image are extracted and each component is halftoned and reverse halftoned. To perform steganography, the region incrementing method uses a random select function that divides the classified message into different regions and in pixels of each region is randomly embedded into the halftoned and reverse halftoned CMY components by performing the pixel preprocessing steps. Two shares are then generated by combing all the halftoned components for one share creation and all the reverse halftoned components for the other share creation. The classified message is then revealed by overlapping both the shares.

The paper organization consists of the following sections: Sect. 2 provides a description of some of the "research works related to visual cryptography and steganography," Sect. 3 provides a detailed view of our "proffered research algorithm," the Sect. 4 shows the "test results generated from the experiments performed on our proffered algorithm," followed by "conclusion" in Sect. 5 and "references."

2 Related Work

In the research paper [5], the authors have proffered the expansion to the LSB method used for steganography and have also proffered an edge-adaptive scheme that performs the selection of regions to be embedded based on criteria which includes the size of the classified message and the dissimilarity among the adjacent pixels of the mask image. According to the authors, regions with sharper edges are selected when low rates for embedding are selected. The embedding rates can be increased by adjusting a few parameters adaptively so as to hide more information. The authors performed the experiments on a total count of 600 images showing how their proffered enhancement provides a better security in comparison with the existing LSB and edge-based methods.

In the paper [6], the authors have proffered a method for processing halftone so as to improve the quality of share images. The researchers construct the share images to incorporate meaning full mask images and providing an opportunity to integrate visual cryptography and biometric safety methods. Their experimental results have shown that the regained classified image is a protracted visual cryptography scheme as the size of the share images and the regained image is the same as for the original halftone classified image. The proffered method eventually preserves the original surveillance as that of the protracted visual cryptography.

3 Proposed Algorithm

This section provides the complete specification of the proffered algorithm. Block diagram in Fig. 1 provides the flow of the proffered method which is followed by the step-by-step description of the algorithm.

3.1 Step-by-Step Functioning of the Proposed Algorithm

Step 1: Input: Mask image and classified image //input by the user is the classified image and the mask image.

Step 2: Extract R, G, B components of the images.

Step 3: Visual cryptography process:

    Step 3.1: Input: mask image.

    Step 3.2: Convert mask image to CMY color model and extract each component of CMY model.

    Step 3.3: Apply Floyd_halftone on each CMY component and obtain the halftone of each C, M, Y components.

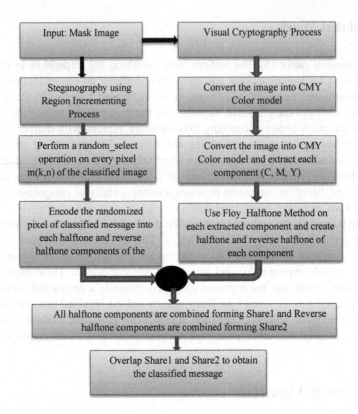

Fig. 1 Proposed information privacy algorithm

Step 3.4: Perform reverse halftone for each of the C, M, Y components.

Step 4: Steganography process using region incrementing method:

Step 4.1: Input: Classified image.

Step 4.2: Design a random_select function //this function processes every black and white pixel.

Step 4.3: For every pixel "m(k,n)" of classified image

Step 4.3.1: Do: call the random_select() and pass the classified image as an argument and go to A1 //encoded classified information.

Step 4.3.2: A1: calculate x_coordinate and y_coordinate values the random positions in the halftoned C components of the mask image and pass the classified image pixel values to the halftoned(C) component and to reverse_halftoned(C) components.

Step 4.4: End of Step4 and goto Step5.

Step 5: Repeat Step4 operations for the M component and the Y components and then goto Step6.

Step 6: Combine the halftoned components into one forming one share1 and then combine the reverse_halftoned images into one forming share2 and then goto Step7.

Step 7: The classified image is retrieved with the overlap of the two created shares.

4 Test Outcomes

The proffered algorithm is materialized by utilizing MATLAB programing. The proffered algorithm was tested on 100 binary images as classified images taken from standard online image-processing database. The implementation results are

Fig. 2 Mask image

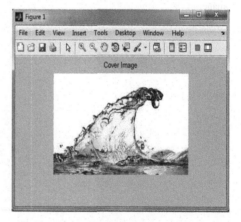

Fig. 3 Classified binary image

Fig. 4 CMY color model

Fig. 5 Halftoned CMY image

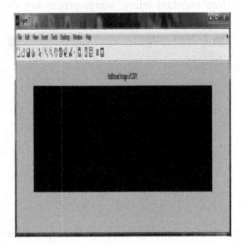

Fig. 6 Share1

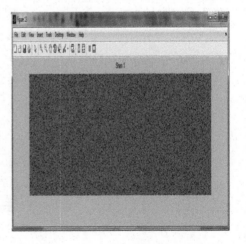

shown below where in the classified image is the image with the text "PRISM" and
the mask image is an arbitrary RGB image (Figs. 2, 3, 4, 5, 6, 7, and 8).

The test outcomes display that the classified image can be retrieved only after the
shares overlap over each other. A single share does not contain all the information
and that both the shares carry the required information which is the classified
message. Thus, to reacquire the classified image, both the shares are needed.

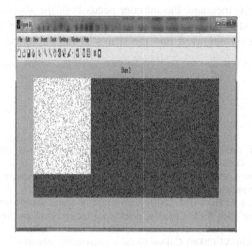

Fig. 7 Share2

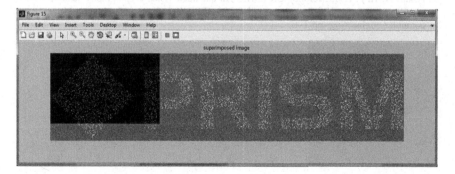

Fig. 8 Overlapped image showing the classified message

5 Conclusion

The proffered algorithm is based on the idea of providing information privacy by using a combinational approach of steganography and visual cryptography. The use of region incrementing method to perform steganography on the classified message adds a higher degree of randomness making it difficult for the infiltrator to retrieve the classified message. The algorithm embeds the pixels into the halftoned and reverse halftoned CMY components of the mask image and then visually encrypting the halftoned and reverse halftoned components to create share; thus, to access the classified message, the intruder needs both the shares as the algorithm avoids any information leak if any classified share is obtained because the classified message is randomly engraved into the shares of the halftoned and reverse halftoned components.

References

1. Chandramouli R (2001) Nasir Memon: analysis of LSB based image Steganography techniques. IEEE Trans Image Process 3:1019–1022
2. Parvez MT, Gutub AAA (2008) RGB intensity based variable-bits image steganography. Asia Pacific services computing conference. IEEE, pp 1322–1327
3. Askari N, Moloney C, Heys HM (2012) A novel visual secret sharing scheme without image size expansion. In: 25th IEEE Canadian conference on electrical and computer engineering, pp 1–4
4. Liu F, Wu CK, Lin XJ (2008) Colour visual cryptography scheme. IEEE Conf IET Inform Secur 2(4):151–165
5. Luo W, Huang F, Huang J (2010) Edge adaptive image Steganography based on LSB matching revisited. IEEE Trans Inform Forensic Secur 5(2):201–214
6. Askari N, Heys HM, Moloney CR (2013) An extended visual cryptography scheme without pixel expansion for HalfTone images. In: 26th IEEE Canadian conference on electrical and computer engineering, pp 1–6

Power-Aware Virtual Machine Consolidation in Data Centers Using Rousseeuw and Croux Estimators

Lincolin Nhapi, Arun Kumar Yadav and Pallavi Khatri

Abstract In order to maximize resource utilization and lessen energy consumption, dynamic virtual machine (VM) consolidation has been found to be effective. The idea here is that energy consumption can be reduced by keeping inactive servers in operating modes that consume less power (such as sleep, hibernation); thus, the power consumed due to idle servers running at all times is eradicated. In this paper, we propose an adaptive virtual machine consolidation mechanism based on Rousseeuw and Croux estimators Sn and Qn, respectively. These two estimators are used to dynamically adjust the upper utilization threshold of server CPU utilization so as to adapt to volatile workloads. When it is necessary, the sleeping nodes are activated so as to be allocated to new requests of VMs or migrating VMs. We have evaluated the proposed mechanism through simulation in CloudSim, and experimental results have shown that our technique can be applied in hyper-scale data centers to effectively utilize resources and in turn lower energy consumption.

Keywords Data center · Virtual machine · Quality of service
Energy-efficient · Dynamic consolidation · Service-level agreement

1 Introduction

Information and Communication Technology (ICT) industry has gone through a revolution since the emergence of cloud computing such that computing resources are now provisioned on-demand, in an elastic fashion and on pay-as-you-go basis.

L. Nhapi (✉) · A.K. Yadav · P. Khatri
Department of Computer Science and Engineering, ITM University,
Gwalior, Madhya Pradesh, India
e-mail: lbnhapi@gmail.com

A.K. Yadav
e-mail: arun26977@gmail.com

P. Khatri
e-mail: pallavi.khatri.cse@itmuniversity.ac.in

© Springer Nature Singapore Pte Ltd. 2018   457
D.K. Mishra et al. (eds.), *Information and Communication Technology
for Sustainable Development*, Lecture Notes in Networks and Systems 9,
https://doi.org/10.1007/978-981-10-3932-4_48

Organizations can now outsource their computational requirements to the cloud and thus they can be able to avoid huge investments in a private computing infrastructure. By outsourcing their computational needs to the cloud, organizations can consequently avoid costs of maintenance and upgrades. They also have the option of constructing private cloud data centers and so as to manage their resources and provisioning processes better. Cloud computing continues to rise on a daily basis; this has seen large corporations around the world to establish large-scale data centers with thousands of servers. However, these servers gobble a significant amount of electrical energy and incidentally causing a rise in operational expenditure and carbon dioxide (CO_2) released into the atmosphere. Inefficient use of computing resources is one of the many sources of energy wastage. As outlined in [1], server utilization in data centers hardly approaches 100% since most of the time they operate in between 10 and 50% of their maximum ability. Furthermore, the narrow dynamic power range of servers compounds the problem of low server utilization: servers still use up to 70% of their peak power even if they are totally idle [2]. Hence, from the energy consumption standpoint, it is highly uneconomical to keep servers underutilized.

The remainder of this paper takes the following organization: Sect. 2 introduces the idea behind energy-aware consolidation of virtual machines while Sect. 3 outlines related works. We present the system model in Sect. 4, give performance metrics in Sect. 5, and propose our methodology in Sect. 6. In Sect. 7, we present our experimental results and analysis and go on to conclude our findings in Sect. 8 as well as give future directions.

2 Power-Aware Dynamic Consolidation of Virtual Machines

One of the technologies which forms the base upon which cloud computing is built is virtualization. Virtualization enables live migration of virtual machines (VMs) between the physical nodes with a minimal downtime usually less than 1 s. The use of live migration [3] enables VMs to be dynamically consolidated on physical servers as fluctuations in the workload occur. This has the advantage of keeping the number of working physical nodes minimal for as long as possible. The process of consolidating VMs is comprised of two basic procedures: identifying servers that are not fully utilized then initiate live migration so as to reduce the number of active hosts and identifying overused hosts then initiate live migration so as to offload VMs from overburdened hosts.

In order to maximize resource utilization and lessen energy consumption, dynamic VM consolidation has been found to be effective [4]. The idea here is that energy consumption can be reduced by keeping inactive servers in operating modes that consume less power (such as sleep, hibernation), thus the power consumed due to idle servers running at all times is eradicated. When necessity arises, the servers can be switched ON to accommodate new requests of VMs or migrating VMs. Nonetheless, modern service applications often experience volatile workloads which results in ever changing resource utilization patterns. This warrants a close attention to the challenge of dynamic or automatic VM consolidation. It is not a petty issue. Therefore, cloud service providers have a challenging task of dealing with the issue of balancing performance and energy consumed. In other words, the aim is to minimize electricity consumption and, on the other hand, fulfill quality of service (QoS) demanded by clients. The QoS requirements are laid out in service-level agreements (SLAs).

3 Motivation and Related Work

This section discusses past approaches that have been found in the literature regarding energy-aware effective dynamic VM consolidation. The writers in [5] formulated the problem under study, dynamic VM consolidation, to resemble a bin-packing scenario. They also applied a forecasting method based on historical data and a genetic algorithm to minimize electricity consumption. In this research, they did not make use of an overload detection algorithm but they focused on VM placement. By applying a forecasting method and genetic algorithm (GA), they found out that the GA-based method outperforms the bin-packing approach in terms of minimization of power consumption. However, by use of forecasting and GA, their method resulted in increased time during training of the genetic algorithm.

Some authors have proposed the use of thresholds, in these kinds of approaches, upper and lower thresholds are manually set. Thus when the current CPU usage goes beyond the maximum threshold, an overload occurs and when the current CPU usage is below the minimum threshold, an underload is said to have occurred. To this effect, the authors in [6] proposed the use of two fixed thresholds, minimum threshold and maximum threshold, so as to enable the identification of overloaded and underloaded. But the inherent problem with fixed thresholds is that they are unfit for an environment with applications that share resources and have variable workload patterns. The ability of the system to change its behavior accordingly is thus a requirement. In another approach by the authors in [7], they put forward a proposal for determining an optimal host overload detection policy based on Markov chain model. They postulated that by exploiting the average intermigration time as specified by the QoS objectives, they can optimally solve the problem of host overload detection. However, the satisfaction of the Markov properties may be untrue for variable types of workloads.

A negative correlation exists between energy conserved and the violation of SLAs [7] since energy consumption is typically reduced at the expense of increased SLA violation. Thus, from another viewpoint, meticulous resource management algorithms in large-scale data centers are required to intrinsically balance the trade-off among the amount of energy consumed and breach of SLA. Consequently, one of the early works to investigate virtualized data centers on a large scale was proposed by the authors in [8]. On top of hardware weighting and consolidation of VMs, they suggested a new method for managing power in virtualized environments. Furthermore, they also suggested division into resident and universal levels, of the resource management task. The power usage by guest VMs is monitored at local level while on the other hand coordination of multiple physical machines is done at global level. Their experimental evaluations show a power consumption of up to 34% without considerable losses in performance.

The authors in [9] took a different angle in which they first conducted a competitive analysis and have shown competitive ratios of optimal online deterministic algorithms. They worked out competitive ratios for the challenging task of dynamic VM consolidation and likewise did the same for VM migration of a single machine. They went further to divide the problem of VM consolidation into four subproblems: (1) detection of overloaded host; (2) detection of underloaded host; (3) identification of VMs to migrate from an over utilized node; (4) implementing another placement scheme for the VMs migrated from overused and underused nodes. They suggested innovative heuristics that can dynamically adapt to changes in the workload by making use of statistical methods, for instance the median absolute deviation (MAD), interquartile range (IQR), linear regression (LR) and among others. They also proposed another algorithm which takes into account the amount of electricity consumed, and they named it PABFD meaning power-aware best-fit decreasing. The PABFD was a follow-up study to their earlier work in which they proposed a modification to the best-fit-decreasing algorithm [6]. Their techniques managed to perform static-threshold-based methods in reducing power consumption. However, when performance is of paramount importance, these methods can be improved.

Finally, the authors in [10] introduced a novel method, namely enhanced optimization (EO) policy for resource management in cloud data centers. This idea is based on a single step for alleviating the VM placement problem rather than doing it in multiple steps. Moreover, for optimization of different objectives in cloud environment, they proposed a solution that forms its basis on Technique for Order of Preference by Similarity to Ideal Solution (TOPSIS). In this methodology, performance measures such as the amount of energy consumed, breaching of SLA, and the frequency of VM migration are all taken into account at the same time. This idea forms the foundation on which the authors put forward new algorithms as such to tackle problems associated with dynamic VM consolidation. The following techniques were proposed: (1) TPSA (TOPSIS Power & SLA Aware Allocation) and (2) TACND (TOPSIS Available Capacity Number of VMs Migration Delay). By implementing this multi-criteria approach, the proposed techniques were able to significantly lower the number of virtual machine migrations and improve service quality. However, electricity usage was still high in comparison with other methods.

4 System Model

In this work, we consider the case of an Infrastructure as a Service (IaaS) cloud provider. To this effect, we have modeled the scenario of an IaaS provider resembled by a huge data center consisting of multiple different physical hosts. The capacity of each host is measured by its CPU performance and the amount of RAM. Users submit VM requests randomly, and each request contains specifications such as 2 processor cores, 4 GB RAM and 100 GB storage. Resource utilization of each VM (footprint) differs according to usage by clients, and so there is a possibility that users may request for more than their required specifications implying that resources are wasted. The customers establish service-level agreements (SLAs) with the IaaS provider to define the QoS provided.

We propose a mechanism which consists of the following 3 modules:

- Server Manager Module
  This module maintains an overview of resource utilization by servers/hosts and does initial assignment of VM requests to servers. It also performs VM replacement in case of server overload or underload. In other words, it periodically does optimization of VM assignment. This enables remaining idle servers to be set to an energy saving mode.
- Load Analyst Module
  This module is responsible for analyzing and predicting when an overload is likely to occur on a host. Prediction will be based on historical demand behavior or resource utilization of virtual machines running on a host. This enables migration to start before an overload occurs and thus avoid violation of SLA.
- Virtual Machine Migrator
  This module decides on which virtual machine(s) to migrate and also performs the actual VM migration. It is also responsible for changing power mode of a host (Fig. 1).

5 Power Models and Performance Metrics

Occasionally, recent studies [8, 11] have come to an agreement that energy consumption by servers can be estimated or represented by a direct proportion between CPU utilization and electricity consumption. In other words, there exists a linear relationship among these two. Nonetheless, we make use of authentic power consumption data as made available through survey findings in [12]. In this study, the energy consumption of servers considered is shown in Table 1.

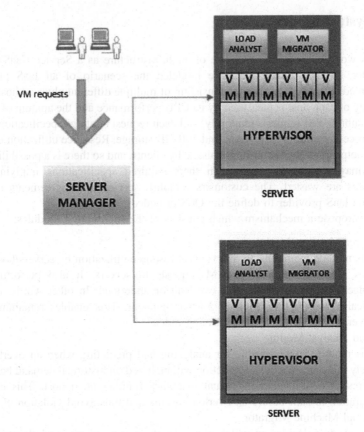

Fig. 1 The system model

Table 1 Electricity usage of servers at variable workloads (W)

Utilization level	Idle	0.1	0.2	0.3	0.4	0.5	0.6	0.7	0.8	0.9	1.0
Dell PE R730	46.9	87.8	107	124	141	156	171	191	215	243	272
Dell PE T630	48.1	89.6	108	126	142	157	172	191	217	247	273

Performance Metrics

(1) Consumed Energy (E)

The overall amount of power consumed by servers in a data center is represented by this metric. Additionally, energy consumption is modeled as the summation of power consumed during a period of time according to Eq. (1).

$$E(t) = \int P(t)dt. \tag{1}$$

where P is power and t is a period of time.

(2) SLA violation (SLAV)

$$SLATAH = \frac{1}{N} \sum_{i=1}^{N} \frac{K_{si}}{K_{ai}}. \tag{2}$$

$$PDM = \frac{1}{V} \sum_{j=1}^{V} \frac{U_{dj}}{U_{rj}}. \tag{3}$$

where N is the number of hosts; V is the number of VMs; K_{ai} is the total time of the host i being in the active state (serving VMs); K_{si} is the total time duration in which utilization levels of host i have reached 100% resulting in performance degradation and thus violation of SLA; U_{rj} represents the overall CPU time demanded during the lifetime of VM; and U_{dj} is a measure that estimates performance bottleneck caused by VM migrations on a particular VM j.

$$SLAV = SLATAH.PDM. \tag{4}$$

(3) SLA Violations (ESV)
This is a joint metric resulting from the calculation of both energy consumption and SLAV.

$$ESV = E.SLAV. \tag{5}$$

(4) Migration Frequency (MF)
This metric is a measure of the overall number of virtual machines migrated. If the frequency of VM migrations is low, then performance degradation is also minimal else it is high. Therefore, it is imperative to reduce the number of VMs migrated.

6 Proposed Methodology

Rousseeuw and Croux Estimators S_n and Q_n

Like the MAD proposed in [6], the two scale estimators S_n and Q_n have the capability to be applied as ancillary or scale estimates. However, unlike the MAD, they exhibit more efficiency and are unbiased with respect to symmetric distributions [13].

The following is a definition of the estimator S_n:

$$S_n = d.med_a\{med_b|y_a - y_b|\}. \tag{6}$$

where for each a the median of $\{y_a - y_b; b = 1 \ldots n\}$ is calculated resulting in n numbers from which their median provides the final estimate of S_n. The value 1.1926 is used as a default value of d, and its purpose is to maintain consistency [13].

In Eq. (5), the inner median is taken as a high median, with its order statistic being of rank $h = [n/2] + 1$ while the outer median is taken as a low median, with its order statistic being of rank $[(n + 1)/2]$.

The estimator Q_n is defined as

$$Q_n = c.\{|y_a - y_b|; a < b\}_{(h)}. \tag{7}$$

where c represents a constant element and $h = \binom{k}{2} \approx \binom{n}{2}/4$ where $k = \frac{n}{2} + 1$ represents approximately 50% of the observations. Thus, out of the $\binom{k}{2}$ interpoint distances, we take the kth-order statistic. This has some similarities to S_n in the sense that by taking the median twice, this yields the interpoint distances which are at minimum as big as their 0.25 quartile.

We therefore define the upper utilization threshold (T_{High}) as in [9], but contrary to them, we apply the two estimators aforementioned.

$$T_{High} = 1 - m.S_n. \tag{8}$$

$$T_{High} = 1 - m.Q_n. \tag{9}$$

where $m \in \mathbb{R}^+$ stands for a factor or parameter of the function controlling the aggression with which the system consolidates VMs. From another perspective, the parameter m enables the fine-tuning of how risky consolidation can be, the lower value of parameter m, the lesser amount of electricity consumed, but on the other hand the SLA violations resulting from aggressive consolidation of VMs increases.

7   Performance Evaluation and Results

The proposed VM consolidation schemes based on the two estimators S_n and Q_n have been implemented, and their performance has been compared to other VM consolidation methods proposed in [9], namely the MAD, IQR, and Threshold (THR). The VM selection policy, minimum migration time (MMT), was the

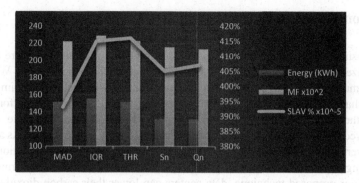

Fig. 2 Comparison of simulation results

Table 2 Simulation results in tabular form

Policy	Energy (KWh)	MF $\times 10^2$	SLAV% $\times 10^{-5}$
MAD	151.59	222.36	393
IQR	155.61	229.03	415
THR	151.74	223.01	416
S_n	132.17	215.76	405
Q_n	131.80	213.56	407

preferred choice in this experiment. In this research, the most important performance parameters of interest are energy consumption and SLAV; however, MF has also been assessed. For each scheme, we have used real-workload data made available as part of the CoMon project [14] (Fig. 2; Table 2).

From the graph and table above, it is evident that our two proposed VM consolidation policies; S_n and Q_n have a better performance than MAD, IQR, and THR policies in both energy conservation and violation of SLAs (SLAV). They were able to achieve a notable reduction in power consumption levels by more than 16%, and the number of VM migrations (i.e., MF) was also reduced marginally.

According to a report by [15], the amount of electricity consumed by US data centers in 2013 was estimated to be 91 billion kilowatt-hours and this is forecasted to reach around 140 billion kilowatt-hours annually by 2020. This will result in American businesses incurring US$13 billion annually in electricity bills and discharging closely 100 million metric tons of carbon pollution per year. If a 16% decrease in electricity consumption is achieved by using our proposed method, this will result in an estimated total savings of up to US$2 billion dollars.

8 Conclusion and Future Works

In this study, we have proposed two VM consolidation methods that were able to minimize energy usage and improve QoS by reduction of VM migrations. We have used simulation and real-world data to evaluate our proposed techniques. Simulation results have shown that our suggested methods S_n and Q_n have a better performance than other methods proposed in the literature. Therefore, our technique can be employed in huge data centers to efficiently and effectively utilize servers. As a result, cloud service providers can bring down their overall electricity consumption, limit operating expenses, and increase their return on investment (ROI). Furthermore, by use of our proposed technique, data centers can lower their carbon dioxide (CO_2) emissions and thus contributing to the green data centers initiative. In future, we plan to investigate the effect of using methods such as linear regression and forecasting of resource usage requests to efficiently allocate resources in cloud data centers.

References

1. Barroso L, Hölzle U (2007) The case for energy-proportional computing. Computer 40:33–37
2. Fan X, Weber W, Barroso L (2007) Power provisioning for a warehouse-sized computer. In: Proceedings of the 34th annual international symposium on Computer architecture—ISCA'07
3. Song X, Shi J, Liu R, Yang J, Chen H (2013) Parallelizing live migration of virtual machines. ACM SIGPLAN Notices 48:85
4. Hermenier F, Lorca X, Menaud J, Muller G, Lawall J (2009) Entropy. In: Proceedings of the 2009 ACM SIGPLAN/SIGOPS international conference on virtual execution environments—VEE '09
5. Hlavacs H, Treutner T (2012) Genetic algorithms for energy efficient virtualized data centers. In: 2012 8th international conference on network and service management (CNSM) and 2012 workshop on systems virtualization management (SVM). IEEE, Las Vegas, NV, pp 422–429
6. Beloglazov A, Abawajy J, Buyya R (2012) Energy-aware resource allocation heuristics for efficient management of data centers for cloud computing. Future Gener Comput Syst 28:755–768
7. Beloglazov A, Buyya R (2013) Managing overloaded hosts for dynamic consolidation of virtual machines in cloud data centers under quality of service constraints. IEEE Trans Parallel Distrib Syst 24:1366–1379
8. Nathuji R, Schwan K (2007) Virtual power. ACM SIGOPS Oper Syst Rev 41:265
9. Beloglazov A, Buyya R (2011) Optimal online deterministic algorithms and adaptive heuristics for energy and performance efficient dynamic consolidation of virtual machines in cloud data centers. Concurr Comput: Pract Exp 24:1397–1420
10. Arianyan E, Taheri H, Sharifian S (2015) Novel energy and SLA efficient resource management heuristics for consolidation of virtual machines in cloud data centers. Comput Electr Eng 47:222–240
11. Kusic D, Kephart J, Hanson J, Kandasamy N, Jiang G (2008) Power and performance management of virtualized computing environments via lookahead control. Cluster Comput 12:1–15
12. Lee Y, Zomaya A (2012) Energy efficient utilization of resources in cloud computing systems. J Supercomput 60:268–280

13. Rousseeuw P, Croux C (1993) Alternatives to the median absolute deviation. J Am Stat Assoc 88:1273
14. Park K, Pai V (2006) CoMon: a mostly scalable monitoring system for PlanetLab. ACM SIGOPS Oper Syst Rev 40:65
15. Delforge P (2015) America's data centers consuming and wasting growing amounts of energy. Natural Resources Defense Council

A Rule-Based Self-Learning Model for Automatic Evaluation and Grading of C++ Programs

Maxwell Christian

Abstract A procedural and formulated evaluation of the practical work or task performed by a student, irrespective of the procedure or method incorporated by him, has always been a challenging job, as perceived by many researchers and practical program evaluators since long [1–3, 5]. So there always has been a continuous process of improvisations in different techniques to properly judge and evaluate the tasks performed by students [5, 6]. Also the time required to solve the trivial and re-occurring common errors, during the initial phase of the student learning of new concepts, leads to more time consumption and hence needs to be reduced. Thus, the model that can evaluate, grade and inform the work accomplished by the student against the actual requirement can smoothen the learning curve and ease the job of the evaluator always comes handy.

Keywords Automatic evaluation · Automatic grading · Rule-based evaluation
Rule-based grading · Self-learning models

1 Introduction

Understanding programing concepts and implementing them have majorly observed as a poles-apart scenario [6]. The student studying a particular programming concept and then implementing them in-line with other concepts that he has studied is observed to be lacking good coupling and also good cohesion in the primary stages of the learning curve [6, 7]. Also, it is observed that the time consumed in coupling the concept and the implementation is quite high in the initial

M. Christian (✉)
MCA Department, GLS University, Ahmedabad, Gujarat, India
e-mail: max4sall@gmail.com

© Springer Nature Singapore Pte Ltd. 2018
D.K. Mishra et al. (eds.), *Information and Communication Technology
for Sustainable Development*, Lecture Notes in Networks and Systems 9,
https://doi.org/10.1007/978-981-10-3932-4_49

stages of learning [4, 6, 7]. The major reason observed behind the time consumption is also the time consumed to solve the trivial and re-occurring common errors, experienced by the students [4–7].

2 The Proposed Model

2.1 The Need

A continuous interaction with students and evaluators alike has derived a few ground rules for assessment. There are multiple methodologies possible to solve the given problem, many a times both student and the evaluator is unaware of the better methodologies than what they are adopting to solve [8, 9]. Many times the implementation is not exactly in-line with the actual requirement. Students waste lot of time doing things not asked for [6, 7]. A dire need exists for using a task and implementation-based dynamic evaluation technique, i.e. evaluation on the basis of the method adopted to solve the concerned problem may be conventional or totally innovative [10, 11].

There also lies a situation where the actual outcome may not be up to the expectation due to the student's lack of proper step-by-step approach towards the problem solving where the student may have wasted fair enough of his time in solving trivial logical error(s), which may be easily identified by the evaluator though his expertise and experience [11, 12]. But on the contrary, solving the same error(s), on a regular basis, by a student/evaluator, again may lead to time crisis towards the overall completion of the entire system evaluation [12, 13, 14].

To narrow down this gap amongst the actual and the expected evaluation, in terms of both time and quality, an intelligent evaluation model which may help on the first hand towards the evaluation process and then backed by the intelligent expertise of the actual evaluator can result in saving the time consumption and also aid the student to understand his own trivial error(s) and the situations that may have lead him towards generating such error(s).

Hence, the automatic evaluation will be a first step towards the evaluation process, still leaving the final verdict in the hands of the evaluator, whose human expertise and experience, as known, can never be out—matched by machine or an algorithm.

2.2   The Architecture

The Fig. 1 demonstrates the architecture of the proposed model expected to be a
self-learning model for rule-based automatic evaluation of C++ programs
   The model as seen in the Fig. 1 is composed of the following major components:

1. *Rule base*: All the rules used to evaluate a developed program will be stored
   here and will keep on increasing gradually with the inputs of the
   trainer/evaluator
2. *Parser*: The parser will basically study the structure of the developed program
   and then direct them to the *code analyzer* (for static analysis like code correctness) and to the *concept analyzer* (for dynamic analysis like member method
   implementations)
3. *Grade Analyzer*: The parsed and analysed code (both by the code and concept
   analyzer) will be studied by the grade analyzer for the grading purpose.
4. *Message base*: It will be used to output the feedback messages, which will be in
   normal plain form rather then the conventional technical format.

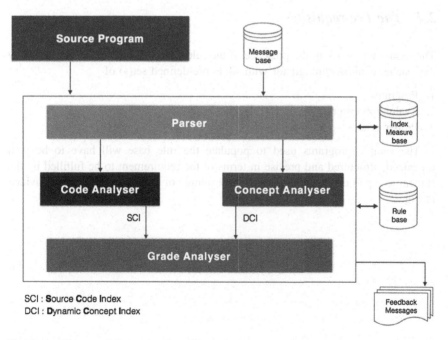

SCI : **S**ource **C**ode **I**ndex
DCI : **D**ynamic **C**oncept **I**ndex

Fig. 1 Architecture of the proposed model

2.3   The Working

To help the student with his developed program and to provide him the feedback on his developed version of the program, the model will work in the following listed steps:

1. The student will first input his program to the model
2. The model will then select the rule set, if specified by the evaluator
3. In case there is no question-specific rule set defined, then the model will use the entire rule base available
4. Then, the program will be evaluated against the selected rule(s)
5. The grading will also be done accordingly
6. The student will be provided the feedback in plain messages
7. When a new rule is encountered, if in case, while evaluating the new version of the implementation, the model will learn it and add it to the knowledge base of rules thus improving itself.

2.4   The Pre-requisites

The model will work in the presence of the rule base will have to be populated by the teacher/evaluator/invigilator with some pre-defined set(s) of

1. Programs
2. Evaluation criteria
3. Proper grades

The sample programs used to populate the rule base will have to be well organized, structured and precise in terms of the requirement to be fulfilled by the student as a part of the development. An example of a sample question is provided in Fig. 2.

1   Develop a C++ program that performs following operations.

[A]   Create two classes StudentList and StudentNode with private data   05
members as mentioned below

StudentNode class:

1.   A member data to store name of the student

2.   A member data to store the total marks of the student

3.   A member next that points next element in the list

StudentList class:

1.   A member head that will point the start of the list.

2.   A member count that will store number of students in the list.

10

[B]   Provide constructors to define the StudentList object in following
ways

1.   StudentList objStudentList
//Creates the list with default data value.

2.   StudentList objStudentList(name, totalMarks);
//Creates the list with first student created with data value as
name and totalMarks.

3.   Provide proper destructor to avoid memory leak

50

[C]   Perform following operations on StudentList

1.   Overload the << operator to display the list along with the
count of total number of Students in the list.
E.g. cout << objStudentList;
// It will display the contents of the objStudentList and also the
total number of Students in the list

2.   Overload the + operator to perform union operation for two
lists.
E.g. objStudentList3 = objStudentList1 + objStudentList2
// It will merge the content of objStudentList1 and
objStudentList2 and store the result into the objList3, duplicate
data is allowed in the list

3.   Overload unary ++ operator to increment all the values of the
specified list.
E.g. objStudentList1++;
// It will add 1 mark to every Student of the objStudentList1

4.   Non member function to reverse the list.
E.g. reverse(objStudentList);
// This function should reverse the objStudentList

Fig. 2 Sample program for the model

3 Conclusions

The expected outcome of my research will be a self-learning system which shall be capable of evaluating the practical assignments related to C++ program(s) based on a rule set selected and defined by the evaluator, and providing grade to the developed program and also provide feedback as and when required in form of plain messages.

References

1. Doshi JC, Christian M, Trivedi BH (2014) Effect of conceptual cue based (CCB) practical exam evaluation of learning and evaluation approaches: a case for use in process-based pedagogy, technology for education (T4E). In 2014 IEEE sixth international conference on technology for education, pp 90–94
2. Forsythe GE, Wirth N (1965) Automatic grading programs. Commun ACM 8:275–278
3. Higgins CA, Gray G, Symeonidis P, Tsintsifas A (2005) Automated assessment and experiences of teaching programming. J Educ Resour Comput (JERIC) 5:5
4. Ihantola P, Ahoniemi T, Karavirta V, Seppälä O (2010) Review of recent systems for automatic asessment of programming assignments. In: Proceedings of the 10th Koli calling international conference on computing education research. pp 86–93
5. Romli R, Sulaiman S, Zamli KZ (2010) Automatic programming assessment and test data generation a review on its approaches. In: 2010 International symposium information technology (ITSim), pp 1186–1192
6. Caiza JC, Del Alamo JM (2013) Programming assignments automatic grading: review of tools and implementations. In: Proceedings of the Inted 2013, pp 5691–5700
7. Rodríguez-del-Pino JC, Rubio-Royo E, Hernández-Figueroa ZJ (2012) A virtual programming lab for moodle with automatic assessment and anti-plagiarism features
8. Queirós RAP, Leal JP (2012) PETCHA: a Programming exercises teaching assistant. In: Proceedings of the 17th ACM annual conference on innovation and technology in computer science education, pp 192–197
9. Spacco J, Hovemeyer D, Pugh W, Emad F, Hollingsworth JK, Padua-Perez N (2006) Experiences with marmoset: designing and using an advanced submission and testing system for programming courses. ACM SIGCSE Bull 38:13–17
10. Enstrom E, Kreitz G, Niemela F, Soderman P, and Kann V (2011) Five years with Kattis—using an automated assess ment system in teaching. In: Frontiers in education conference (FIE), 2011. Institute of Electrical and Electronics Engineers, Piscataway, NJ, T3 J–1
11. Malmi L, Korhonen A, Saikkonen R (2002) Experiences in automatic assessment on mass courses and issues for designing virtual courses. ACM SIGCSE Bull 34(3):55–59
12. Zanden BV, Anderson D, Taylor C, Davis W, Berry MW (2012) CodeAssessor: an interactive, web-based tool for introductory programming. J Comput Sci Coll 28(2):73–80
13. Yu YT, Poon CK, Choy M (2006) Experiences with PASS: developing and using a programming assignment assessment system. In Sixth international conference on quality software, 2006. QSIC 2006. Institute of Electrical and Electronics Engineers, Piscataway, NJ, pp 360–368. doi:10.1109/QSIC.2006.28
14. Douce C, Livingstone D, Orwell J (2005) Automatic test-based assessment of programming: a review. J Educ Resour Comput (JERIC) 5:4

Study and Research on Raspberry PI 2 Model B Game Design and Development

Nishant Sahni, Kailash Srinivasan, Karan Vala and Saurabh Malgaonkar

Abstract This paper gives a detailed account of the steps required to develop a multimedia system, which marries the software component—an arcade-style game developed using Pygame libraries and the hardware component—a portable arcade game machine developed using a Raspberry Pi processor. It also describes the various functionalities Pygame libraries offer to game developers and emphasizes its ease of use along with the benchmark statistics.

Keywords Raspberry Pi · Pygame · Python · Games · Gaming
Game development

1 Introduction

Having a computer as a gaming machine is one of the most common and finest forms of playing games. Due to the rapid increment in technology, evolving hardware and software requirements, the way games are being viewed by one and all is changing. For instance, arcade games have been the center of attraction for the past 30 years. Year by year, it is becoming very economical for game developers, attracting a huge group of customers, and as a result, 2D games are still being

N. Sahni (✉) · K. Srinivasan · K. Vala
Computer Engineering Department, Mukesh Patel School of Technology
Management & Engineering, NMIMS University, Mumbai, India
e-mail: nishantsahni1994@gmail.com

K. Srinivasan
e-mail: kailashs1802@gmail.com

K. Vala
e-mail: karanvala22@gmail.com

S. Malgaonkar
Department of Computer Science, Whitacre College of Engineering,
Texas Tech University, Lubbock, TX, USA
e-mail: saurabhmalgaonkar@gmail.com

© Springer Nature Singapore Pte Ltd. 2018   475
D.K. Mishra et al. (eds.), *Information and Communication Technology
for Sustainable Development*, Lecture Notes in Networks and Systems 9,
https://doi.org/10.1007/978-981-10-3932-4_50

developed and sold. One of the main advantages is that 2D graphics are easy and relatively less expensive to implement as compared to 3D graphics. As a result, 2D games are enjoying a great comeback and have reinvigorated interest in the arcade style of gaming. The main objective of this project is to understand and create a developer-friendly portable gaming machine which runs Python-based games developed using Pygame libraries [1]. The various hardware components required to construct the portable gaming machine are highlighted. The developer-oriented aspects with respect to game development with Python and Pygame are also explored [2]. The hardware components that put together to build this gaming machine [3] are as follows:

1.1   Raspberry Pi

This was found to be the most suitable processor for the gaming system. The following table gives its precise specifications (Table 1).

1.2   Display Device

The main purpose of a display device in our system is to output the game content. The display device used is a LCD display with an HDMI port. As the display is capacitive touch-enabled, touch can be used as another form of input for a game running on the system.

1.3   Controls

These are required to carry out the actions in the game and to interact with the system. We are using a controller comprising of joysticks and buttons for input.

Table 1 Raspberry Pi 2 Model B specifications

Parameters	Specifications
RAM	1 GB
CPU	900 MHz
Operating system	Linux operating system
Video output	HDMI (rev 1.3 and 1.4)
Power	Micro USB socket
Dimensions	85 × 56 × 17 mm
GPU	Dual core 1Gpixel/s to 1.5Gtexel/s
Audio output	3.5 mm jack, HDMI
USB	4 × USB 2.5 connector

A wired or wireless keyboard and mouse can also be used for testing. The physical controls need to be wired. We have identified the ports that each button is wired to and then edited the control configuration file to map the keys to the right in-game buttons.

2 PyGame Design and Development

Pygame is a platform on which the actual games are written using the high-level Python programming language as Pygame exclusively contains Python modules [4] (Fig. 1).

Program files contain information about sprites and other data. Along with this data, image as well as input data is stored in the database. Whenever any of these functions are called by the user, i.e., when the user starts the game, the object handler is initiated. This object handler controls various functions such as sound, text, and displaying text messages. The array contains sprite information that communicates directly with the object handler. For example, you have the sound handler to control the sound and the text handler to display the text messages, and so on. The object handler retrieves sprite information as soon as other data requests are received from the other handlers. Depending on what function is called, the handler performs that in particular.

2.1 Library Support

Simple DirectMedia Layer (SDL) is a game creation library on which Pygame is built. SDL simplifies the task of porting games from one platform to another. SDL

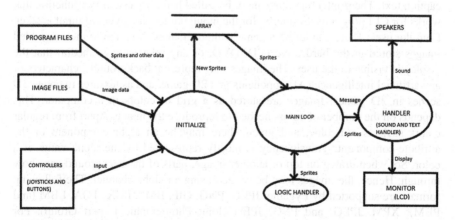

Fig. 1 System design for a touch screen game

provided a common way to create a display on multiple platforms. As it became very easy to work with, game developers have been using this for a long time for developing commercial games. It is written in C, which is a language commonly used for writing games. It has the ability to work with the hardware at a low level. Pygame allows Python programmers to use the very powerful SDL library.

2.2 Visualization Using Pygame

Pygame consists of many subsections for a variety of game-related tasks. Pygame provides functions only for creating programs with a graphical user interface (GUI). Programs with graphical user interface (GUI) can display a window with colors and images. The following 'Hello World!' script demonstrates the basics of initializing Pygame, creating displays, event handling, and drawing on the screen.

Input:

```
import pygame, sys
from pygame.locals import *
pygame.init() #initializing pygame
DISPLAYSURF = pygame.display.set_mode((600,400))
pygame.display.set_caption('Hello World!')
while True: #main loop
for event in pygame.event.get():
if event.type == QUIT:
pygame.quit()
sys.exit()
pygame.display.update()
```

For the above python code, the corresponding output is a blank window with the name 'Hello World!' at the top of the window, i.e., the title bar which holds the caption text. The print() function cannot be called here, because it is a function that works for CLI programs. Similarly, for the input() function, keyboard input is taken from the user. There cannot be a game without images. It is due to the assembled images stored in the hard drive (CD, DVD, or any other storage media) that the display is visible to the user. The images may represent backgrounds, characters or any artificial intelligence (AI) opponents in a 2D game, and as textures to create 3D scenes in 3D games. Images are stored as a grid of colors in a computer. This depends on the number of colors needed to reproduce the image. Apart from regular colors such as red, blue, and green, there may be an alpha-component or the attribute-component. Translucency is mostly represented by the alpha value of a color. So, when drawn on top of another image, parts of the background can show through. Hence, the images may be created using an alpha channel. The following formats are supported by Pygame: JPEG, PNG, GIF, BMP, PCX, TGA, LBM (and PBM), XPM. JPEG, and PNG. JPEG (Joint Photographic Expert Group): For shrinking the file size, they use a process known as lossy compression. This can reduce the quality of the image. PNG (Portable Network Graphics) is the most versatile image format because it has the capacity to store a huge variety of image types. It compresses the image well and also supports alpha channels. Thus, JPEG is used only for larger images which have a lot of color variations, otherwise PNG is used.

2.3   Game Designing

The game code must draw lines, circles, and other geometric figures to represent in-game figures on the screen. An entire game can be created just by using these draw functions and without loading any images. Some of the functions are described as:

pygame.draw.rect: This function is used to draw a rectangle on the surface. Using 'pygame.rect,' rectangular dimensions can be drawn along with the width of the line.

pygame.draw.polygon: While calling 'pygame.draw.polygon,' lists of points are taken as parameters and a polygonal shape is drawn between them. An optional width value can also be taken. For example, if width is excluded or set to zero, the polygon will be filled; otherwise, only the edges will be drawn.

pygame.draw.circle: This function draws a circle on the surface. The center point and the radius of the circle are taken as parameters along with a value for the width of the line. If width is excluded or zero, the circle will be drawn along with a line; else, a solid circle will be drawn.

pygame.draw.ellipse: A squashed version of a circle may be called an ellipse. This function takes a rectangle-style object that the ellipse must fit into. Width is also taken as a parameter.

pygame.draw.line: This function is used to draw a line between two points, for example, the start and the end point.

pygame.draw.aaline: The jagged appearance of lines occurs due to the aliasing effect. But Pygame draws anti-aliased lines that are much smoother than the lines that are drawn using the function 'pygame.draw.line.' Even though anti-aliased lines have better visual quality, they are slower to draw in comparison with the lines drawn using the 'pygame.draw.line' function.

Surfaces can be represented by various images, but most of these details are hidden by Pygame. As soon as the surface is stored in the memory, it can be drawn, transformed or copied to another image. For instance, a surface object can be represented as the screen. The function 'pygame.display.set_mode' returns a surface object used to represent the display. There are two main ways of creating surfaces in Pygame. A surface can be created by calling the 'pygame.image.load' function. This surface will match the dimensions and colors of the image file. Another way to create a surface is to call the 'pygame.Surface' function passing the surface size as parameters. The following line of code creates a surface of size 256 by 256 pixels using the 'pygame.Surface' function: blank_surface = pygame.Surface((256,256)).

When surface objects are used, Pygame will hide this detail from the user. The type of Timage used does not matter, the code will run anyway. If all the images are of the same format, then there will be not much load on Pygame. The code will be executed with ease. If the convert function is called, avoiding all the parameters, then the current surface will be converted such that it matches the display surface. In conclusion, if the source and the destination of the surfaces are of the same type, it will be copied much faster. The problem arises when the image contains an alpha channel (the convert function may neglect it). In response to this, Pygame provides a method (method alpha) that helps converting the surface to a certain format. This information (alpha) is preserved within the image. Consider the following lines of code:

background = pygame.image.load(background_image_filename).convert()

mouse_cursor = pygame.image.load(mouse_image_filename).convert_alpha()

The load function reads a file from your hard drive and returns a surface containing the image data. The call to pygame.image.load reads the background image and then immediately calls convert. This function converts the image to the same format as our display. The mouse cursor is loaded in a similar way (convert_alpha). The mouse cursor image contains alpha information, which means that portions of the image could be completely invisible.

Rectangular Objects A rectangle can be defined using the x and the y coordinates with respect to the top left corner of the screen followed by the height and the width of the rectangle. Similarly, x and y coordinates can be used as individual tuples followed by the height and the width. For example, the following code can define a rectangle having the same dimensions:

my_rect1 = (120,120,250,160)

my_rect2 = ((120,120), (250,160))

The main benefit of Rect objects is that once a Rect object is created, its position can be adjusted, points whether inside or outside can be detected, and intersection of other rectangles can be found.

The following two lines of code construct the Rect objects that are equal to the two rectangle tuples in the previous example:

my_rect3 = Rect(120, 120, 250, 160)

my_rect4 = Rect((120, 120), (250, 160))

Clipping is the process of removing or cutting a part of the region that is irrelevant. To clip the unwanted region, the 'set_clip' function is called along with a Rect-style object. In case a portion that you require is clipped by some unforced error, it can be retrieved by the 'get_clip' function. The following is an example showing how clipping is implemented

```
screen.set_clip(0, 0, 550, 400) #sets the region
draw_map() #draws on to the top half of the screen
screen.set_clip(0, 400, 550, 170) #sets the clipping area
draw_panel()
```

Usually, the entire screen must be filled while creating an image on the display; otherwise, parts of the previous screen will be visible. A simple way to avoid this is to call the fill method that takes a color as a parameter. The following code will fill the screen:

screen.fill((0,0,0))

Here, the three parameters are the red, green, and blue values of the color.

Blitting literally means bit block transfer.

This method is very useful if relevant image data have to be copied from one surface to the other. This method can be used to draw backgrounds, characters, fonts, etc., in the game. One of the ways to use this method is:

screen.blit(font((0,0))

The above code is an example stating that blitting of the surface is possible by calling the screen.blit function.

Sprites are a very useful feature inside of a Pygame program. They are nothing but two-dimensional images that can be manipulated within a graphical scene. Sprites usually support 2D games. They basically come in handy with objects that support collision. The attributes that sprites consist of are: Name (name of the selected sprite), Image (name of image file), Frame rate, Random (whether the selected sprite has random movement or not), Bounces (whether there is any change in the direction of the selected sprite after collision), Visible (visibility of the sprite on the display/screen), etc.

Sounds, Pygame can load files of the format MP3 or WAV. Sound files are imported as soundObj = pygame.mixer.Sound('bloop.mp3') soundObj.play()

import time

time.sleep(1) # stop and sound plays for 1 s soundObj.stop()

Above is a sample code for running an audio file that may be saved in another file. Calling the play() method starts the program execution. As Pygame can load only one music file at a time, call the pygame.mixer.music.load() function to load a single audio file every time you want another audio file to be played. Call the pygame.mixer.music.stop() function to immediately stop the current running audio file.

3 Inferences

The ongoing work explains why Pygame is appropriate for two-dimensional game development. It provides a developer-friendly platform for shorter and more cost-effective development cycles. In terms of hardware compatibility, it provides entities to perform tasks for handling external input using keyboards, mouse, and controllers (joysticks). Also, the Pygame library supports in-game physics through an Open Dynamics Engine (ODE), typically referred to as PyODE. This engine has two main features—rigid body simulation and collision detection. Rigid body simulation involves incorporating real-world physical properties such as mass distribution, mass, elasticity, etc., to in-game objects. Collision detection helps in determining the force with which two in-game objects or bodies collide, and their reaction after the collision [5]. This engine exclusively works with Pygame libraries. Thus, physics-based games can also be implemented using Pygame [3]. The future work consists of comparing the existing system with the new upcoming equivalent gaming consoles and focus more on optimization of performance with respect to many available parameters in game designing and programming.

4 Results

Basic objects and scenarios were designed and executed on the system, and the following are some benchmarking results documented. After the source file is compiled, it is copied to a removable storage device such as pen drive and then loaded through the terminal for the execution on a Linux OS and the benchmarks could be obtained (Tables 2, 3, and 4).

Power Consumption was 1.037 W (max) for the above-mentioned operations.

With Raspberry Pi as the processor, which supports quad-thread programming and handling, transportation is much easier and faster due to its size, providing

Table 2 System memory access for game processing benchmarking

Memory utilization (KB)	Scaling (MB/s)
8	1011
32	1014
512	679
1024	391

Table 3 Vector graphics rendering with CPU, GPU, and memory utilization (MB/sec)

Object/Scenario (Memory KB)	A = A + B * A	A = A + A + B	A = A + A
16	1975	2290	2793
32	1685	1866	2141
512	775	811	839
1024	425	438	436

Table 4 Request response through threading for the processing operations

One thread	Four threads
Total time: 657.536 s	**Total time**: 180.786 s
min 65.31 ms per request response	**min** 65.27 ms per request
avg 65.75 ms per request response	**avg** 65.37 ms per request
max 71.02 ms per request response	**max** 99.27 ms per request

high-performance and basic PC functionalities like accessing the web browser, etc. Also, due to the HDMI ports, the display quality will be very good. This paper gives an account of the Pygame framework's capabilities, which include several entities with functions for manipulating graphics, colors, creating objects, playing sounds, and so on. The light-hearted nature of these games appeals to hardcore and casual gamers alike. Pygame enables the development of these games with minimum effort and expense to developers.

References

1. Sweigart A (2012) Making games with Python and Pygame
2. Mcgugan W (2007) Beginning game development with Python and Pygame
3. Sahni N, Srinivasan K, Savla H (2015) A review on developing an Arcade game machine and an Arcade game using Raspberry Pi and Pygame. Int J Comput Appl (0975 – 8887) 120(17)
4. Jones R (2010) Rapid game development in Python
5. Noufal Ibrahim KV (2010) Creating physics aware games using PyGame and PyODE: the Python papers monograph 2: 20. In: Proceedings of PyCon Asia-Pacific 2010
6. Raspberry Pi 2 Model B (2015). https://www.raspberrypi.org/products/raspberry-pi-2-model-b/. Accessed 2 May 2015
7. Notepad++ (2015). https://notepad-plus-plus.org/. Accessed 2 May 2015

Improved Churn Prediction Based on Supervised and Unsupervised Hybrid Data Mining System

J. Vijaya and E. Sivasankar

Abstract Retaining a customer plays a vital role in success of every sales firm. Not only it increases the profit margin of the firm but also it maintains the ranking of the firm in the competitive market. Every organization competes for the success of itself in the competitive market, so it aims to retain the existing customers through customer preservation techniques. Because preserving an existing customer is less costly than adding new customer. Hence, customer association management (CAM) and customer maintenance (CM) are two important parameters in determining the success of the organization. Also quality service is another parameter which reduces the customer churn to a greater extent. Hence, every organization conducts a customer churn forecast as a valuable step because it aims at customer maintenance and organization success. In this work instead of single classifier resulting in low efficiency, hybrid supervised and unsupervised techniques are deployed to achieve improved churn prediction. In stage one, data cleaning is carried out to eliminate deviations in data set. In stage two, clustering algorithms such as K-Means and K-Medoids are used to group customers with similar trends. In stage three, hold-out methods based training and testing data sets are obtained from the above clusters. In the next stage, the training and testing are carried out by algorithms such as decision tree (DT), k-nearest neighbor (KNN), support vector machine (SVM), linear discriminant analysis (LDA), and naive bayes (NB). In the final stage, sensitivity, specificity, and accuracy are measured to evaluate the efficiency of the hybrid system.

Keywords CAM · Churn · CM · K-Means · K-Medoids
Decision tree · Support vector machine · Linear discriminant analysis
Naive bayes · K-nearest neighbor

J. Vijaya (✉) · E. Sivasankar
National Institute of Technology, Tiruchirapalli 620015, Tamilnadu, India
e-mail: vijayacsedept@gmail.com

E. Sivasankar
e-mail: sivasankar@nitt.edu

© Springer Nature Singapore Pte Ltd. 2018
D.K. Mishra et al. (eds.), *Information and Communication Technology
for Sustainable Development*, Lecture Notes in Networks and Systems 9,
https://doi.org/10.1007/978-981-10-3932-4_51

485

1 Introduction

There is a tremendous growth in telecom sector year by year. Telecommunication companies are growing due to the invention of various high end technologies such as cloud computing, Internet of Things, m-payments, and due to evolving communication technologies. In day-to-day life, mobile communication and broadband connectivity have become part of the society and also the number of connected devices is increasing at a high rate. So telecom sector has become an area of innovation, growth, challenges, and competitions. It is known from Global Mobile Consumer Survey report that customers of USA are looking at their mobile devices about 8 million times per day on an average [1]. Hence, there is always continuous demand from consumers for various digital technologies that are offered by the telecommunication companies. The companies should be able to provide both voice and data with high quality, reliability, and also at affordable cost. Also the customer maintenance and customer churn depend upon various factors such as the quality of service, customer geographic location, cost per service, and customers' need. In front, customer churn forecast plays a vital role in the success of every telecom firm [2]. Hence, various intelligent data mining methods such as decision tree, naive bayes, logistic regression, random forest, artificial neural networks, inductive rule learning, and support vector machines are used to predict the customer churn ahead [3–8]. These techniques are applied not only in telecom but also in bank, newspaper, pay-TV, supermarket, credit cards, and automobile industries' churn prediction [3, 9, 10]. Nowadays, instead of using single classifier, hybrid supervised and unsupervised combined data mining techniques are deployed for improved churn prophecy [11, 12]. In this work, KDD Cup 2009: French Telecom Company Orange data set is deployed to analyze customer churn forecast. In stage one, data cleaning is carried out to eliminate deviations in data set. In stage two, clustering algorithms such as K-Means and K-Medoids are used to group customers with similar trends. In stage three, hold-out methods based training and testing sets are obtained from the above clusters. In the next stage, the training and testing are carried out by algorithms such as decision tree (DT), k-nearest neighbor (KNN), linear discriminant analysis (LDA), support vector machine (SVM), and naive bayes (NB). In the final stage, sensitivity, specificity, and accuracy are measured to evaluate the efficiency of the hybrid system.

2 Literature Survey

Churn prediction problem is solved by number of data mining techniques. In earlier days, a lot of researchers utilized single-model-based classification algorithms such as decision tree, logistic regression, artificial neural networks, inductive rule learning, random forest, naive bayes, and support vector machines to predict the customer churn [3–6, 8]. Nowadays, researchers concentrate on hybrid model which is a combination of two or more classifiers or a combination of clustering and classification. Indranil Bose and Xi Chen (2009) selected 14 attributes from a collection on 271 attributes related to revenue contribution and minutes of usage of the customer. He has used five clustering techniques, namely fuzzy c-means, K-Means, BIRCH, self-organizing map, and K-Medoids, to segment the customers and the result of segmented customers is given as an input to the decision tree, and performance is evaluated using top decile lift [13]. Some of the researchers worked on landline mobile customer churn prediction, but landline customer data set is not available in proper format. Huang, Kechadi, and Buckley (2012) extracted and modified new set of features such as aggregated call details, account details, bill, and payment from the entire set, which was evaluated using seven traditional classification methods. The accuracy is amplified by the newly extracted features. Performance was measured in terms of true-positive and false-positive rates [4]. Khashei, Hamadani, and Bijari (2012) implemented a hybrid model by combining artificial neural network and multiple linear regression model. Their proposed artificial neural network with multiple linear regression gives better result compared to single artificial neural network as well as traditional classification techniques such as support vector machine, linear discriminant analysis, k-nearest neighbor, and quadratic discriminant analysis [14]. Kyoungok Kim, Chi-Hyuk Jun, and Jaewook Lee (2014) concentrated on network patterns within customers because within a network, the churn information is passed from one customer to another. He identified who was the initial churner, thus targeting these people to reduce customer churn. Louvain community detection algorithm used to find structure of community [15]. Churn prediction problem is also applicable for various domains such as automobiles, gaming, banking, and credit card forecasting. Kyoungok Kim, Chi-Hyuk Jun, Jaewook Lee (2014), proposed modified Fisher's discriminant algorithm which was based on Fisher's linear discriminant function for credit card churn forecasting. The output of the modified Fisher's discriminant algorithm produces better profitability than discriminant algorithm [2]. Torsten J. Gerpott and Nima Ahmadi (2015) divided the entire contract period for landline customers into three stages such as earlier winback, late winback, and proactive period. Winback period is the activity where the service providers try to retain the customer after he submits his termination notification. They added winback reasons to the original data set and analyzed it using survival analysis. The results show that winback produces improved result compared to the previous [16]. Ozden Gur Ali and Umut Arıturk (2014) proposed dynamic churn prediction which means churn prediction is based on the time as well as on the customer. In normal churn prediction, each customer has only a single observation, but in dynamic prediction, each customer is observed during multiple times. They formulated three ways of churn prediction, namely single-period training data, multiple-period training data, and single-period data with addition lags. They built the model using logistic regression and decision tree for the above banking data set [17]. Adnan Idris, Muhammad Rizwan, and Asifullah Khan (2012) proposed a PSO-based undersampling method as the number of non-churners is very high compared to the number of churners. The various reduction techniques were evaluated. His data set is modeled using random forest and k-nearest neighbor. His approach produced better AUC value [18].

3 Churn Prediction Models and Methods

The block diagram of the proposed model is given in Fig. 1. The orange data set is preprocessed using effective data cleaning methods. After the cleaning process, the clustering approach is carried out using unsupervised techniques such as K-Means and K-Medoids algorithms. The various clusters obtained from the above-mentioned method are divided into training and testing sets using hold-out method. And each of the cluster training and testing data sets is handled by various supervised data mining techniques. The hybrid mechanism efficiency is measured in terms of sensitivity, specificity, and accuracy.

3.1 Data set

French Telecom Company Orange has given two versions of data sets. One is of smaller data set and other is of larger. For our study, we have taken training part of small version. The data set involved in this work consists of 50,000 instances with 230 features. Of the total features, available 190 are of numerical type and 40 are of nominal features. The names of the attributes are hidden inside the data set in order to maintain privacy of the customer. Majority of the customers about 46,328 are non-churn customers and about 3672 customers are predicated to get churn.

Fig. 1 Block diagram of hybrid churn prediction model

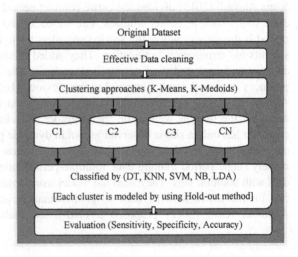

3.2 Data set Preprocessing

In the above data set, 60% of the data consist of missing values. Hence, preprocessing plays a vital role in increasing the consistency and accuracy of the system. From the above, data set features with 25% of the missing values are eliminated to reduce the system with 67 useful features. From the above-obtained data set, all the missing values of numerical and nominal features are replaced with mean and mode of the attribute values. Some of the nominal features are showing higher deviation in the attribute values. Hence, those features with higher deviations are removed from the data set, reducing the features count to 49. Since the clustering algorithm handles only numerical values, all the string values are converted to numeric. In the next step, normalization is carried out using min-max normalization.

3.3 Clustering Algorithm

Many data mining techniques such as image segmentation, customer segmentation, market analysis use clustering technique. Clustering technique is a process of grouping-related data items without much knowledge of group definition. The main aim of the clustering is to identify highly efficient clusters with high similarity within a cluster. From the clustering, homogeneous groups are formed to construct supervised classification models.

3.3.1 K-Means

In 1957, Stuart Lloyd proposed a partition-based clustering technique called K-Means algorithms. N input tuples are divided into K partition, and K should be less than N. It must satisfy the following criteria: (i) Each tuples must be a member of at least one cluster and (ii) each cluster should have at least one tuple. The K-Means algorithm consists of the following steps as shown below.

1. Initially, K value is initialized and it should be less than N.
2. In the next step, K tuples are randomly chosen as initial cluster points.
3. Next, we measure the distance from each tuples to the cluster center of all the clusters.
4. The cluster center which is at minimum distance is chosen as the cluster of the tuples considered.
5. After adding the tuples to the cluster, a new cluster mean is computed and it is taken as a new cluster center.

6. The above steps are repeated and the cluster is fixed until all the tuples are checked and there is no change in the mean value of the cluster.

3.3.2 K-Medoids

In 1987, Kaufman and Rousseeuw proposed a partition-based clustering technique called K-Medoids algorithm. It constructs K partition of N input tuples, where K less than or equal to N. The K-Medoids algorithm consists of the following steps as shown below.

1. Initially, K value is initialized and it should be less than N.
2. In the next step, K tuples are randomly chosen as initial cluster points.
3. Next, we measure the distance from each tuples to the entire chosen cluster center.
4. The cluster center which is at minimum distance is chosen as the cluster of the tuples considered.
5. After adding the tuples to the cluster, new cluster means are taken as a combination of already known medoids and randomly select any non-medoids object present in the data set.
6. The above steps are repeated and the cluster is fixed until all the tuples are checked and if there is less cost function of the cluster in the consecutive steps.

3.4 Classification Algorithm

Classification technique is a process of grouping-related data items which is known in advance. The main objective of the classification is to assign class label to the unidentified data items based on the model created from the training samples.

3.4.1 Decision Tree

In 1993, Quinlan proposed a divide and conquer method-based classification technique called decision tree. The decision tree classification algorithm consists of the steps as shown below.

1. Find the information gain for all the attributes.
2. A binary tree starts with highest information gain attribute.
3. The tree construction stops until there is no further attributes.
4. After constructing a tree, the classification rules are built based on each path from the root to leaf.

3.4.2 K-Nearest Neighbor

The k-nearest neighbor is one of the easiest classification techniques in data mining. The k-nearest neighbor classification algorithm consists of the steps as shown below.

1. Initialize the number of nearest neighbor, which is K.
2. Find the distance between each unknown test case and all known training cases.
3. Each unknown test case value is predicted by a majority vote by its K neighbors.

In this work, Euclidean distance measure is used for finding distance between data point and nearest neighbor = 3.

3.4.3 Support Vector Machine

In 1992, Boser, Guyon, and Vapnik proposed Gaussian radial-basis kernel functions-based SVM. SVM supports both linear and nonlinear data. Our churn prediction is linearly separable because it is a binary classification problem, namely churners or non-churners. The customers are separated by a hyperplane that separates the two classes. The classification function of SVM is calculated in Eq. 1 as follows:

$$d(X^T) = \sum_{i=1}^{l} y_i a_i X_i X^T + b_0 \tag{1}$$

where X^T is a test instance; y_i is the class label of support vector Xi; α_i and b_0 are numeric parameters; and l is the number of support vectors.

3.4.4 Naive Bayes

Naive bayes is a simple statistical classifier based on Bayes' theorem. Bayes' theorem is used to find the posterior probability $P(H/X)$ from prior probability $[P(H)]$, posterior probability conditioned on X $[P(X/H)]$, and $[P(X)]$ prior probability of X.

$$P(H/X) = \frac{P(X/H)P(H)}{P(X)} \tag{2}$$

Here, we go for kernel density estimation used to estimate probability density function for data values, and it overcomes the data smoothing for our proposed data set.

3.4.5 Linear Discriminant Analysis

In 1936, R.A. Fisher proposed a classification technique called linear discriminant analysis. Linear discriminant analysis uses continuous independent variable and categorical dependent variables similar to logistic regression. LDA is used to solve binary classification problem like churn prediction. Variant of LDA is Fishier linear discriminant, multiclass LDA, Quadratic discriminant LDA, etc. Face recognition and bankruptcy prediction problem are also solved by LDA. In this work, result is based on Quadratic discriminant LDA.

4 Experiments and Results

4.1 Experiment Setup

This work comprises a set of experiments. Initial set of experiment is conducted to evaluate the performance of the single-classification model such as decision tree (DT), k-nearest neighbor (KNN), linear discriminant analysis (LDA), support vector machine (SVM), and naive bayes (NB). Next set of experiment is carried out to appraise the hybridization performance on the K-Means algorithm and K-Medoids clustering algorithms along with classification algorithms decision tree (DT), k-nearest neighbor (KNN), linear discriminant analysis (LDA), support vector machine (SVM), and naive bayes (NB). Finally, we compare the effectiveness of single-classification techniques and hybrid learning system.

4.2 Performance Measures

In our study, we have used accuracy, sensitivity, specificity as performance measures to analysis the performance of the single classifier and hybrid model-based classifier. The objective of the churn prediction model is to acquire high sensitivity with low specificity. Table 1 shows a confusion matrix (Japkowicz 2006) where D_{11} is the number of the correct predicted churn customers, D_{12} is the number of the incorrect predicted churn customers, D_{21} is the number of the incorrect predicted non-churn customers, and D_{22} is the number of the correct predicted no-churner customers. From the confusion matrix, sensitivity is defined as the fraction of churn test cases that were correctly classified as churn, calculated by Eq. 3.

$$Sensitivity = \frac{D_{11}}{D_{11} + D_{12}} \tag{3}$$

Specificity is the fraction of non-churn test cases that were incorrectly classified as churn, calculated by Eq. 4.

$$Specificity = \frac{D_{21}}{D_{21} + D_{22}} \tag{4}$$

4.2.1 Performance Based on Single Classifiers

In single classifier model, our proposed data set is divided into training and test sets based on hold-out method and finds the performance calculated during the experimentation using accuracy, sensitivity, and specificity as tabulated in Table 2. It shows that our proposed data set works well for the entire classification algorithm compared to normal data set and also SVM reaches our objective function of 100 percentage sensitivity and maximum accuracy 92.55. Figure 2 picturing the accuracy of the various classification algorithms performs individually.

4.2.2 Performance Based on K-Means Clustering Along with Classifiers

In hybrid model, our proposed data set is segmented using K-Means clustering method; then, each of the clusters divided into training and test data sets based on hold-out method. Training data set is modeled using classifiers, and test data set is predicted based on the model designed by classifiers. Finally, output of all cluster value is aggregated. In this, we go for various classification algorithms,

Table 1 Churn prediction confusion matrix

Actual	Predicted	
	Churn	Non-churn
Churn	D_{11}	D_{12}
Non-churn	D_{21}	D_{22}

Table 2 The performance of classifiers

Classifiers	DT	KNN	SVM	NB	LDA
Accuracy	87.61	90.91	92.55	91.39	89.52
Sensitivity	93.68	98.00	100.0	99.00	97.96
Specificity	12.23	02.84	00.00	01.00	00.16

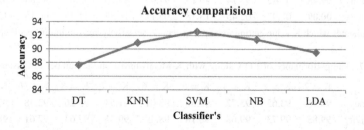

Fig. 2 Accuracy curve for various classifiers

such as decision tree (DT), k-nearest neighbor (KNN), linear discriminant analysis (LDA), support vector machine (SVM), and naive bayes (NB), and find the performance calculated during the experimentation using accuracy, sensitivity, and specificity as tabulated in Tables 3, 4, 5, 6, and 7, Here K represent the no of clusters. It's varies from 2 to 10. Bold indicates which K value getting maximum accurcy, sensitivity and minimum specificity value. The graphical view of accuracy is shown in Fig. 3. The result shows that (1) all the classifier decision tree (DT), k-nearest neighbor (KNN), linear discriminant analysis (LDA), support vector

Table 3 The performance of DT along with K-Means based on the number of clusters

#k	K = 2	K = 3	K = 4	K = 5	K = 6	K = 7	K = 8	K = 9	K = 10
Accuracy	88.85	89.34	87.66	88.38	88.00	**90.32**	88.71	88.84	88.83
Sensitivity	94.43	94.42	94.32	94.21	93.71	**95.67**	94.27	94.45	93.92
Specificity	9.82	08.57	13.03	10.84	08.28	**07.77**	10.35	10.53	10.07

Bold indicates which K value getting maximum accurcy, sensitivity and minimum specificity value

Table 4 The performance of KNN along with K-Means based on the number of clusters

#k	K = 2	K = 3	K = 4	K = 5	K = 6	K = 7	K = 8	K = 9	K = 10
Accuracy	89.84	90.85	89.74	89.52	90.57	91.18	**94.72**	92.15	92.25
Sensitivity	97.61	97.94	97.15	97.57	98.03	98.15	**98.49**	98.40	98.36
Specificity	03.25	02.89	03.92	02.63	02.22	02.16	**01.94**	02.29	02.28

Bold indicates which K value getting maximum accurcy, sensitivity and minimum specificity value

Table 5 The performance of SVM along with K-Means based on the number of clusters

#k	K = 2	K = 3	K = 4	K = 5	K = 6	K = 7	K = 8	K = 9	K = 10
Accuracy	93.28	91.71	91.41	92.29	91.72	91.74	**93.77**	91.68	92.72
Sensitivity	100.0	100.0	100.0	100.0	100.0	100.0	**100.0**	100.0	100.0
Specificity	00.00	00.00	00.00	00.00	00.00	00.00	**00.00**	00.00	00.00

Bold indicates which K value getting maximum accurcy, sensitivity and minimum specificity value

Table 6 The performance of NB along with K-Means based on the number of clusters

#k	K = 2	K = 3	K = 4	K = 5	K = 6	K = 7	K = 8	K = 9	K = 10
Accuracy	93.13	92.15	**94.14**	90.50	93.72	84.67	87.74	87.23	81.05
Sensitivity	**99.48**	97.45	97.74	89.15	97.09	89.70	93.07	92.84	85.36
Specificity	**00.09**	02.97	02.45	13.18	05.11	11.53	11.35	06.18	12.95

Bold indicates which K value getting maximum accurcy, sensitivity and minimum specificity value

Table 7 The performance of LDA along with K-Means based on the number of clusters

#k	K = 2	K = 3	K = 4	K = 5	K = 6	K = 7	K = 8	K = 9	K = 10
Accuracy	93.16	93.65	**93.72**	92.56	93.01	91.54	93.26	92.38	92.89
Sensitivity	**99.85**	99.73	99.68	98.18	98.49	99.45	97.63	97.61	98.74
Specificity	**00.06**	00.09	00.14	00.62	01.94	00.89	02.95	00.37	01.59

Bold indicates which K value getting maximum accurcy, sensitivity and minimum specificity value

machine (SVM), naive bayes (NB) along with K-Means produce improved result than their individual performance. (2) Out of the five hybrid models, SVM along with K-Means produces better accuracy, sensitivity, and lower specificity.

4.2.3 Performance Based on K-Medoids Clustering Along with Classifiers

In this hybrid model, instead of K-Means algorithm we go for K-Medoids algorithm along with similar classification algorithms and find the performance calculated during the experimentation using Accuracy, Sensitivity, and Specificity is tabulated in Tables 8, 9, 10, 11, and 12, Here K represent the no of clusters. It's varies from 2 to 10. Bold indicates which K value getting maximum accurcy, sensitivity and minimum specificity value. The graphical view of Accuracy shown in Fig. 4. The result show that (1) all the classifier decision tree (DT), k-nearest neighbor (KNN),

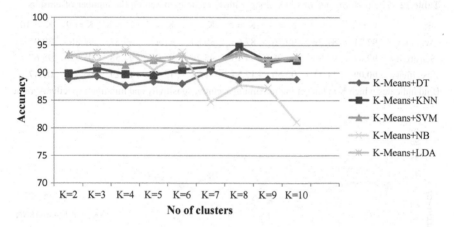

Fig. 3 Accuracy curve for various classifiers along with K-Means

Table 8 The performance of DT along with K-Medoids based on the number of clusters

#k	K = 2	K = 3	K = 4	K = 5	K = 6	K = 7	K = 8	K = 9	K = 10
Accuracy	88.78	87.62	88.27	88.29	**90.43**	85.52	85.58	87.10	87.08
Sensitivity	94.32	93.53	93.27	94.20	**95.24**	92.10	92.63	92.95	93.61
Specificity	9.83	10.75	11.30	10.69	**09.15**	11.87	10.95	10.44	09.20

Bold indicates which K value getting maximum accurcy, sensitivity and minimum specificity value

Table 9 The performance of KNN along with K-Medoids based on the number of clusters

#k	K = 2	K = 3	K = 4	K = 5	K = 6	K = 7	K = 8	K = 9	K = 10
Accuracy	90.46	90.90	89.96	89.27	88.53	**94.41**	91.43	90.15	91.59
Sensitivity	97.79	97.95	97.78	96.98	96.86	**98.99**	98.23	97.41	97.92
Specificity	03.28	02.78	02.55	03.48	03.13	**01.99**	02.96	03.20	02.23

Bold indicates which K value getting maximum accurcy, sensitivity and minimum specificity value

Table 10 The performance of SVM along with K-Medoids based on the number of clusters

#k	K = 2	K = 3	K = 4	K = 5	K = 6	K = 7	K = 8	K = 9	K = 10
Accuracy	93.29	91.30	93.22	92.63	92.39	92.38	**93.32**	92.40	92.55
Sensitivity	100.0	100.0	100.0	100.0	100.0	100.0	**100.0**	100.0	100.0
Specificity	00.00	00.00	00.00	00.00	00.00	00.00	**00.00**	00.00	00.00

Bold indicates which K value getting maximum accucry, sensitivity and minimum specificity value

Table 11 The performance of NB along with K-Medoids based on the number of clusters

#k	K = 2	K = 3	K = 4	K = 5	K = 6	K = 7	K = 8	K = 9	K = 10
Accuracy	**93.14**	92.50	92.32	89.09	91.60	86.49	88.47	86.89	83.99
Sensitivity	**99.48**	99.30	98.61	94.34	98.14	91.85	93.45	92.02	88.64
Specificity	**00.14**	00.40	01.02	05.31	02.38	06.92	07.83	08.00	13.43

Bold indicates which K value getting maximum accucry, sensitivity and minimum specificity value

Table 12 The performance of LDA along with K-Medoids based on the number of clusters

#k	K = 2	K = 3	K = 4	K = 5	K = 6	K = 7	K = 8	K = 9	K = 10
Accuracy	**92.21**	87.29	90.76	87.64	89.84	87.44	89.37	91.86	89.95
Sensitivity	**99.01**	92.80	94.98	94.15	95.24	94.08	94.58	95.95	95.62
Specificity	**00.09**	11.11	02.50	17.54	11.21	12.93	10.29	06.41	09.47

Bold indicates which K value getting maximum accucry, sensitivity and minimum specificity value

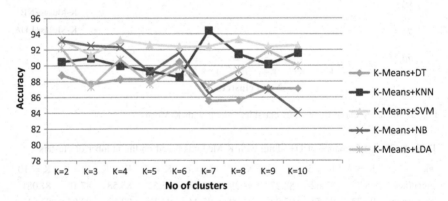

Fig. 4 Accuracy curve for various classifiers along with K-Medoids

support vector machine (SVM), linear discriminant analysis (LDA), naive bayes (NB) along with K-Medoids produce improved result than their individual performance. (2) Out of the five hybrid models, SVM along with K-Means produces better accuracy, sensitivity, and lower specificity. (3) Both the clustering algorithms (K-Means and K-Medoids) are performing well with the slight variation in their efficiency.

Table 14 shows peak performance (maximum accuracy, sensitivity, and specificity) among the various numbers of clusters explored in clustering algorithm.

4.2.4 Performance Comparison with Other Existing Approaches

The proposed supervised and unsupervised hybrid learning system shows better performance than various hybrid learning system done by previous research work tabulated in Table 13. Indranil Bose and Xi Chen (2009) selected 14 attributes from a collection on 271 attributes related to revenue contribution and minutes of usage of the customer. He has used five clustering techniques, namely fuzzy c-means, K-Means, BIRCH, self-organizing map, and K-Medoids, to segment the customers, and the result of segmented customers is given as an input to the decision tree, and performance is evaluated using top decile lift [13]. Adnan Idris, Muhammad Rizwan, Asifullah Khan (2012) proposed a PSO-based undersampling method as the number of non-churners is very high compared to the number of churners. The various reduction techniques were evaluated. His data set is modeled using random forest and k-nearest neighbor. His approach produced better AUC value [18]. Ying Huang and Tahar Kechadi (2013) proposed weighted K-Means algorithm hybrid with First Order Inductive learning classification algorithm for prediction [12] (Table 14).

Table 13 The performance comparison of proposed method with existing approaches

Performance comparison with existing hybrid models	Accuracy
Our proposed model K-Means + KNN	94.72
Chr-PmRf by Adnan Idris [18]	66.59
Weighted K-Means +FOIL by Ying Huang [12]	87.24
KM-boosted C5.0 by Indranil Bose [13]	92.84

Table 14 Performance comparison of highest accuracy, sensitivity, and lowest specificity values

Performance measure	Classifiers	Single classifiers	K-Means hybrid	K-Medoids hybrid
Accuracy	DT	87.61	90.32	**90.43**
	KNN	90.91	**94.72**	94.41
	SVM	92.55	**93.77**	93.32
	NB	91.39	**94.14**	93.14
	LDA	89.52	**93.72**	92.21
Sensitivity	DT	93.68	**95.67**	98.24
	KNN	98.00	98.49	**98.99**
	SVM	**100.0**	100.0	100.0
	NB	99.00	**99.48**	99.01
	LDA	97.96	**99.85**	99.48
Specificity	DT	12.23	**07.77**	09.15
	KNN	02.84	**01.94**	01.99
	SVM	**00.00**	**00.00**	**00.00**
	NB	01.00	**00.09**	00.14
	LDA	00.16	**00.06**	00.09

Bold indicates which K value getting maximum accurcy, sensitivity and minimum specificity value

5 Conclusion

In this work, the orange data set is preprocessed using effective data cleaning methods. After cleaning process the clustering approach is carried out using unsupervised techniques such as K-Means and weighted K-Means algorithms. The various clusters obtained from the above method are divided into training and testing sets using hold-out method. And each of the cluster training and testing data sets is handled by various supervised data mining techniques. The hybrid mechanism efficiency is measured in terms of sensitivity, specificity, and accuracy. The proposed hybrid models are very useful for success of every firm.

References

1. Web page reference. http://www2.deloitte.com/us/en/pages/technology-media-and-telecommunications/articles/telecommunications-industry-outlook.html
2. Mahmoudi N, Duman E (2015) Detecting credit card fraud by modified Fisher discriminant analysis. J Exp Syst Appl 42:2510–2516
3. Burez J, Van den Poel D (2009) Handling class imbalance in customer churn prediction. J Exp Syst Appl 36:4626–4636
4. Huang BQ, Kechadi MT, Buckley B (2012) Customer churns prediction in telecommunications. J Exp Syst Appl 39(1):1414–1425
5. Hwang H, Jung T, Suh E (2004) An LTV model and customer segmentation based on customer value: a case study on the wireless telecommunication industry. J Exp Syst Appl 26(2):181–188
6. Larivire B, Poel DVD (2014) Predicting customer retention and profitability by using random forests and regression forests techniques. J Exp Syst Appl 29:472–484
7. Neslin SA et al (2006) Defection detection: measuring and understanding the predictive accuracy of customer churn models. J Market Res
8. Xia GE, Jin WD (2008) Model of customer churn prediction on support vector machine. J Syst Eng Theory Pract 28(1):71–77
9. Anil Kumar D, Ravi V (2008) Predicting credit card customer churn in banks using data mining. Int J Data Anal Tech Strateg
10. Iturriaga FJL, Sanz IP (2015) Bankruptcy visualization and prediction using neural networks: a study of U.S. commercial banks. J Exp Syst Appl 42:2857–2869
11. Tsai CF, Lu YH (2009) Customer churns prediction by hybrid neural networks. J Exp Syst Appl 36(10): 12547–12553
12. Huang Y, Kechadi T (2013) An effective hybrid learning system for telecommunication churns prediction. J Exp Syst Appl 40:5635–5647
13. Bose I, Chen X (2009) Hybrid models using unsupervised clustering for prediction of customer churn. J Organ Comput Electr Comm 19:133–151. Copyright: Taylor & Francis Group, LLC
14. Khashei M, Hamadani AZ, Bijari M (2012) A novel hybrid classification model of artificial neural networks and multiple linear regression models. J Exp Syst Appl 39(3):2606–2620
15. Kim K, Jun C-H, Lee J (2014) Improved churn prediction in telecommunication industry by analyzing large network. J Exp Syst Appl 41(1):6575–6584

16. Gerpott TJ, Ahmadi N (2015) Regaining drifting mobile communication customers: predicting the odds of success of winback efforts with competing risks regression. J Exp Syst Appl 42:7917–7928
17. Ali OG, Arıturk U (2014) Dynamic churn prediction framework with more effective use of rare event data: the case of private banking. J Exp Syst Appl 41:7889–7903
18. Idris A, Rizwan M, Khan A (2012) Churn prediction in telecom using random forest and PSO based data balancing in combination with various feature selection strategies. J Exp Syst Appl 38:1808–1819

Simulation-Based Comparison of Vampire Attacks on Traditional Manet Routing Protocols

Joshua Reginald Pullagura and Dhulipalla Venkata Rao

Abstract A mobile ad hoc network is wireless network without infrastructure and is self-configuring network of mobile nodes where nodes move to their places randomly, leave around or join the network. Over the many years, simulation work has been carried out to achieve a new reliable routing protocol. This paper seeks to rationalize Vampire attacks on the network layer that deals with route establishment. During route establishment and packet forwarding, Vampire attacks perpetually weaken the network by draining nodes' energy. The Vampire attacks degrade the performance of routing protocol. All the basic traditional routing protocols those are examined here are open to Vampire attacks, which are damaging, and is time-consuming to troubleshoot. Simulation results show DSDV performs better than its counter parts.

Keyword AODV · DSR · DSDV · Vampire

1 Introduction

Mobile ad hoc networks (MANET) is defined as network without any permanent infrastructure. It consists of nodes distributed randomly and communicates with each other in radio environment. MANETs even though mainly designed for military applications can also be deployed in commercial sector. In recent times, there is a rapid growth in personal area networking. Routing plays an important role so that packets are delivered successfully to the destination.

In general, every protocol consists of static nodes. The main objective of a good protocol is to identify the path to destination and send the encrypted data from

J.R. Pullagura (✉)
Vignan University Guntur, Guntur, Andhra Pradesh, India
e-mail: pjreginald@gmail.com

D. Venkata Rao
Narasaraopet Institute of Technology, Narasaraopet, Guntur, Andhra Pradesh, India
e-mail: dv2002in@yahoo.co.in

© Springer Nature Singapore Pte Ltd. 2018
D.K. Mishra et al. (eds.), *Information and Communication Technology for Sustainable Development*, Lecture Notes in Networks and Systems 9, https://doi.org/10.1007/978-981-10-3932-4_52

source to destination. Since the network has no fixed infrastructure, each node should genuinely participate in packet forwarding process. This can be concluded through simulation, which can be modeled hypothetically for further investigation. Here, simulation is carried out using NS2 simulator, NS2 plays a role of both emulator and simulator [1]. NS2 provides default implementations for creation of network nodes and creating links between nodes and dealing with routing algorithms. It is also used for the implementation of transport layer protocols such as UDP and TCP and some of traffic generators [2]. The motivational work carried out here is to compare the traditional routing protocols with Vampire attacks with the help of a simulator. AODV and DSR are categorized as on-demand protocols. DSR protocol [3] allows the nodes to establish the route and they react when there is need, whereas DSDV [4] falls under proactive category where table updation provides better route. The performance comparison of AODV, DSR and DSDV routing protocols with Vampire attacks is evaluated and monitored and simulation is carried out on NS-2.35 simulator. DSDV routing protocol exhibits enhanced performance in comparison with the other two protocols. Average energy and total energy are computed and evaluated for analysis purpose, and the number of nodes utilized for the simulation are varied from 20 nodes to 30 nodes and increased to 40 nodes with one vampire.

2 Vampire Attacks

Security is the salient feature of the ad hoc networks. It is the most essential factor for network functions which include route discovery, forwarding the packets and route maintenance. Ad hoc networks are open to threats and hence counter measures have to be incorporated in the initial stage of design. Security in wireless networks is much more challenging when compared to wired network [5]. In mobile networks, the majority of the functions are shared by all the nodes. This scenario creates the core security threat that is specific to the network. The attacks on networks are mainly due to selfishness nature of the node or due to malicious nature of the node and other reasons are not properly defined. The attacks are categorized as active and passive attacks, respectively. Active attacks range from message replay, injecting erroneous messages, impersonation of nodes, thus threatening the availability, integrity, authentication, and non-repudiation. Almost six layers of the OSI model are prone to attacks. The worst hit layers are network and data link layers. Network layer deals with the routing concept which is the most essential phase in source to destination packet delivery.

Vampire attacks are not protocol specific. This type of attack degrades the performance of a network. At the network layer, the source node inserts entire path in the packet header. The intermediate nodes will not make any decision about the route but follow the route given by source node in packet header and it does not depend on the physical design. The active Vampire attacks are further classified into two categories.

2.1 Stretch Attack

Here, an infected node creates an alternate longer source routes, causing packets to travel through more number of nodes in the network affecting the network lifetime. A non-malicious source would select the route which is short and stable to the destination, but the malicious node selects very longer and unstable route. In sensor and adhoc networks nodes, battery is the one of the main concerns, as the route with high energy is drained more. The route stability also decreases as one or few nodes with low energy may drain out. The stretch attack is more challenging when compared to the other attacks.

Figure 1 shows randomly placed nodes where malicious (colored) node is causing stretch attack, thereby draining the nodes' power in the network.

2.2 Carousel Attack

Here, a harmful node transfers a packet in a route with more number of loops. This causes the infected node to appear several times in the route, increasing route length beyond the established quantity.

In this paper, we consider the vampire attacks both stretch and carousel on the three basic routing protocols. The stretch attack expands route distance travelled by packets, causing them to be processed by more number of nodes. The harmful node does not consider hop count and shortest route to the destination [6]. Carousel attack [7] causes immoderate battery drain for few numbers of nodes, whereas

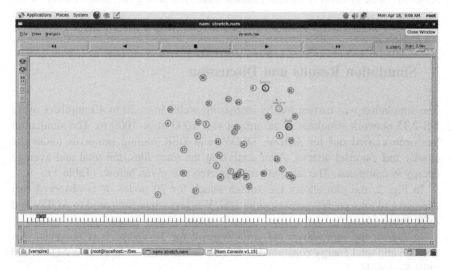

Fig. 1 Stretch attack

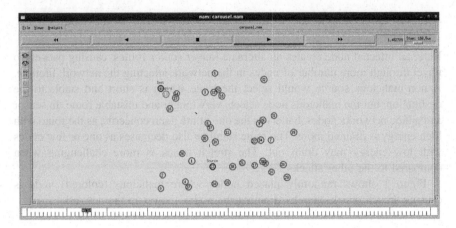

Fig. 2 Carousel attack

Table 1 Simulation parameters

Number of nodes	20–40
Routing protocol	AODV, DSDV, DSR
Traffic source	CBR
Area	1000 m × 1000 m
Tool	NS-2.35

stretch attack causes precise energy drain for all the nodes in the network [6]. These attacks are causing serious threat to network layer.

Figure 2 shows randomly placed nodes in the simulation where malicious (colored) node is causing carousel attack thereby draining the nodes power in the network.

3 Simulation Results and Discussion

The simulation was carried out by increasing nodes from 20 to 40 numbers using NS-2.35 network simulator in an area of size 1000 m × 1000 m. The simulation has been carried out for AODV, DSDV, and DSR routing protocols under the stretch and carousel attacks. After analyzing the trace files, the total and average energy is computed. The simulation parameters are given below: (Table 1)

In Fig. 3, the plot shows the stretch attack for 20 nodes. It is observed that average and total energy consumed by DSDV is less when compared to AODV and DSR Protocols.

In Fig. 4, the plot shows the stretch attack for 30 nodes. It is observed that average and total energy consumed by DSDV is less when compared to AODV and DSR Protocols.

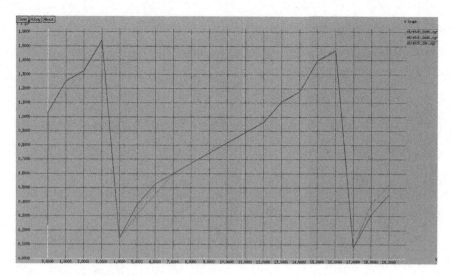

Fig. 3 Stretch attack 20 nodes

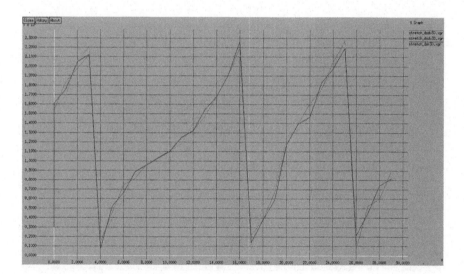

Fig. 4 Stretch attack 30 nodes

In Fig. 5, plot shows the stretch attack for 40 nodes. It is observed that average and total energy consumed by DSDV is less when compared to AODV and DSR protocols which are tabulated in the table given below.

When nodes are linearly increased from 20 to 40 numbers, the energy drained due to stretch attacks in DSDV protocol is comparatively less than AODV and DSR protocols. There is very less variation noticed when the nodes are dynamically increased in the three traditional protocols.

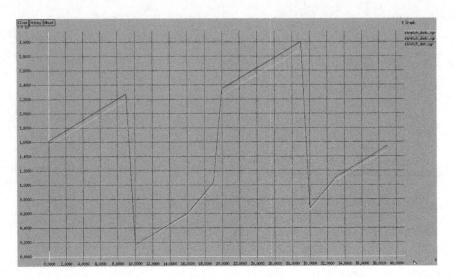

Fig. 5 Stretch attack 40 nodes

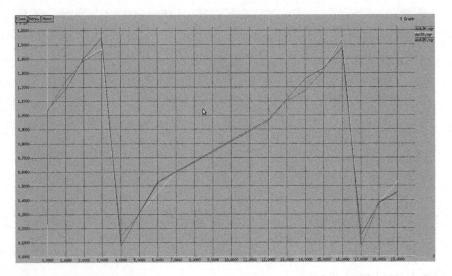

Fig. 6 Carousel attack 20 nodes

From Figs. 6, 7 and 8 we conclude that that DSDV protocol takes lesser average and total energy than AODV and DSR routing Protocols which were subjected to Carousel attack. When the number of nodes is less, simulation results exhibit distinguishable variations and when number of nodes are increased the variation are less noticeable. For the three traditional protocols the average energy and the total energy are also increased with the increase in number of nodes (Table 2).

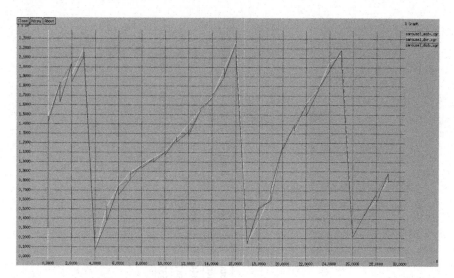

Fig. 7 Carousel attack 30 nodes

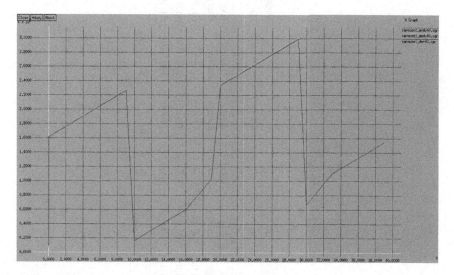

Fig. 8 Carousel attack 40 nodes

Table 2 Energy table

Vampire attack													
Protocol	Carousel attack						Stretch attack						
No of nodes	20		30		40		20		30		40		
	Avg energy	Total energy	Avg energy	Total energy	Avg energy	Total energy	Avg energy	Total energy	Avg energy	Total energy	Avg energy	Total energy
AODV	61.9577	1548.94	90.1494	2253.74	119.53	2988.3	61.791	1544.72	90.6991	2267.48	120	2999.9
DSR	61.7954	1544.89	91.03	2275.96	120.232	3005.81	61.9122	1547.27	91.0597	2276.49	120.19	3004.75
DSDV	61.3363	1533.41	87.6067	2190.17	119.82	2995.49	61.4908	1537.27	90.7174	2267.93	115.612	2890.3

4 Conclusion

Vampire attacks disrupt the working of the ad hoc mobile networks immediately. The average and total energy consumed by the nodes due to Vampire attacks is less in DSDV protocol when compared to the AODV and DSR protocol because all the nodes in the network will be knowing all possible destination details. When the Vampire attacks are occurred in the network, it not only consumes higher power, but also consumes more time. There are many solutions and techniques proposed to prevent from these attacks, but are not so effective; hence, there is a need for a better solution which avoids these kinds of attacks.

References

1. Siraj S, Kumar Gupta A, Badgujar R (2012) Network simulation tools survey. Int J Adv Res Comput Commun Eng 1
2. Weingartner E, vom Lehn H, Wehrle K (2009) A performance comparison of recent network simulators. In: IEEE International Communications (ICC'09), Dresden, Germany
3. Johnson DB, Maltz DA, Dynamic source routing in ad hoc wireless networks. In: Imielinski T, Korth H (eds) Mobile computing. Kluwer Academic Publishers
4. Perkins CE, Bhagwat P, Highly dynamic destination sequence-vector routing (DSDV) for mobile computers. Comput Commun Rev
5. de Morias C, Agrawal DP, Ad hoc and sensor networks theory and applications. World Scientific Publishing Co. Pte. Ltd
6. Vasserman EY, Hopper N (2013) Vampire attacks: draining life from wireless ad hoc sensor networks. IEEE Trans Mobile Comput 12(2)
7. Yang H, Luo H (2004) Security in mobile ad hoc networks: challenges and solutions. IEEE Wirel Commun 11(1)

Author Index

© Springer Nature Singapore Pte Ltd. 2018
D.K. Mishra et al. (eds.), *Information and Communication Technology for Sustainable Development*, Lecture Notes in Networks and Systems 9, http://doi.org/10.1007/978-981-10-3932-4